TRENCHLESS CONSTRUCTION for UTILITIES

DIG

O-DIG
CONFER

IPHE
DIG 85

NO-DIG 85

TRENCHLESS CONSTRUCTION for UTILITIES

Proceedings of the
First International Conference

London
16–18 April 1985

Proceedings Edited by
F. E. Bruce
Emeritus Reader in Public Health Engineering
University of London

Conference
Organized by:
Institution of Public Health Engineers
Grosvenor Gardens House
35-37 Grosvenor Gardens
London, SW1W 0BS
Tel: 01-834 8785

Published 1985

ISBN 0 905188 07 1

Produced in England by **Inprint of Luton Ltd.**
95-115 Windmill Road, Luton, Beds.

CONTENTS

ORGANIZING COMMITTEE

Chairman	
E W Flaxman	Past-President, IPHE, and Partner, Binnie & Partners
Members	
P A Banks	President, IPHE, and Partner, John Taylor & Sons
J C Clancey, OBE, BEM	Vice-President Emeritus, IPHE
K J Flemons	Vice-President, IPHE, and Associate, Sir Frederick Snow & Partners
C Prescot	Exhibition Manager, Westrade Fairs Ltd.
P B Rumsey	Manager, Water Research Centre
R West	Managing Director, Westrade Fairs Ltd
J L Wilson	Senior Engineer, National Joint Utilities Group
J Wishart	Binnie & Partners
Secretary	
Cdr D J Dacam, OBE	Secretary, IPHE

Exhibition organized by:-

Westrade Fairs Ltd.,
P.O. Box 53,
Rickmansworth,
Herts.
WD3 2AG
Telephone: 0923 778311

MESSAGE FROM MRS. LYNDA CHALKER, MP,

Minister of State for Transport

I am pleased that the Institution's Conference NO-DIG 85 was such a success. It is also gratifying to note the increasing use of trenchless techniques for the replacement of services in streets.

I am sure that there is still considerable potential to extend the use of trenchless techniques. I hope that the momentum generated by the Conference will help equipment manufacturers to overcome problems such as sensing the presence of other apparatus; and that such improvements will encourage the utilities to make greater use of these types of equipment.

There is no doubt that trenchless techniques can offer wider advantages to society through reducing traffic disruption and avoiding the risk of poor reinstatement of trenches. These are subjects which are obviously very relevant to Professor Horne's review of the Public Utilities Street Works Act. I am delighted to say that we have now received Professor Horne's report, and we intend to make an early decision about its publication. The Government will certainly consider very carefully what the review report has to say on the subject of trenchless excavation.

WELCOME AND INTRODUCTION TO NO-DIG 85

PETER A. BANKS, President, Institution of Public Health Engineers

Lord Nugent, Ladies and Gentlemen,

As President of the Institution of Public Health Engineers I offer you a very warm welcome to this First International Conference and Exhibition on Trenchless Construction for Utilities.

The Institution has always tried to provide a forum for the promotion and advancement of new technologies in its field. Judging from the response we have had to this Conference by way of numbers of papers offered, delegates registering (nearly 400) and exhibitors (67) exhibiting in the accompanying Exhibition, we have certainly identified the new technology of current interest at this time.

The concern with ageing existing infrastructure, both in this country and in many other older industrial nations, and with its installation in the newer industrialized nations, represents a challenge for the use of the techniques we are here to discuss. It is a chance for us as engineers and practitioners to ease the burden, the social cost that is borne by society at large, of installing and renewing that infrastructure.

We have deliberately sought papers on this subject by experts from all over the world. It is a technology where developments are taking place not just in one country alone, and it is important that we hear of each other's work.

The Institution with its many overseas members is particularly glad to welcome so many delegates from outside this country here today. Over 100 overseas delegates from 20 different countries is very gratifying, as it confirms that a wide interest exists.

A symptom of the awareness of the public in the United Kingdom of the importance of our subject is to be found in the Review of the Workings of the Public Utilities Street Works Act commissioned by Government which Professor Michael Horne has led. I hope that the discussions at this Conference will assist in the debate and the decision-making that will emerge from the Review.

We will, I think, Ladies and Gentlemen, be demonstrating that if society wishes to be relieved of a great deal of the cost and inconvenience of conventional installation methods for underground utilities, we have the means to do so.

The germ for the idea of this NO-DIG 85 lay in the Presidential Address given to this Institution by Ted Flaxman in 1980. His task in his concluding Address on Thursday will be to identify the major issues that we will have exposed in our discussion for others to take note of and act upon. He will also be speaking of the next steps that the Institution will be taking to further this subject. Indeed, he will be making an announcement about the Institution's plans for a special award to spur on this technology.

When we designed this Conference we decided to concentrate on trenchless construction below 900mm diameter, and we have excluded the techniques of mole ploughing because of course this necessitates some surface disturbance and obviously cannot be used where other services exist.

We believe carefully aligned specialist conferences are, as your response indicates, better value than those which are too diffuse, but next time it may be appropriate to adjust our definition and widen the scope a little more

And now it is my privilege to introduce Lord Nugent who has so kindly agreed to deliver an opening Address for us. Lord Nugent is a member of the House of Lords Select Committee on Science and Technology and has been associated with water and environmental matters for many years — as Chairman of Thames Conservancy, Chairman of the National Water Council and, more particularly in the context of the Conference, he is an Honorary Fellow of the Institution of Public Health Engineers. His contribution to the field in which we all operate has been very considerable and we are particularly pleased that he is able to undertake this task for us.

Ladies and Gentlemen, I have great pleasure in asking Lord Nugent to present his opening Address.

OPENING ADDRESS

The Rt Hon the LORD NUGENT OF GUILDFORD, PC

I must first congratulate the Institution of Public Health Engineers on the promotion of this NO-DIG 85 Conference and Exhibition. Together with last month's Symposium on Plastics for Pipeline Renovation, this forms the most important expositon of modern pipeline technology ever to be put on in Britain. It is bound to be of great interest to the water industry, as well as to the gas and electricity industries.

I am very pleased to hear that the Conference is being attended by over 100 overseas visitors from more than 20 countries because this new technology is a common need, not only in all advanced countries, but also in the developing countries, where the need is for new public utilities.

But to return to the acute needs of the water industry. Two years ago, the House of Lords Select Committee on Science and Technology reported on the water industry and called attention to the huge cost of replacing Britain's sewers and water mains, many of which, after 100 to 150 years of life, were now showing every sign of collapsing. It was estimated that to carry out this replacement by the traditional methods of digging and backfilling would cost £45 billion.

Britain has the most comprehensive system of water distribution and sewerage in the world, with 280 000km of water mains, 240 000km of service pipes and 210 000km of sewers. At the time of the Select Committee's report, the water industry's current expenditure on renovation and repair was £350 million per annum, but the Select Committee estimated that this left a shortfall of at least one-third of what was required. So the progressive deterioration continues. Under the existing strategy of repair, the industry copes with 80 000 burst water mains and 5000 sewer collapses per year.

On top of the engineering costs is the enormous social cost of traffic dislocation and inconvenience to the public.

We have here a grim scenario, worsened by the Government's tightening squeeze on Regional Water Authorities' finances. There seems to be some corrosion in the corridors of power — a blockage requiring engineering attention! The problem is to raise the level of Parliament. The only man who made a serious try to do this was Guy Fawkes! He failed — but with the benefit of your modern tunnelling techniques, who knows what he might have achieved? That market for NO-DIG disappeared 400 years ago. But an active new market has appeared in Whitehall — for Mole-ing. No national newspaper can afford to be without an Unclemole, or a Telemole. How they would like to be able to drill a shaft direct into the offices of the Prime Minister and all Secretaries of State!

The water industry's crisis of crumbling sewers and water mains cannot be solved by the traditional methods of open cut and backfill. But these modern techniques of trenchless renovation and NO-DIG pipelaying offer a practical solution at half the cost, which *is* within our resources.

The ten Regional Water Authorities have begun to use these techniques, but I was concerned to hear at the Symposium last month that at present only £8 million per annum is being spent on renovation. The figure should be at least £80 million p.a. This indicates a reluctance to change from traditional methods, using the agent Local Authority's direct labour force. Of course, this system must continue for a few years yet, until the renovation programme gets into full stride, and eliminates bursts and collapses. I hope the Water Authorities will get the message from this Conference.

My friends in the water industry have done a magnificent job in the past eleven years, modernizing the mixed bag of assets which they inherited from the Local Authorities — stretching from top class to no class at all. After giving priority to the sewage treatment works, they are now turning attention to their hidden assets of sewers and water mains.

In many cases, records are innaccurate, or even non-existent. Thus the advanced technology described in the papers to be presented in Session 5 is particularly apposite. The first task is to locate the water mains and sewers, and then examine them *in situ*. The radar technique described by Mr Daniels in Paper 5.1. is particularly promising, while Mr Garnett's use of alternating magnetic field technology, as discussed in Paper 5.2, allows subsequent examination. Having located the buried pipes and cables, the second task is to proceed to the renovation of the corroded or damaged pipe by one of the new techniques.

Undoubtedly, digital mapping, as described in Paper 5.3 has an essential part to play in ensuring that all underground utilities work together and do not cut across each other — literally! Here the importance of the National Joint Utilities Group as co-ordinator cannot be overestimated.

I must say a special word of welcome to our overseas contributors, who have added so much to this Exhibition and Conference. We are especially pleased to see our friends from Japan, who have done such brilliant developmental work in recent years to perfect microtunnelling techniques. Their Paper 4.4 on the Telemole and Telemouse makes fascinating reading. Japan's annual programme for sewerage and sewage treatment of over £2000 million fills me with envy!

In conclusion, I repeat my congratulations to the Institution of Public Health Engineers and to all the distinguished contributors. All who attend this Exhibition and Conference will receive a unique introduction to the most advanced modern technology for NO-DIG pipeline construction, which has an invaluable benefit to confer on our country and on all advanced civilization throughout the world.

TUESDAY 16 APRIL

SESSION 1. Chairman: P. A. Banks, President, Institution of Public Health Engineers

1.1 Is there a NO-DIG future?

Donald F Rees OBE, FIPHE *Donald Rees Partnership*

The combination of attending to existing underground services and installing new services has considerable impact upon the movement of surface traffic if the traditional methods of construction are employed.

The paper discusses the effects that surface traffic disruption can have upon the nation's industrial efficiency at a time when it faces severe competition in both the domestic and overseas markets. It then considers the effect of scale on the costs of carrying out the construction operations by non-disruptive methods and how such market scale may be developed.

Finally the paper speculates upon the path which non-disruptive techniques may follow based upon present day technology, with a brief look further into the future.

INTRODUCTION

It is indisputable that NO-DIG 85 is timely. A few simple facts underline the need to provide new services and maintain and refurbish the old ones with the minimum overall expenditure and with the maximum overall effect — and the term overall is used advisedly for that is the gist of the NO-DIG case.

Within a century we have moved from the horseless carriages — counted in their hundreds — to a traffic load of over twenty million private and commercial vehicles. A figure which is increasing at a rate of around 2 per cent per year.[1]

Industrial shift has concentrated the population into vast urban conurbations which must be served with gas, electricity, water, sewerage, oil, telecommunications, district heating — and who can predict what other demands will arise in our technically advancing society? Perhaps a return to the pneumatic transport systems or their modern equivalent moving manufactured goods from factory to dockside; home deliveries from centralized macro-markets — all of course with automatic address tracking and direct account charging. Refuse disposal (that which is not already re-cycled) may be pelletized for tube transportation to processing stations. After toll roads and toll bridges why not toll holes? Perhaps 'Rent-a-hole' could become a viable commercial proposition.

Certainly underground space will continue to be exploited in every conceivable manner thereby escalating the problems of plotting a free route for each new service facility.

COMPETITIVE BRITAIN

To achieve or retain a competitive position in the world markets Great Britain Limited must look very seriously at its operating overheads which are already blunting its trading efforts. Britain's market share is under constant attack from highly efficient industrialized nations and more recently from those emerging countries where populations are moving from the fields to the factories in their determined and relentless striving for a better standard of living, if not for survival itself.

We know full well so many of the factors upon which our national prosperity depends:

Greater efficiency in the employment of capital by more professional management.

The fuller acceptance of robotics with which we already compete.

A radical change in employment relationships.

Greater realism in union philosophy.

We appear, however, to tolerate overhead costs which any competent business would have reduced severely as an essential commercial requirement.

One such overhead which concerns industry generally is brought into focus at NO-DIG 85. It is the enormous burden which the nation carries in the form of disruptive construction works which are not only costly in their direct execution but which do untold harm to the economy by slowing up transport and bringing inefficiency to trade and industry.

Our transport systems are already inadequate and our main artery to our European markets, in the form of the Channel Tunnel, remains a politically dominated myth, and yet these systems form the essential inter-link of our industrial economy.

Our water and sewerage facilities have become enmeshed in local and national politics to such an extent that the serious problems of maintaining and renewing such vital underground services are being lost under sweeping dogmatic policy decisions for unrelated reasons. With the local authorities further strictured by rate-capping and Treasury pressures, there seems little hope that a serious programme of renovation and renewal will be forthcoming on a scale which would encourage the investment to support further technological advance.

Unless there is a greater market security we will lose the opportunity of applying our technology for the purposes of reducing installed costs and of improving the speed and quality of operation. Sadly industry will turn its back for good financial reasons and we will lose yet another opportunity to exploit our skills in a field of increasing worldwide need.

Water waste is excessively costly and deplored socially when water rationing is the consequence. Sewers collapse daily, endangering the public and costing them dearly in emergency repairs and environmental nuisance.

Government is of course well aware of the situation through the House of Lords Report on the subject and the National Development Council's report from the civil engineering development committee entitled *Construction to 1990*. The problems are even more critically emphasized in the Council's later report on the state of the country's public sector infrastructures. Ironically the 1990 report states "Government should give unstinted support to projects of all sizes which would help the construction industry to compete more effectively in foreign markets, even where they may not be a top priority from a purely domestic viewpoint."

NO-DIG interests would wholly endorse that view and would confirm that they desperately need that 'unstinted support' in order to enable them to head the export markets with the confidence born of home market experience. The priority from that domestic viewpoint is surely high enough to strengthen the resolve and the costs of continuing neglect will provide future generations with an almost insurmountable financial and environmental problem.

The House of Lords Select Committee on Science and Technology[2] spent a great deal of time investigating all aspects of our sewer and water problems and their Lordships appeared to have little hesitation in expressing their views as to the seriousness of the situation and were anxious to establish the extent to which modern technology could contribute to resolving the problems. A Department of the Environment Memorandum on that report confirmed that "the Government have been concerned about the situation which exists in some parts of the older towns. The Government is giving encouragement to the WRC-led programme of survey work and research in conjunction with the construction industry into new approaches to renovation which are being adopted by water authorities and local authorities."

The sentiments are admirable although directed to only one section of the underground services industry. NO-DIG on the other hand sets out to encompass all of those services which, to maintain or replace them, involve surface disruption if outdated methods are used.

WATER RESEARCH CENTRE RESPONSIBILITIES

The WRC is a unique organization with an Engineering Division at Swindon devoted particularly to the renovation of sewers. They assess materials, systems, survey and operational equipment and aim to guide industry in the recognition of the water authority requirements. They also advise the authorities as to the developments which are taking place and assist industry in such developments when requested. Indeed some developments are the result of Joint Venture between the WRC and commercial organizations.

The role of WRC has evolved somewhat in the same manner as the renovation industry as a whole. Because renovation is not a subject "in the book" WRC has found itself the focal point of reference. "What method should be used" — "How should I specify the work" — "How long will it last" — "What load will it carry" etc, etc. The questions are endless and the answers to them must acknowledge their own limited knowledge and very heavy responsibility. The result — born of a natural caution — must have effectively damped down much of the pace of innovative development as more and more attention has been given to quality of the product and the ability to produce quality in the field (as against the laboratory).

Some may well be grateful for this damping down effect as it has in some measure coincided with the damping down of renovation as a growth market.

In order that more engineers should make their own engineering decisions WRC have produced many publications providing guidance in so far as the state of the art has allowed. The *Sewerage Rehabilitation Manual* will be followed this year by the recommendations on Conditions of Contract and Specifications.

In their full appreciation of the social and commercial impact of surface traffic disruption arising from conventional construction operations the WRC have paid particular attention to the need to minimize such disruption in undertaking the renovation or replacement work.

THE EFFECTS OF DISRUPTION

To continue to apply traditional techniques to the installation and rehabilitation of services is inconceivable. The negotiation through the network of existing services precludes the use of economical mechanical equipment in most cases. Existing pipes and cables must be accurately located and carefully supported as work proceeds and the crucial backfilling operations must be undertaken with equal care if later settlement damage is to be avoided. Altogether a labour-intensive job which is time-consuming and therefore excessively disruptive and excessively expensive.

Already disruption arises on a significant scale due to the decay of existing services. The direct costs of repair are increased by up to a factor of perhaps four times by the social and other consequential costs of the work, particularly when emergency collapse repairs are involved. Traffic diversions may be required and in any case traffic congestion will always force frustrated drivers into diversionary side streets where the increased loading can compound the problem through further service failures.

Add together the direct costs of repair, the cost of additional policing, the loss of taxable profit for shopkeepers, commerce and industry, increased accident rate with medical and hospital services for the injured, the loss of production through injury, consequential damage to other services, inefficient employment of capital in vehicle delays — and numerous other factors which together place an incalculable burden on the economy. That other straw of an overhead which we can ill afford to carry.

A paper from WRC deals specifically with the subject wherein they endeavour to estimate some of the costs involved.

In one way and another all of these costs must be reflected in the prices we charge for our goods and services — that blunting of our competitive edge. These overheads can be hidden or clouded but there is no escape from the final impact.

A QUESTION OF NEED

We may look with envy at the Japanese 'miracle'. It is no miracle, any more than was the case of post-war Germany. It is the result of a disciplined nation acting effectively in the face of world competition to produce reliable quality of good design at saleable prices. They seem to have got the measure of those factors upon which their prosperity depends and it is interesting therefore to note that with so much of the country unsewered, Japan has been a leader in developing non-disruptive methods for the installation of services beneath city streets overburdened with surface traffic. They had the need and the determination to meet it with the best of modern technology. In so doing they could well take the lead in the NO-DIG business; furthermore they have a healthy economy, a strong domestic market and justifiable confidence in their exporting competences.

When the costs of disruption are properly considered it becomes abundantly clear that to dig holes in the time-honoured way has little to commend it, for the shovel has become an expensive tool — some say a sign of impending bankruptcy. Surely few will argue the case for the shovel against the modernization of the operation on lines which will help to keep the surface unmolested.

WHERE DO WE GO FROM HERE?

NO-DIG 85 could be our watershed. As the first conference of its kind it is hoped that the debates in this forum will motivate Government, local government and industry to act with more noticeable enthusiasm.

Many systems already exist, some proven, some embryonic. Some of them are inherently cheaper than those disruptive methods which have in any case reached the limits of their potential. Some are more expensive and will remain so until the scale of demand brings in its train economy of operation and furthermore encourages that investment which can improve and enhance such developments in cost efficiency and quality of performance.

In order to encourage such investment it has been suggested that the employment of non-disruptive techniques should carry a form of subsidy which acknowledges national benefits. The calculations would be complex and no doubt the basis of endless discord. It is not a worthy proposal in that it could prove to be a disincentive to the achievement of high operational efficiency.

The only effective solution is to develop systems which are cheaper and more efficient in every way. The benefits of lower costs and lower national overheads would then be a bonus to the country without hidden tax-supported subsidies. If an incentive was required it could be interesting to consider legislation whereby, with due notice, surface excavation for certain services in clearly defined 'sensitive' locations would be prohibited. Such a bold approach, by generating a positive market for NO-DIG techniques, should give the industry the confidence which is fading so rapidly.

MAN'S INGENUITY

Man's ingenuity is not lacking. What is lacking is that commercial stimulation which still does not exist on a sufficiently realistic scale in spite of the wide publicity given to the relentless decay which is affecting one of the nation's largest capital assets.

Ironically war has done more to stimulate technological advance than has the battle for peaceful survival. The wartime battle had to be won regardless of cost. Today's battles between nations are industrial with modern technology in the forefront.

Some of our most promising NO-DIG systems have emanated not from the water industry but from the highly commercialized gas industry where decayed gas mains were not only a danger to the public and a financial drain through expensive leakage but also these factors had a serious effect upon their public image.

The gas industry co-operated with private industry to produce the revolutionary pipe-bursting system and the water industry will be a beneficiary in the application of pipe-bursting to sewers and water mains. It is a clear example of the results which can be achieved by presenting industry with a problem to test man's ingenuity and showing that industry that the reward for investment and success is a substantial domestic and export market.

What has been achieved overall in the water industry brings great credit, but little or no profit, to the pathfinders who have speculated considerable resources of money and manpower in the belief that demand would follow the successful matching of need. Sadly it was an illusion. With a few outstanding exceptions — mostly by large companies with substantial financial backing — innovative developments have tended to 'freeze' at a 'nuts and bolts' stage whereas many of them were at the very threshold of moving ahead into the world of high technology, exploiting the advantages of robotics and high production techniques which would produce low installed prices against major market supply.

It is this next stage of development which is so important if the NO-DIG objectives are going to be met. It is astonishing that a government which claims to encourage such private enterprise should move the target to a more distant horizon just as the industry was getting the bull's-eye in the sights and was ready to squeeze the trigger.

THE TECHNOLOGICAL REQUIREMENTS

We do not in fact have to take a great leap forward into a technological world of fantasies. To undertake service installation and renovation work in a non-disruptive manner and at economic prices is mainly a matter of drawing together the essential elements of existing technology and adapting them to today's needs.

Certainly it may mean pushing technology to its limits and perhaps beyond those limits as time expands our understanding. Most of what we need is, however, here on the shelf.

It is inappropriate to project a scene of centuries hence when communities may be protected from their own ecological errors, living in vast protected environments where our crude services of the 20th century are unwarranted as science has run more of its course (but such statement made a century or so ago would have been incorrect as we continue to use our ancient structures!).

Our projection must be confined to "today and tomorrow", helping to wrestle with the problems arising from our decaying underground heritage and the shock of living in this age of scientific discovery.

What indeed does this "today and tomorrow" bring within the bounds of possibility, probability, and, with the right support, certainty?

To "know your enemy" is a sound first principle and so we must identify those things which may frustrate a smooth NO-DIG operation. The 'enemy' which is in the form of the multitude of cables, pipes, ducts and culverts, old timber piles, abandoned timber water mains — the services of our modern society and the dross of bygone days.

We have our methods of searching out many of these underground hazards — our 'mine detectors' of the utilities industries — our modern hazel twigs. As yet however they do not cover the full spectrum of materials which may be encountered.

An ideal concept is in the form of a continuously recording, full road width 'sweep', which in one pass would identify and plot each service by type, material, alignment, depth, diameter — live or abandoned.

Into this complex plot would then be fed the requirements of the new service specifying the gradient, size, access frequency, limitations on directional variations, clearance required etc.

The final product would be a 3-D representation defining route options together with the estimated cost and timescale assessments relating to the various suitable techniques which are available.

CCTV will be advanced far beyond its present simple visual inspection. It will confirm such factors as alignment variations, pipe deformation, crack intensity as the now outdated vidicon tube gives way to the measurable grid of the charge-coupled device — the indestructible 'chip'. A future 'pass' through a piped service will also recognize the installed design, externally, and the presence of the now infamous and destructive void.

Engineers will carry portable, battery-operated, survey probes, as a matter of course to give them access to the previously inaccessible and thereby to verify the quality of new and remedial work, with video recordings as their 'notebook'.

But even with all of these devices the NO-DIG systems will be at the mercy of some unexpected and unidentified obstructions which could cause line abandonment, surface chaos and costly recovery work.

AUTOMATIC GUIDANCE

It is for such circumstances that we require the mole with the self-seeking guidance system which can identify those freak conditions sufficiently well ahead to take avoiding action, but within specific and pre-programmed parameters according to the sensitivity of the particular service to directional adjustment.

The pre-programming of a microtunnel installation on the basis of the 3-D survey may seem to be beyond our present technological limits. However it has taken Dr Peter Zollman[3] a mere decade to produce tunnel guidance systems which today are capable of slaving pipe-jacked conduits around complex curves by using pre-programmed offsets which are fed into a computer memory module which then provides offset data, without human intervention, comparing the actual measured values of the tunnel drive with the design data.

Dr Zollman concludes that "Instrumental tunnel guidance is an essential part of small-bore tunnelling. It achieves major savings of time and operating costs, enhances accuracy, improves safety and has made precise remote control of tunnelling machines practical. It leads the way to further automation for unmanned operation in inaccessible conditions and helps the tunnelling industry to remain competitive."

With such guidance near to hand, how far from the scientific mind lies the self-seeking mole?

SCALE: COST RATIOS

Unfortunately the economics of systems are being judged by the prices sought today. Obviously the prices have to take into account the considerable costs of development. Not only is this bound to be a conservative calculation in the present circumstances but the whole subject of scale runs through the cost equation.

Great strides have been made in the transportation of excavated material by pneumatic and hydraulic means. From Japan, again, we witness the miniaturization of tunnelling wherein the water used for high-pressure face excavation is then employed as the carrier of the waste material.

James Thomson, pioneer of UK pipe jacking, in reporting upon the Japanese microtunnelling scene[4] concludes "There is little doubt that the era of microtunnelling has arrived and that new cheaper rigs and mass production of pipe will bring down the cost of this type of work. As authorities take a more realistic view of the true costs of open cut, an ever increasing share of the market will be carried by microtunnelling."

A clear endorsement of the view that competitive prices, rather than subsidies will promote the demise of disruptive trench warfare. The future trend is spelt out quite clearly in the last sentence of Mr Thomson's conclusions. We have the classic case of price being dependent upon client demand and client demand being dependent upon price. Industry can no longer speculate upon a healthy market — such speculation has been too consistently wrong — it is now up to the client to break this cyclic situation by producing the unstinted support which although clearly intimated is still lacking.

SIZE-FOR-SIZE AND ENLARGEMENT BREAKTHROUGH

Certainly one of the most dramatic recent developments in the pipeline replacement market has been the gas industry's initiated development which the WRC in co-operation with industry have now applied to the sewer and water main business.

The system involves the fracturing of the existing service by the progression of a pneumatic hammer mole which then forces the broken fragments into the adjacent ground as a new pipe is placed within the tail of the mole. The replacement pipe may be of equal diameter to the original and current trials indicate that the replacement can be with a pipe of significantly greater diameter.

Such techniques dispose of the justifiable criticism that most renovation work results in the reduction of pipe cross-sectional area, so affecting hydraulic flow efficiency and reservoir capacity.

Surely such work as this must be a candidate for that 'unstinted support' so that experience in the field can be fast and effective with the export potential for this British development exploited before such opportunities are lost, as happened with the other British first, the bentonite shield.

WHAT WILL THE FUTURE BRING?

We are now on the verge of being able to exploit a whole array of developments which should play a vital part in our NO-DIG future.

The 'chip' commanding pre-programmed tunnel drives and mastering tiny robots in the confines of the service pipe. The laser beam as a datum for the slaving of directional control. The laser beam as a cutting medium for dealing with obstructions and cutting connection entries — and even perhaps converting, *in situ*, silica sands into glass pipes and clay into clayware pipes.

In situ conversion may be far-fetched, but the *in situ* placement of a permanent conduit lining using 'instant set' materials is a close target for innovative development.

Micromoles, minimoles and tunnels, be they jacked, face-built or formed *in situ*; put them to work with self-seeking heads incorporating soil analysis probes and CCTV; computer program them for pre-planned guidance control through a 3-D plot. Can one then foresee the Contracts Manager in his control room monitoring a dozen or so drives for accuracy, rate of progress, geological variations, drive pressures and to keep him awake he could check the directional changes slaved by the tracking head to make sure that the failsafe system hasn't failed. With such records the submission of the final account takes just a few minutes of computer time!

Water ingress during construction has been a dominant factor in tunnel construction. Compressed air, closed-face diaphragms, chemical treatment, freezing and of course the British bentonite shield — they have all had their applications in the costly job of overcoming water pressure dominated drives.

In our new era we have dispensed with the need to provide a suitable underground environment for our tunnel miners and so there would seem to be no good reason why we should not work by remote control under conditions of equalized pressure within the watertable.

THE WORLD OF NATURE

We have only to glance at the world of nature to realize the extent to which we lack the abilities which exist in nature.

Robert Gibbings said that "he liked to feel that we were not superior to other forms of creation but equal to them, each in our own sphere" "—we are inclined to flatter ourselves on our intelligence, assuming superiority because other creatures seemed lacking in powers that we have developed. We seem to forget that they have powers of which our imagination has no conception."

There are in nature senses of smell, location, sight and sound which baffle the imagination and which in human terms would be enormous attributes. In our own simple way we have used scientific means of aiding our own natural senses to provide us with optical devices, directional aids, ultrasonics, radio sound and picture transmission and instrumentation to identify various noxious gases.

We still marvel at the ultrasonics of the bat, the migratory powers of the birds, the homing instincts of animals, the ability of some butterflies to detect sugar in a 1:300,000 solution against our meagre average of 1:200, the sensitivity of bees to ultraviolet light, the moths which can home on a female from a mile or more away when we cannot detect the faintest odour at an inch distance.

Perhaps as time unfolds these miracles of nature, science may eventually simulate such phenomena and extend our own limited dimensions.

EXPORTABILITY

We have an opportunity to lead the world in this new field of engineering, much of which is not in the textbook.

We are now fairly clear as to the need and we appreciate that we have in this country the challenge of widely mixed geology supporting what is most certainly the most complex mass of underground services in the world. We have an exceptional surface traffic density in a compact country where disruption has immediate and serious consequences. We have ingenuity and the ability to design and engineer appropriate products on the basis of the very latest technology.

We have an excellent reputation in public health engineering and indeed have been responsible for founding major underground service networks in numerous cities overseas.

We know that as a nation we must export strongly in an increasingly competitive world market.

What we lack is the foundation of a genuine home market within which we can build our experience and demonstrate our abilities. Without such a facility we cannot develop our technology responsibly and successfully for the justification and the incentive will be lacking.

We can commit resources on the basis of such support as is implied in specious government statements which would appear to be set aside in the pursuance of the broad brush policies to discipline local government expenditure.

If acceptable installed costs are required they will only be produced through the economy of scale. Cox[5] has quoted tendered prices for identical inversion lining contract work on lengths varying between 40 and 400 metres. The long run was approximately one-third of the cost per metre of the shortest run.

We are considering the requirements of quite a substantial proportion of our extensive network of sewers and water mains alone and with economic operation being mainly a function of scale it does not require much imagination to appreciate why so much importance must be placed upon breaking through the current semi-sterile market situation.

THE BENEFITS OF GOVERNMENT SUPPORT

Government, in cooperation with local authorities and utility organizations could produce a fourfold benefit to the country in this new market sector:

It could accelerate the rectification of dilapidated services before more of them reach a point of no return.

By establishing market confidence it could promote methods of renovation and replacement which will be cheaper and more efficient than the current outworn systems.

It could reduce the country's overhead burden by reducing surface disruption.

In total Exchequer terms the cost of getting the NO-DIG concept more widely operational could not only serve Britain here at home but could ensure that Britain became the established leader in a business which has worldwide needs and for which we have an abundance of skills.

IS THERE A NO-DIG FUTURE?

Certainly there is a NO-DIG future.

The question is "Which country is going to accept the challenge of leadership?"

REFERENCES

[1]Department of Transport, *Transport Statistics Great Britain 1972-1982.* HMSO 1983 2.15 *Motor Vehicles Currently Licensed 1946-1982.*

[2]House of Lords Select Committee on Science & Technology, *Inquiry into the Water Industry Vol. 1, Report,* HMSO.

[3]Zollman, Peter, Guidance Systems for Smallbore Tunnelling. *Tunnels & Tunnelling,* Morgan-Grampian Plc.

[4]Thomson, James, Micro Tunnelling, *Tunnels & Tunnelling,* Morgan-Grampian Plc.

[5]Cox, Graham, *New Life for London's Ancient Sewers,* Lecture to ALGS, December 1984, Institution of Civil Engineers.

1.2 The size of the market

James Thomson *Consultant*

In this review of the size of the market initially consideration is given to the existing UK pipeline and duct infrastructure and to the value and volume of all work currently being undertaken. Each of the statutory bodies' activities is considered with a view to building up a more detailed picture of the nature of the work. The international market for underground pipe and duct installation is reviewed with specific information on aspects of work in Japan and Germany. The second section considers the demand for trenchless pipelaying and divides and describes the current techniques into four broad categories. The two preceding sections are related in the final section by considering the current application, including restrictions and drawbacks, of equipment and techniques and makes some broad assessment of the potential scope for trenchless pipelaying.

INTRODUCTION

It is my belief that we today are on the threshold of what is likely to be one of the most significant developments in the civil engineering industry, a development that will compare with the replacement of the navvy by the mechanical excavator. It will radically affect all of us whether we work within the statutory authorities, for consultants, for contractors, for pipe suppliers or for manufacturers of equipment.

But it is the public who will gain the most by not being affected if trenchless pipe methods take over.

In trying to build up a picture of the size of the market, it has been necessary to make a number of definitions and assumptions, all of which are spelt out in the text.

It has been surprisingly difficult on some occasions to obtain reliable information; in fact some fundamental items such as the lengths of sewers and water mains laid in UK annually proved to be not available.

To do justice to the subject on a fully international basis requires a major study which neither time nor funds permit.

It is to be hoped during the conference that further and better information will come to light and that alternative assumptions may be propounded. The challenge of evolving a whole new philosophy of design and construction for pipelaying is going to require the combined wisdom, ingenuity and faith of a great many people.

HISTORICAL DEVELOPMENT OF THE PIPE INFRASTRUCTURE UK

Although this is an international conference, not only in deference to our host country, but because the United Kingdom has virtually a fully developed pipe infrastructure, it is worth looking in closer detail at the state of this system today.

The United Kingdom by virtue of its density of population, the industrial revolution and the consequent growth and prosperity has installed a comprehensive service network. Over 95 per cent of the population is served by the sewer and potable water system with comparably high percentages for electricity, telecommunications and gas.

By comparison with a number of other western countries there is a strong public demand to protect the visual environment so that the percentage of wires and cables installed underground is perhaps higher.

The system as we know it today has been constructed from about 1840 onwards with many of the early pipelines now reaching or exceeding their design life.

Probably 20 to 25 per cent of the sewerage system was installed more than 70 years ago.

Table 1.2.1 sets out a summary of the length of underground mains and their current estimated replacement cost as of 1983.

This table does not take into account the service connections which it is estimated would add a further £27,000 million on to the replacement cost total (based on £1,650 per dwelling).

These essential services are provided and maintained by the statutory undertakers who have both obligations to provide services and rights to lay their pipes and ducts in public highways, etc.

In addition to these utilities, pipelines and ducts are constructed for the oil industry and for the pumping of a variety of other materials, although the length of these lines is comparatively small. Most recently the advent of cable TV has brought a newcomer into the already crowded underground scene.

CURRENT LEVELS OF WORK

Table 1.2.2 gives estimated current annual work load for the various authorities.

Although this workload will fluctuate for each authority as major capital spending schemes are carried out, the figures give a reasonable picture of what is undertaken annually and projections indicate that with the need for replacements, improvements and new developments, the basic load as indicated in the table of work should probably be maintained into the immediate future.

With nearly 16,000km of pipes and ducts plus 1,800,000 services being laid annually, there would appear to be some scope. It is estimated that the total value of this work is approximately one billion pounds and that represents about 30 per cent of the civil engineering construction market in the UK.

About £250 million is probably the value of pipes and ducts installed into the ground which means that £750 million is accounted for by excavation, backfilling, temporary and permanent reinstatement. Figure 1.2.1 which comes from the IPHE report on *Trenchless Pipelaying* illustrates graphically the relative costs of the components in constructing a sewer.

It is worth looking at the individual undertakers' activities to try and get better understanding of the nature of work carried out.

Gas

In the gas industry pipelines consist of 18,500km of transmission pipelines in the 600mm to 1,050mm range and 214,000km of distribution system in sizes ranging from 50mm to over 1 metre in diameter.

It is projected that over the next five-year period, 10,000km of new distribution mains will be installed and 12,000km of mains replaced. In addition 30,000km of new and replacement services will be provided.

Although historically cast iron and steel were the main pipe materials and represent about 87 per cent of the existing network, today over 90 per cent of pipework in both mains and services is in polyethylene.

More than 50 per cent of the work carried out is for the renewal of existing mains and services.

30 per cent of new mains and 58 per cent of replacement work are located in the carriageway.

The differential costs for laying of polyethylene pipe show the significant influence of location.

180mm Polyethylene Pipe	Installed Average Price
Laid in open ground	£22
Laid in footpath	£29
Laid in carriageway	£54

Water

The water resources and sewerage and sewage treatment resources are in England and Wales vested in ten Water Authorities who also have responsibilities for land drainage and flood protection.

The overall annual UK capital expenditure on water resources and supply by the statutory authorities is in the order of £350 million of which probably 50 to 60 per cent is spent on mains and services. In addition private developers for new housing etc. also add a significant amount each year to the network.

It is unusual for water mains to exceed 1,000mm in diameter and about 75 per cent of new work lies in the sizes up to 200mm. Over 50 per cent of work is in the renewal and reinforcement area. Substantial expenditure is also made by the use of *in situ* lining methods for rehabilitation.

The predominant pipe materials are ductile iron and PVC.

Sewers

It is estimated that in the UK annual capital expenditure by the statutory authorities on sewerage, sewage disposal and pollution control is £420 million with about 40 per cent being spent on sewer schemes and of this amount the greatest percentage being allocated to rehabilitation, renewals and reinforcement. In 1982 only about £20 million was spent by the water authorities on truly new work. Probably several times this figure was invested by the private sector on new developments.

About 50 per cent of the existing network is less than 225mm i.d. and 75 per cent of pipes are clayware.

For the current work about 70 per cent of all sewers laid are below 300mm in i.d. and probably over 90 per cent of these are laid in clayware.

Above this size concrete, both reinforced and unreinforced, predominates.

About 5 per cent of the existing network is man-entry size (950mm and above), although new work in the larger sizes probably does not average out as high as 5 per cent. The figure varies from year to year but about 4 per cent or 100 km a year is the average.

About two-thirds of all sewers lie under roads, with the remaining one-third being beneath footpaths, gardens, fields, etc.

It is estimated that about 2 per cent of the sewers are in critical need of repair and something over 3,000 collapses per year are reported.

It would appear that about 8 per cent of sewers are laid below 3 metres deep. There are large regional variations due to local geography.

Electricity

The underground duct system covers both the transmission system, where about 14,500km of duct carry mains from 66kV to 400kV, and the distribution system, which has a length of 360,000km, the largest length, 270,000km, being utilized for the smallest classification of cable, that is 650V and less. Current work follows a similar pattern with two-thirds of the ducts being for the 650V cable or less category.

The majority of electric cables are laid in plastic duct. Occasional large diameter man-entry cable tunnels are required but these do not represent a large volume of work.

For technical reasons it is established practice to lay electric and telephone lines at shallower depths than the pipe services. Consequently these services are more prone to damage.

A survey carried out by the National Joint Utility Group estimated that the annual rate of incidents for all services was 73,000 incidents per annum with repair costs of £15 million, without any customer inconvenience cost or social cost being added into the calculation.

Telecom

Although it is feasible to lay cables directly in the ground, the majority of telephone cables are laid in ducts. It is policy to provide duct space for long term needs and to facilitate repair and replacement.

For many years the standard 3⅝″ (90mm) earthenware duct in single, 2, 4, 6 and 9 way configurations was used. Due to emission problems in manufacture, production ceased in 1968.

Increasingly PVC or polyethylene is now being used. A 25mm duct is used for customer connections.

On main duct work 90mm plastic is used singly or in formation, and an intermediate 55mm size has been developed.

A specially developed clayware duct is now being manufactured and used as an alternative in the 90mm size.

THE INTERNATIONAL MARKET

In looking at the market in more global terms, it is necessary to restrict our viewpoint somewhat to those areas on which we can forecast a positive need for such equipment and techniques based on the level of development, and secondly exclude some areas of the world, such as the Eastern Block where demands and trading can be distorted by political issues and make it difficult to forecast.

It has also been decided to exclude developing and less-developed countries and concentrate on what is broadly known as the Western market. It is fully appreciated this is a very arbitrary definition and before anyone takes me to task I too can cite the work carried out in places like Singapore, Thailand, India, Egypt and the Middle East. That a substantial market exists outside the areas I have delineated is for sure, and it is a market that will have an increasing demand for such techniques in the future years. To reiterate I am talking in terms of Western Europe, N. America, Japan, Australia and New Zealand.

Before considering the likely overall demands it is worth highlighting some specific information from various parts of the world.

Japan

By comparison with European countries the pipeline infrastructure is less well developed in Japan. Roughly 1½ times the area of Britain, its 120 million population live in 18 per cent of the area, mainly along coastal plains. Many of the cities have been built in the area of alluvial river deltas which add to the problems of the underground infrastructure.

It is stated that 70 per cent of Japanese homes still lack main drainage connections despite the massive programme of soil construction work promoted by the Japanese government. In the period 1978-82 the average length of sewer laid per annum was 7,000km and 75 per cent of the total is in diameters less than 300mm. 10 per cent is in diameters over 1,000mm. This high percentage of large diameters is now falling as the bigger schemes in cities are completed. The recurrent annual expenditure on sewer pipelines is in the order of £3,700 million which alone is five times what is spent in UK on all pipelines. Pipeline materials in the sewerage field are 50 per cent concrete with clay accounting for 15 per cent, PVC 26 per cent.

Germany

A country with a similar population to the UK, with a long tradition of infrastructure development and with a high percentage of population connected to public utilities.

Despite having a population and geographical area which is about 10 per cent greater than the UK, the annual investment in civil engineering work is something like four times that of the UK.

A programme of sewer installation to bring up to 90 per cent of the population on to the public sewer system involves a total of 30,000 to 40,000km of new sewer which has to be laid before the year 2000. The total cost is budgeted in the order of £15,000 million which is £1,000 million per annum. Again this is a greater figure for sewerage than is spent in UK per annum for all pipes and ducts.

Europe

With the sparse information at our disposal it would appear that the markets in Germany, France and Italy are all likely to be substantially bigger than that of the UK.

Looking at the overall picture for Europe and Scandinavia, I estimate based on demographic figures and values of civil engineering construction that the total annual length of pipelines to be installed for all utilities is not less than 150,000km per annum with a current annual investment of £7.5 billion with probably £4.5 billion being spent on the water and sewerage.

The Western World — the overall market

Taking the European figures and adding Japan at its present level of infrastructure development and with the work being undertaken in N. America and to a lesser degree in the Antipodes, the annual market is probably 450,000 to 500,000km a year of all types of pipeline and duct with an overall expenditure of between 20 and 25 billion pounds.

THE DEMAND FOR TRENCHLESS PIPELAYING

From the information on the UK one begins to obtain a picture of the needs and demands of the authorities and of the public.

With over 16,000km of pipes and ducts being laid a year, and from the location patterns, probably at least 8,000km (50 per cent) will be located in highways. This is bound to lead to a great deal of inconvenience, delays and damage.

When an authority has to cross a rail track or a motorway, the disturbance, social costs and damage virtually eliminate using open cut methods.

As there is no penalty or convenient mechanism for charging by the road authority to the undertaker who wishes to construct the pipe, the incentive to find an alternative route, use a trenchless method or even to minimize the work period using open cut, does not exist.

It has been calculated that with trench construction, in the roads of inner cities and towns, the social disruption costs are on average three times the construction costs and this multiplier can be as high as 10 times.

The government are now encouraging authorities to take into account social costs in their comparative budgeting when considering such work. At the risk of antagonizing my friends in the statutory authorities, I would propose that the government be encouraged to introduce legislation to enable road authorities to charge statutory undertakers a fee based on traffic densities, number of disturbance days and the extent of damage. Apart from anything else, it would give encouragement to the efficient contractor to put forward non-disturbance or at least minimum-disturbance methods.

It should also be pointed out that the open-cut system of pipelaying does lead to an excessive number of instances of damage to other services which to a great degree could be avoided by the use of trenchless methods.

A third point that has already been touched upon is the sheer illogicality of undertaking trenching methods to lay a pipe in the ground. With something like £1 billion spent a year in UK on pipes, ducts and services and with only £250 million spent on providing and laying the permanent hardware, £750 million is devoted to breaking up the surface, excavating the soil, backfill and reinstating the surface.

If one applies Pareto's theory to new and replacement market then 20 per cent of the length will be critical and absorb about 80 per cent of the cost.

This means that we should concentrate out thoughts on the most difficult 3,200km in UK terms which is likely to be deeper, more difficult, in congested locations and probably with unfavourable ground conditions. The cost of this work using our theory will be 80 per cent of £490 million (that figure is after deduction for the provision and laying of the pipes) which will result in a sum of £390 million. It is within this critical 20 per cent that the majority of problems lie and which offers the greatest opportunity to trenchless pipelaying techniques.

The same demand is mirrored in the rest of the western world where society is becoming increasingly intolerant of the delay, costs and inefficiencies of digging up the road.

Applying Pareto's theory to the total market we could be considering a total critical length of around 100,000km a year with a potential cost in the region of £15 billion.

TRENCHLESS PIPELAYING

The time has come to now look at the other side of the equation and loosely define what is meant by trenchless pipelaying techniques. I am conscious that a number of papers follow this one setting out in detail many aspects of special techniques, but in order that we may consider the potential market, I have taken four broad divisions.

1. Techniques to lay pipes and ducts up to 250mm

Typical techniques of trenchless methods in this category include mole boring (percussion boring), horizontal drilling, auger boring, etc.

Both cased and uncased boring can be undertaken and lengths tend to be under 50 metres.

Accuracy requires to be reasonable rather than precise.

A relatively low value of the work being undertaken in this division demands simple, less sophisticated equipment which can be rapidly installed and operated from small pits and can operate economically using local labour rather than specialist operators.

2. Techniques to lay pipes and ducts from 350mm to 950mm

Typical techniques which exist today are:

(a) The pilot pipe method, i.e. the Komatsu Iron Mole
(b) Remote control slurry shield-tunnelling techniques, i.e. Iseki, Telemole and Telemouse.
(c) Auger bore — KA-MO, American Auger, etc.

A watertight corrosion-free pipeline is a necessity and some techniques employ temporary liners which are replaced with the permanent pipe or duct whilst others use a single-pass system.

Within this division there will be a demand for equipment capable of undertaking lengths of 100 metres and above.

The machine should be capable of being directionally controlled during boring and achieving an accuracy of 25mm radius of the target.

The machine should be capable of working within a variety of ground conditions and water tables including dealing with obstructions of differing types.

There may well be subdivisions within the range of equipment as many occasions will arise where the full sophistication is not required and the economies of installation and benefits of lower cost will be the most important factor.

3. Techniques to lay pipes and ducts from 950mm upwards

These techniques for trenchless pipelaying are well known and established and include all the variations of tunnel driving and pipe jacking.

4. On-line replacement

In the field of pipe rehabilitation we have also the possibility of replacing a defective pipe or duct by use of mains bursting techniques which tow in behind a plastic replacement sleeve.

This technique has already found great use in the gas industry where it was pioneered and it is expected to find increasing use in the water industry and perhaps in other areas.

It is understood that the current range of applications is up to 200mm.

APPLICATIONS OF TRENCHLESS PIPELAYING TO THE MARKET

In order that we can determine what share of this vast volume of work can be undertaken by trenchless pipe methods, it is necessary to determine what part is being played at the present time.

The picture varies substantially from country to country. One area in which trenchless pipelaying has been used universally and is overall the dominant method is in pipes over 950mm.

The techniques of tunnelling and jacking are chosen for their economy, convenience and lack of disturbance.

In Japan for example it would appear that about 500km a year or around 80 per cent of the large diameter pipe and duct is laid by trenchless techniques at this time.

In UK about 50km a year of segments and pipe jacking are installed but with the revitalization of pipe jacking this figure could be increasing at the expense of open cut.

Looking in overall terms at the market it is likely that the annual length of pipe and ducts laid over 950mm for all purposes could be in the order of 5,000km and that a guestimate of the use of trenchless methods would be 50 per cent of this length or say 2,500km, which represents an existing market of something like £1,500 million or more a year.

With improving techniques and hopefully more attractive prices to the client it can be envisaged that an increasing share of the market, probably up to something like 75 per cent of all pipes laid, could be achieved.

In the middle range from 350mm to 950mm lie the most uncharted waters. The two countries which have undertaken the greatest amount of work are Japan and Germany.

In Japan it is stated that 200km of small pipe is laid annually and there are 20 types of microtunnelling equipment. In Germany the techniques are still very much in the trial stages but about 15 systems are in use or are currently being developed.

The estimated total of all pipes and ducts laid in this range of sizes is currently around 50,000km a year and for a variety of reasons, but mainly cost, about 25 per cent of these could lie in the critical area and thus be favourable to trenchless pipelaying.

At the present time there appear to be two major restrictions to increasing the share of the market by microtunnelling methods.

Firstly the cost which at this time appears excessive due to the high capital cost of machines, expensive special pipes and shaft constructions. All these costs will with competition, rationalization, and experience reduce substantially for the benefit of the customer, thus opening up an increasing share of the market.

Another problem is more technical and is in finding a method of installing lateral connections economically and effectively.

Other technical problems such as dealing with stones, obstructions, the control of line and level, the advance warning of obstructions all appear to be either solved or well on the way to a satisfactory solution.

The final range of sizes is for pipes up to 250mm and it is virtually impossible to come up with any meaningful lengths or values as so much of this work is in new or replacement service connections to properties. Records both of lengths and costs are either sparse or non-existent.

What is for sure is that the volume of work is immense albeit the individual cost of each job is probably well under the cost of one metre of trenchless piping of a larger size.

The UK Gas Board have been leaders in adopting these techniques together with pipe bursting, which is our last category of trenchless methods.

They estimate that in 1982/83 33 per cent of mains replacement and 39 per cent of service replacement was undertaken by these methods. Applying these figures to their quoted statistics this indicates 1,000km of replacement mains and over 2,000km of service replacement. Furthermore, they estimate that by 1988 50 per cent of mains and service replacement will be done by using trenchless methods.

It is also of great interest that British Gas have over 800 pneumatic earth-boring machines.

This pattern in British Gas is confirmed in equivalent US gas utilities where a survey in 1984 showed that 30 per cent of installations for gas distribution pipes were undertaken by trenchless methods.

Presumably what is good for gas could be applicable to all the other utilities albeit with adaptation to suit their specific needs and problems.

As the overwhelming majority of pipes to be installed lie in this category, the potential market probably runs into hundreds of thousands of km annually in the western world, but it is also obvious that systems will have to be evolved which allow local contractors and maintenance men to undertake the work rather than specialists.

CONCLUSIONS

A number of conclusions result from the consideration of the information on the market.

1. Whatever assessments one uses in determining the size of the market, potentially a huge volume of work is ripe for the application of trenchless methods.
2. The 'Trenchless Revolution' has already started, which is evidenced by the work underway in Japan and Germany in the middle sizes and the interestingly large percentage of work undertaken by the Gas Authorities in UK and USA, this work being in the smaller sizes and by on-line replacement methods.
3. There is little doubt that when the true costs to the community of traditional trenched pipelaying are taken into account then the viability as well as the desirability extends to many more situations.
4. Although a great deal of pioneering work has been done and great progress made in the development of equipment and techniques, much remains to be done particularly in the reduction of costs.
5. It is significant that the major advances in the field have been made in those regions where close co-operation exists between contractors, consulting engineers, public officials, equipment manufacturers, pipe suppliers and elected representatives.

It is regrettable that in so many cases the decision-making process is often based on short term self-interest, politics, prejudice, hidebound tradition or sometimes downright ignorance.

REFERENCES

A wide range of sources were utilized including:

[1]Institution of Public Health Engineers, *Trenchless Pipelaying,* Report of the Institution's Working Party.

[2]Tunnels & Tunnelling, *Small Bore Tunnelling and Pipe Jacking*, Supplement October 1984, various articles.

[3]Nitze, R T, The Public Utilities Street Works Act 1950, *Public Health Engineer, 13*, 2, 63-66, April 1985.

[4]Cox, Graham, Myths & Mythologies of the Sewerage System, *Water and Waste Treatment,* December 1984.

[5]Glennie & Robinson, *The Sewers of England and Wales — The System and its Structural Problems.* WRC Report 138E, June 1984.

[6]Smith, B J, Developments in Trench Excavation and Reinstatement *Public Health Engineer, 13,* 2, 73-76, April 1985.

[7]Water Authorities Association *The Water Industry in Figures,* leaflet.

[8]Clow & Stenton, External Plant *P.O. EE Journal,* October 1981.

[9]Electricity Board, *Handbook of Electrical Supply 1982/83.*

[10]Stein, Bielecki & Maass, *Small Diameter Jacked Pipes. Developments and Experiences from Hamburg,* Europipe Conference Basle 1983.

[11]Hale & Dean, 'Time to Renew' *Pipeline and Gas Journal,* July 1984.

[12]European Community Contractors, *Construction in Europe. Statistics 1970-1983.*

[13]Yorkshire Water Authority, *Statistics on Sewer and Water Main Construction 1983/84,* Internal Document.

TABLE 1.2.1. Length of underground mains and current replacement costs

Authority	Length of Mains km	Replacement Cost £M
Gas	235,000	15,000
Water	320,000	27,000
Sewers	250,000	60,000
Electricity	375,000	15,000
Telecom	400,000	12,000
TOTAL	1,580,000	129,000

TABLE 1.2.2. Estimated annual work load, UK

	Gas	Water	Sewers	Elect.	Tele.	Total
New & Replacement Mains km	4,500	3,000	2,500	3,200	3,000	16,200
Approx. Annual Value £M	140	150	225	100	90	705
New & Replacement Services 1000's	700	240	200	210	450	1,800
Estimated Value £M based on £200 per average service						360

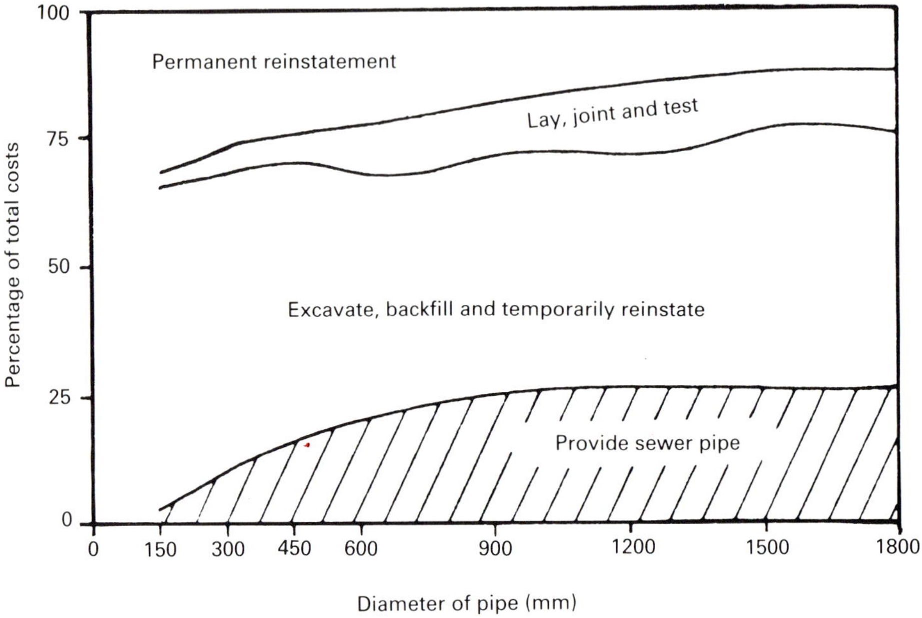

Figure 1.2.1. Costs of provision of pipe as a proportion of total cost.

1.3 Microtunnelling in Japan

S Tohyama, Dr-Eng *Vice President, Japan Sewage Works Agency (Nippon Gesuidoo Jigyoo-Dan)*

The Japanese Government is continuing to place a major priority on sewer construction as part of its policy for improved amenities. This has generated a rapid increase in the use of microtunnelling methods for the installation of small diameter (250 to 350mm) sewers. Over twenty microtunnelling machines have already been developed and the highly competitive market ensures continuous process improvement.

Each of these systems eliminates workers from the in-pipe space by using remote controls. The systems can be classified into five broad categories, according to their mode of excavation and earth/rock/muck removal, namely 1) Thrust 2) Auger 3) Rotating Pipe 4) Waterjet-aided and 5) Other.

In addition to their application to sewer construction, microtunnelling techniques show great future potential for the installation of conduit systems to carry Japan's planned high speed fibre-optics communications network.

INTRODUCTION

In Japan, for the past twenty years the construction of sewers and sewage treatment facilities has been strongly promoted as an important national policy. (Figure 1.3.1). However, for the past several years the Japanese national government has been suffering from large budget deficits, and every effort is being made to set its finances in order. Therefore, spending on public works has been reduced, entailing a corresponding fall in spending on sewage works. Nonetheless, investments in sewage works amounted to Y1,500,000 million (approximately £5,000 million) for FY 1984, in which Y1,100,000 million is for sewer construction and Y400,000 million for construction of sewage treatment plants.

In recent years the demand for sewer pipes in Japan (Tables 1.3.1 and 1.3.2) has switched from large to small diameters for two main reasons. Firstly there has been a trend towards construction of separate collection networks for sanitary sewers and storm drains with ninety per cent of current expenditure concentrated on the sanitary sector. Secondly the focus of attention is shifting to small and medium sized cities now that the largest urban areas such as Tokyo, Osaka, Nagoya and Sapporo have reached reasonable levels of sewer construction.

Sewer pipes are laid under roads to facilitate their use by the premises along both sides. In heavily urbanized areas, water supply pipes, gas pipes, telephone and electric cables have already been laid in place. To carry out sewer projects, which necessarily involve laying sewer pipes in deep locations, it is necessary to fully protect existing underground utilities and sometimes to temporarily relocate them. When the cut and cover method is used, adverse effects may occur depending on soil and other conditions; pumping up of ground water could cause nearby premises to settle; pile driving and extracting could cause vibration; or construction could impede traffic flow — all of which could easily arouse the opposition of local residents. In the case of large sewers in major urban areas, construction has been possible by using shield-tunnelling methods to dig tunnels into which workers can enter to carry out excavation and undertake survey techniques. As experience has grown techniques such as constructing sewers with an inside diameter up to 8,000mm (o.d. 9,700mm) by pressurized slurry shield have been developed.

However, in the construction of small-diameter sewers in smaller urban areas, it has been necessary to develop tunnelling methods to excavate, remove excavated materials, and lay pipes by means of remote control, with a great degree of precision. In Japan, for safety reasons it is not permitted to enter tunnels unless internal diameters are above 800mm. Contractors and equipment manufacturers who have had a great deal of success with the shield-tunnelling and pipe-jacking methods have co-operated fully with each other in attacking this problem. A number of new small-diameter jacking methods have been developed over the past ten years. Currently about twenty types of method are in use and about 200 km of small diameter pipes are laid by these methods annually.

Some of the typical small-diameter pipe-jacking methods currently in use in Japan are discussed below.

SMALL DIAMETER PIPE-JACKING METHODS

Small-diameter pipe-jacking methods can be classified into the following five types, depending on whether they use the excavation or thrust-boring methods. (See Table 1.3.3).

'Jack-in' method

The 'jack-in' method is generally used in soft ground conditions such as soft soil and sandy strata. Two types, the 'one-process' type and the 'two-process' type, are in use. The 'one-process' type is used to install steel pipes or small-diameter concrete pipes directly by a hydraulic jack. The 'two-process' type involves inserting a pilot or guide pipe at first followed by steel or small-diameter concrete pipes, and is the most commonly used method.

Each is provided with its own automatic control equipment for direction; thus correction of direction is possible. The details of the methods may differ depending on equipment manufacturers, but the pipe diameters for use are from 250mm to 700mm, and boring lengths of up to 50m are possible.

Two typical methods within the category of the 'two process' type jack-in method are discussed below in more detail.

Pilot pipe method

A pilot pipe is driven through either by jacking or excavation. The reinforced concrete pipe which follows the pilot pipe is driven using screw or slurry (balanced water pressure) mucking systems.

As a pilot pipe, steel pipe with an outside diameter of 216mm is used in the case of the jack-driving type and a steel pipe with an outside diameter of 230mm in the case of the excavation type. The degree of accuracy of the pipelaying depends on how well the pilot pipe is thrust forward. The pilot head has a structure allowing a maximum direction shift of 3 to 3.5 degrees, and with every 40cm of progress, a correction of direction can be made.

Since the 'pilot head with excavation' type was developed in 1982, construction has become possible even under soil conditions where previously supplementary methods were required. It is particularly effective under conditions of mixed soil strata or high N-value.

The balanced water pressure mucking system is most effective for excavation in sand strata with high water levels. Recently, a type for use in gravel layers (maximum diameter of gravel 80mm) has been developed, bringing the total number of possible types of the method to five as shown in Table 1.3.4. In practice, it is important to select a suitable type to meet the actual soil conditions.

The advantages of this method are as follows:

(1) Because equipment can be disassembled, shaft area can be small. Even if there are obstacles in the upper portion of the available shaft space, equipment is still relatively easy to install.

(2) The method is applicable under a broad range of soil conditions, with N-values from 0 to 50.
(3) By simply changing adaptors, work is possible with concrete pipes of inside diameters from 250mm to 900mm and with steel pipes from 200mm to 1000mm, which saves equipment costs.
(4) Work is possible even under high water levels from 4m to 10m.
(5) Good driving accuracy is possible over distances of 65m or more.

Lead pipe method

Dual lead pipes (inner pipe with 150mm diameter and outer pipe with 200mm diameter) capable of fine directional adjustment are used. The inner pipe is linked to the central axis of the jacking machine and rotates. The head installed at the end of the lead pipe has a built-in target. The jacking machine drives the pipes and corrects direction by surveying the target from the transit installed in the launching shaft. The head can be easily changed from percussion to pressure-cutting depending on soil conditions. After the lead pipe has made the breakthrough, the head for laying concrete pipes is mounted at its rear, and pipes are thrust forward one after another. In the case of pressure-cutting, the muck is carried to the launching shaft between the external and the internal pipes by the rotation of the screw of the internal pipe shaft. During cutting, pressurized water is jetted into the interior of the head and lubricant is injected around the circumference of the rear portion of the pipe.

The advantages of this method can be summarized as follows:

(1) Pipe diameters may range from a steel pipe with a diameter of 150mm to a concrete pipe with a diameter of 600mm, and a thrust length of more than 50m can be made.
(2) The lead pipe moves forward within a ±2mm range of deviation from the centre line given by the installed transit, and keeps a straight line without wiggle, thus ensuring high accuracy of the trailing pipes.
(3) Correcting the direction of the lead pipe and operating the thrust machine is easy, which facilitates operator training.
(4) A number of different heads are available for soils ranging from hard to soft. Even if soil conditions change abruptly during thrusting, the internal pipe together with head can be retracted and a suitable head substituted.
(5) This method permits the smallest shaft area.

Auger method

In this method the interior of the pipe is equipped with a screw head and auger screw, and these rotate to excavate the face.

The auger method includes two types. In the 'one-process' type the lead pipe is thrust forward, immediately followed by and connected to the main pipe (steel pipe or concrete pipe). In the 'two-process' type, after thrusting the lead pipe including the guide pipe through to the arrival shaft, it serves as a guide for the advance of the main pipe. The latter method is commonly applied for laying sewer pipes.

These methods are applicable under soil conditions including relatively hard cohesive and sandy soils, and sand-gravel composites. Pipe diameters can range from 250mm to 700mm and thrust lengths of 60m are possible. Corrections of direction can be achieved by means of attached automatic direction controls.

This method includes the lead pipe thrust method and the sheath pipe thrust method. These methods can be said to be horizontal auger methods with steering control by means of screw augers. This equipment is designed for use under all soil conditions and by switching cutting edges, thrusting is possible in sand-gravel strata and hard soils. The standard thrust length is 50m. Diameters of steel pipes can range from 350mm to 800mm and concrete pipes from 250mm to 700mm. When the sheath pipe type is used, a manhole of 1.5m diameter is sufficient for the arrival shaft.

Advantages of this method are as follows:

(1) Its simple composition facilitates operations. Double pipe operations for advancing the lead and the auger make this a system in which jointing is easy.

(2) It is applicable in sand-gravel strata as well as in hard soils. Auger cutting edges can be replaced or repaired during operations. It can be used under difficult soil conditions by selecting suitable attachments.

(3) During thrusting, constant checking for wiggle is possible and its correction can easily be achieved by using a power wrench to change the position of the cutter.

(4) The problem posed by waterbearing strata can be resolved by using an auger cutting edge covered by a bulkhead at the face, or by using a short pitch auger. When the hydrostatic pressure of waterbearing strata is too great for the use of the above types, then tunnelling is still possible by using the water balance method to equalize the water pressure within the thrust pipe to the underground water pressure.

(5) In rocky ground, auger cutting edges can be replaced with 'toricon' bits or with air hammer bits.

Horizontal boring method

This method includes two types. One is the single casing rotating system in which the pipe to be laid is equipped with cutting edges at its front and thrust forward with jacks while being rotated.

This method can be thought of as horizontal boring and can cope with hard grounds and sand-gravel mixed grounds with pipe diameters of 1500mm or less. Only steel pipe can be used as the pipe to be laid must rotate. Boring length is approximately 50m and this method is used for sheath pipes for road crossing of underground utilities as well as the pipe roof method.

The other is the double casing rotating system in which the rotating casing rod is installed in the pipe to be laid. The pipe is thrust forward while the rotating casing rod excavates the ground. Steel pipe is used for the main pipe, and after completing the excavation the internal casing rod is extracted.

Although these methods lack steering ability, they are highly appropriate for the pipe roof method.

Slurry shield-tunnelling method

This is a method to drive a tunnel by remote control of a small-diameter slurry shield machine. This method uses a rotating cutter for excavation, and the muck is converted into slurry and conveyed to the surface by a pipeline. The muck is then separated by treatment facilities and the slurry is recycled.

Advantages of this method are as follows:

(1) Construction is fast due to the high speed of thrusting and use of the 'one-process' system.

(2) It is suitable for sandy soils with high water tables.

(3) It offers high tunnelling accuracy due to constant monitoring of direction by its television camera.

(4) Operating techniques are relatively simple and require only a short training period. One-man operation is possible.

(5) The combination of pressurized slurry and mechanical earth pressure balanced types minimizes effects on the surface.

CONCLUSION AND FURTHER STUDY NEEDS

The reason why Japan's microtunnelling technology has advanced very fast and is gaining worldwide attention is the huge demand for this kind of technology. The construction of smaller-diameter sewer systems is proceeding intensively, particularly since the introduction of separate systems and the switch of emphasis to smaller and medium-sized cities. As shown in Figure 1.3.2 more than 500km of sewers are laid annually by the jacking method and a breakdown of the construction of small-diameter sewer pipes by microtunnelling for the last five years is shown in Figure 1.3.3. It can be said that microtunnelling technology has been born out of our successful results.

As mentioned above, microtunnelling includes a large variety of types and by combining applications and types, even more varieties can be created. However, the first condition for developing microtunnelling methods is that since within pipes no human operations are possible, the equipment used must have the highest reliability and must be easy for operating personnel to handle. Most equipment used currently has automatic direction control equipment installed. However, direction correction differs depending on soil conditions and needs certain levels of skill. Thus, results are influenced by the quality of the operators. Therefore, some sort of operator qualifications would be necessary as well as establishing suitable training programmes.

Further study needs are summarized as follows:

(1) It is hoped that simplified equipment will be developed in order to prevent excessive intake of muck during excavation in water-bearing soft sand strata.
(2) Excavators for soft rock and gravel strata are under development. However, there is a need to complete research in order to develop better direction correction equipment.
(3) It is hoped to develop more powerful equipment that would permit boring distances of up to 100m under various soil conditions.
(4) There is a need to reconsider the standards on pipe strength and thickness as well as joint structure to prevent water leakage.
(5) There is a need also for research on shaft equipment that can be installed in a short period of time, because at present the boring distance is rather short, making the number of shafts excessive.
(6) It is hoped to develop boring machines that do not depend on supplementary methods of chemical injection.
(7) In 1979 standards for concrete pipes for microtunnelling were stipulated for diameters of 250mm-500mm. It is desirable that these standards be extended to pipes with diameters up to 700mm.

There is little hope that from now on the environment for construction of sewer projects will improve. It will therefore be necessary to improve efficiency of the work through the continuing effort to improve and develop special technologies of this kind.

The development of methods depends on co-operative research efforts by equipment manufacturers, constructors and pipe manufacturers, in response to the needs of users. We are happy to note that adequate joint efforts have been and are being made.

TABLE 1.3.1. Lengths of pipes used (classified by kinds)

(unit: km)

Year	Clay Pipe	Vinyl Chloride Pipe	Hume[1] Pipe	Jacking Pipe	Steel Segment	Concrete Segment	Prefabricated Box Culvert	Others	Total
1978	1,115.1	1,333.0	2,703.8	394.2	141.1	86.9	82.1	109.5	5,965.7
1979	982.2	1,398.6	3,282.2	417.8	177.4	81.2	111.4	129.1	6,579.9
1980	1,027.9	1,705.6	2,913.0	387.3	176.6	56.2	89.4	474.2	6,829.8
1981	1,125.2	2,077.0	3,192.9	409.1	134.8	49.2	108.2	345.9	7,442.3
1982	1,152.6	2,519.1	3,399.8	521.0	139.0	46.8	97.9	320.9	8,197.1
Total	5,403.0	9,033.3	15,491.7	2,129.4	768.9	320.3	488.6	1,379.6	35,014.8

[1]Steel reinforced spun concrete

TABLE 1.3.2. Lengths of pipes used (classified by diameters)

(unit: km)

Dia. of Pipe Year	Less than 200mm	250-300mm	350-450mm	500-900mm	1,000-1,850mm	2,000-2,800mm	3,000-4,800mm	More than 5,000mm	Total
1979	1,497.6	3,151.5	496.1	673.5	527.2	141.6	85.5	6.9	6,579.9
1980	1,872.4	3,151.7	445.5	663.5	501.1	148.6	38.0	8.9	6,829.7
1981	1,980.5	3,682.2	523.2	622.0	452.6	140.6	31.5	9.7	7,442.3
1982	2,210.7	4,115.7	503.4	691.4	493.4	140.5	37.5	4.5	8,197.1

TABLE 1.3.3. Classification of microtunnelling

Method	Type
Jack-in method	One-process type (Direct insert)
	Two-process type (Pilot pipes insert)
Auger method	One-process type (Direct insert)
	Two-process type (Pilot pipes insert)
Horizontal boring method	Single layer casing rotating type
	Double layer casing rotating type
Slurry shield method	Pressurized slurry type
	Hydraulic water balance type
Others	Pulling type
	Others

TABLE 1.3.4. Soil conditions and combinations of methods

System	Pilot pipe	Concrete pipe		Soil conditions								
				Soil viscosity		Silt		Sand/sandy soils				Gravel
			N-value	0–20	20–50	0–20	20–50	3–13		13–50		30–50
			Water Pressure	—	—	—	—	kg/cm^2 0–0.4	0.4 –0.6	kg/cm^2 0–0.4	0.4 –0.6	0.4 –0.6
Type I	Jacking type	Screw mucking		†	X	†	X	†	X	X	X	X
II	Excavation type	Slurry mucking		0	0	0	0	0	0	0	†	0
III	Excavation type	Screw mucking		0	†	0	†	0	X	†	X	X
IV	Jacking type	Slurry mucking		0	X	0	X	0	†	X	X	X
V	Excavation type	Slurry mucking		X	X	X	X	X	X	X	X	†

Constructability: † Most suitable 0 Suitable X Not suitable

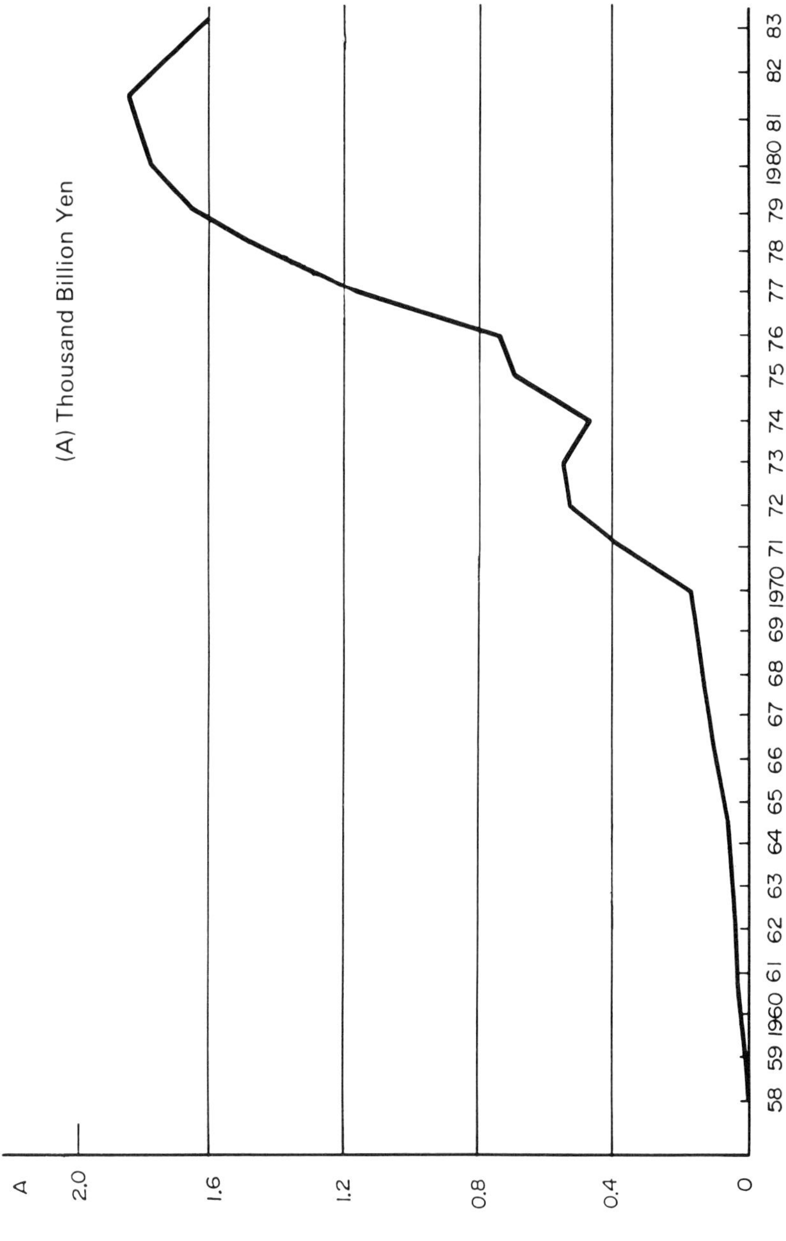

Figure 1.3.1. Investment in sewage works in Japan.

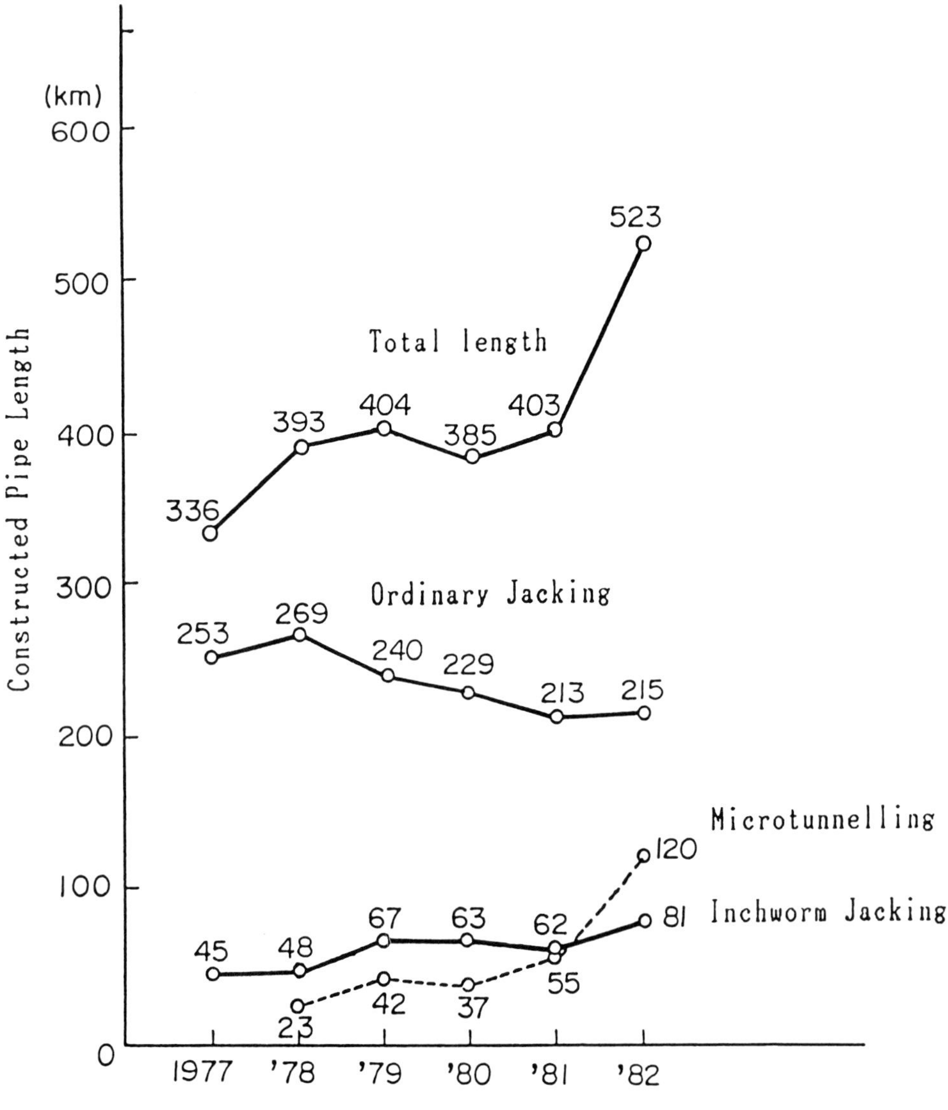

Figure 1.3.2. Construction of sewers by jacking methods in Japan.

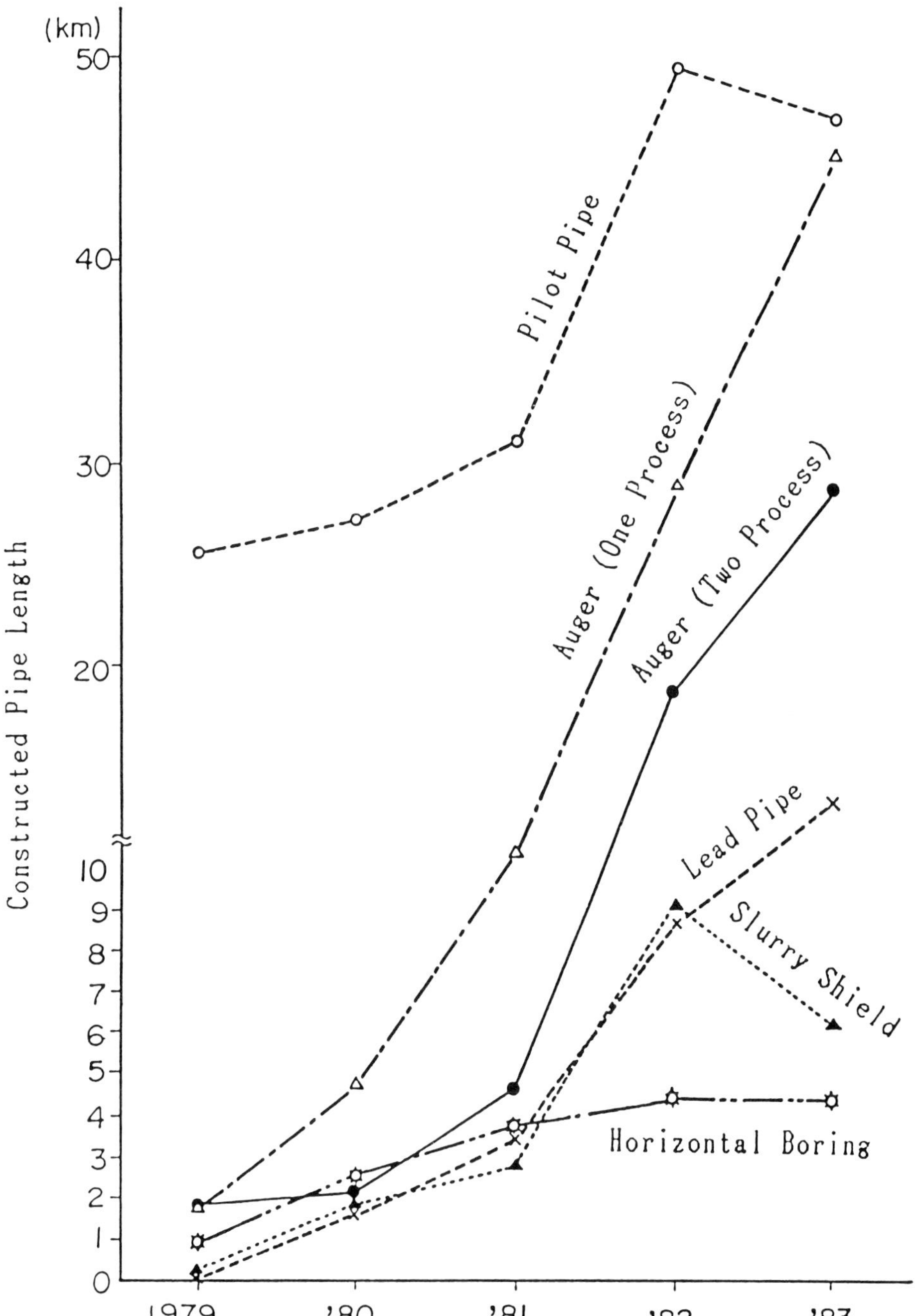

Figure 1.3.3 Lengths of pipes constructed by microtunnelling, 1979-83.

Authors' introductions

Mr. Donald F Rees: Is there a NO-DIG future?

The written paper emphasizes the NO-DIG theme as a realistic endeavour to protect and expand our underground infrastructure in an elegant and non-disruptive manner. The need is there — the technology is there (in fact or potentially) and some very brave investment has been committed in support of some very remarkable developments.

Specious statements have been made implying strong support for new technology, but below ground such support has been more vocal than practical, and far too nominal to justify the considerable investment already made by many somewhat disillusioned enterprises.

Industry *must* be able to recover its investment within a reasonable period and must see a realistic profit opportunity — that's what business is all about!

Developments of the kind which are presented here this week, both from the platform and in the Exhibition Hall, require the support of a healthy home market, and we are fortunate in having the finest proving grounds in the world in which to carry out our trials and perfect our techniques. We have the benefit of Water Research Centre's technical and practical concentration of effort and we welcome the close relationships which they are developing with industry.

But then, I wonder, how much of this technical and industrial effort will be allowed to fall by the wayside due to market frustrations?

I have suggested that the subject of underground assets should be a matter for all-party agreement, for a century of neglect cannot be remedied within the lifetime of a governing party.

The All-Party House of Lords Select Committee on Science and Technology examined our water problems two years ago, when they said that there was a significant risk of decay getting beyond the Water Authorities' control and that the current recession made the task more daunting but did not remove the need to tackle it.

To progress effectively we must have a long-term policy and a long-term strategy with secure funding alongside. History has shown us the extent to which day-to-day political pressures have resulted in the relentless decay of our vast underground assets.

If the political parties cannot agree such a route in the overall national interests, then there could be a case for distancing this vital sector from the Town Hall battles which are raging and place the responsibilities in an entirely separate pocket, empowered to provide that long-term security which would inspire confidence within the whole of the 'underground' industry.

I have expressed concern about the manner in which innovative enterprise can be seriously damped down by the inherent caution of the employer or his consultant — for they are not paid to take risks.

There is also the danger that the ambition of the Water Research Centre to achieve technical perfection will have a similar effect, for they in turn are the focal point of guidance to the industry, in which role they carry a considerable burden of responsibility. It is an unenviable position — it is a cleft stick. They must give sound advice and at the same time encourage new innovative developments — where risks abound! All within a strictured market lacking a contract approach which recognizes the risks involved.

Of course, there are risks associated with the learning curve of new developments. They are risks which need to be taken — and they are risks which need to be shared.

It is interesting to note that the most dramatic solutions have been spawned through the commercially motivated Gas Industry, which had the Need — produced the Market Scale —

accepted the Risk Factor — and got on with the job at an impressive pace. Surely there is a moral there!

I believe that there is scope for our consulting engineers to team up with commercial enterprises to provide a competent Design/Construct approach to this industry, working to Performance Specifications. This has the advantage of combining technical design discipline with the initiative and thrust of responsible commercial enterprise. This could remove the tendency of being drawn into a rather stereotyped approach to our remedial and construction techniques.

IF we are to accelerate the rectification of dilapidated services and expand them to accommodate demand;

IF we are to promote investment in methods which are both economic and efficient;

IF we are to reduce the enormous burden of traffic disruption and social costs;

IF we are to fully exploit the inherent abilities which exist within all of the facets of this industry;

THEN we must seriously review our governmental policies and our current practices of designing, specifying, documenting and executing contract works.

Let Engineers engineer — and give them the authority so to do.

Let the Contractors and Manufacturers exploit their initiative and enterprise.

Let the Employer associate himself with the Risk elements and let his Engineer regain *his* authority which has been eroded over the years.

To summarize my paper:

I would applaud the development of a rolling programme of productive work with adequate support funding to produce investment confidence.

I would like to see our underground activities extracted from the tug-of-war of the political forum and drawn together under an imaginative long-term policy and strategy.

I would plead for a sufficiency of market scale which will ensure a genuine economy of scale.

I would ask that every underground project be associated with a calculation of social disruption costs — even if only to drive home the message that installation alone is only part of the economic equation.

I would recommend that greater freedom of design be promoted through Performance Specifications with Contractors, Consultants and Industrialists working in concert to submit Design/Construct propositions — and I would like to see WRC being involved to their justifiable economic benefit.

I would like to believe that Britain could take its rightful role in projecting its expertise into the world export markets and not let its ideas go by default through inadequate support as was the case with the bentonite shield and with so many inventions from this country.

I would like to see a far stronger attack by every component of the Water Industry to get more done at the 'sharp end' of construction.

As to my question "IS THERE A NO-DIG FUTURE?" Of course there is. There *has* to be! The question remaining is "WHO IS GOING TO LEAD THE WORLD INTO THAT FUTURE?"

BRITAIN COULD. WILL IT??

Mr James Thomson: The size of the market

The writing of a paper on the size of the market is a thankless task in that it is not only difficult to obtain any information about the work carried out (the Government could learn a thing or two from the Statutory Authorities about keeping information classified), but any predictions of what is likely to happen in the future are bound to be held up to ridicule, either at this stage by somebody with a different set of prejudices, or in the future when the benefit of hindsight will come into play.

Considering the market for trenchless pipelaying is reminiscent of the two shoe salesmen who at the turn of the century were sent by their respective firms to darkest Africa. Within days they both cabled back, having made the same observation. The first man, perhaps rather conservative, said, "Am returning immediately. No prospects as nobody wears shoes". The second shoe salesman's cable was rather different. "Send additional help. Huge untapped market, nobody has shoes." The market for trenchless pipelaying is a little like that — we are virtually at the start of what is an untapped market, and in many ways attitudes will determine the rate of progress.

In my opinion a huge potential market does exist, and if I did have any doubts, then reading the papers and talking to the people here make it clear that the revolution is already under way and that a great number of individuals and organizations are already in, or about to get in, the business.

I think it is worth making clear that there is not just one market, but several with their own needs and criteria. I believe that microtunnelling most rapidly springs to mind when we talk of trenchless pipelaying. Designed to undertake sizes from 250 to 950mm, these machines are highly sophisticated miniaturized versions of the bigger machines. Able to work in a wide range of ground conditions and to lay lengths of sewer accurately up to 200 metres, they are, as you would expect, expensive animals, and cause the doubting Thomases to say "trenchless pipelaying won't happen because of the cost". But as can be seen, we are increasingly being offered a wide range of options with varying degrees of sophistication. The tendency will be for techniques to improve and prices to drop, and increasingly it will be viable and convenient to go under.

In terms of length, the most important range is in diameters up to 250mm, where the continued development of various forms of moles will substantially change the work pattern in the coming years with tens of thousands of kilometres being undertaken in the West by this technique. The already remarkably high percentages undertaken in the UK and USA by the gas authorities is a clear indicator.

The third segment of the market which potentially is vast is the "on-line replacement". With the completion of the infrastructure in many of the developed countries, the emphasis will shift to replacement and renewal. Pipe bursting, another development from the gas industry, has achieved considerable success, and no doubt alternative techniques will be developed to match the demands.

The fourth segment of the market is for pipes above 950mm, and trenchless methods in the form of segment tunnelling and pipe jacking hold a substantial percentage of the total length of work undertaken. Particularly in the field of pipe jacking with improved and mechanized techniques, there is every likelihood it will command an increasing share of the market. Prices relative to deep open cut should appear more attractive. Experience in other parts of the world already shows the trend.

To close, I adopt the words appropriate to the coming revolution, and say to you comrades, "Engineers of the world — go under. You have nothing to lose but your prejudices, your shovels and your holes in the road."

Dr S Tohyama: Microtunnelling in Japan

In my paper I have presented the current status of trenchless construction methods for sewerage in Japan, particularly microtunnelling methods.

About twenty years ago, sewerage and sewage treatment facilities had been constructed in only a very limited number of cities in Japan. However, in the 1960s there arose an increasing public demand to control water pollution and to improve the living environment. The Government therefore adopted a positive policy to improve sewerage and sewage treatment facilities. In order to carry out a systematic programme, they made successive five-year plans, which were co-ordinated with the economic development plans, and has endeavoured to implement those plans.

In 1980 and 1981, approximately 1.8 trillion yen (£6000 million) were invested in each year, but at present about 1.5 trillion yen (£5000 million) are put into these works each year, the reduction being caused by the financial difficulty of the national Government. Of these amounts, about 75 per cent was spent on sanitary and storm sewers and the rest on treatment works.

In the past priority had been given to the provision of wastewater treatment facilities for urgent control of water pollution, but in recent years priority has gone to laying sewers, especially sanitary sewers, because the capacity of the treatment works had caught up with that of the sewers. The annual length of sewers constructed has more than doubled in the last eight years. PVC pipes are increasingly used for the small-diameter sewers which are needed by the smaller cities.

About 550km of sewers are constructed each year by jacking methods. The most typical method is the ordinary jacking method in which pipes are thrust by jacks installed in the starting shaft. Recently it has become more difficult to find suitable sites for shaft construction, and other systems of construction are therefore being increasingly used. One is the semi-shield method, in which the front pipe is equipped with a shield machine to excavate the face. Another is the inchworm jacking method, where one or two sets of intermediate jacks are provided to increase the length of drive.

However, microtunnelling for pipes up to 700mm inside diameter is the most rapidly growing method of construction. Five categories of microtunnelling are in use, as listed in Table 1.3.3 and described in the paper. In fiscal year 1984, the length of sewer installed by microtunnelling reached about 180km, and the method will be used more and more in the future. The main recent developments in microtunnelling have been (1) correction of directional wiggle, using either a wedge-shaped pilot head, or steering jacks; (2) target measuring techniques, using a transit or, more recently, a laser beam; (3) remote control of operation; (4) excavation in sand-gravel strata, employing a hydraulic pressure balance type of machine.

As I have indicated, the development of microtunnelling has been rapid, but there is still a need for further development, e.g., to increase the length of drive, to develop pipe materials to prevent joint corrosion and infiltration of ground water, and to apply the method to smaller-diameter pipes such as lateral sewers, and to optical fibre communication networks.

As far as pipe materials are concerned, reinforced concrete is most widely used, but other materials have also been tried, including GRC (glass fibre-reinforced concrete), clay, combined clay and reinforced concrete, PVC and FRP (fibre-reinforced plastic). Collars for FRP pipes are usually steel, but, to avoid corrosion problems, FRP collars are now available.

[Dr Tohyama illustrated his presentation with a number of slides.]

Discussion on Papers 1.1, 1.2 and 1.3

Dr J W Heavens (ARC Slimline Ltd.): As a representative of a company supplying pipes to the industry, I should like to know what role the authors think the pipe manufacturers can play in ensuring a bright and profitable future for microtunnelling. In particular, could Dr Tohyama give examples of developments in pipe design and materials in Japan?

Dr S Tohyama: One example of contributions which the pipe manufacturers could make would be to develop suitable pipes for curved pipe-jacking. Most of the roads in Japan, especially in the smaller towns, are narrow and winding, and, therefore the necessity for curved jacking is very high. They should also set up a very good maintenance and after-care system for their products.

D G Clow (External Plant Division, British Telecom): Certain comments in Mr Thomson's paper require correction, and I have available for any interested delegates a paper which outlines the characteristics of British Telecom's underground system, both trenched and trenchless. [*Underground Construction in British Telecom with special reference to Trenchless Methods,* by D G Clow, April 1985.]

Mr Thomson suggests that trenching is illogical. I would question this in the context of telecommunications duct, which is typically buried at a shallow depth. Fluid-tightness is not essential, so the ducts can accept some degree of ground movement from heavy traffic and adjacent construction work. Shallow depth means that construction is quick, temporary support minimal and reinstatement easier, as only a small amount of subsurface material is disturbed in trenching. Trenching in nearly all cases is much cheaper than trenchless construction; most work is done by contractors in a highly cost-competitive market, and if trenchless methods were more cost-effective to contractors, and hence to the utility, they would be adopted more extensively. The majority of present BT duct installation consists of two or three ducts side by side. It is difficult to install such a formation cheaply by a trenchless method.

Mr Thomson also suggests that utilities should be charged a fee when they open a road, based on traffic density, number of disturbance days and extent of damage. Who pays? The public at the end of the day!

Mr Thomson says that, for pipes under 250mm diameter, 'reasonable' rather than 'precise' accuracy is needed from a trenchless method. In fact, the very situations where trenchless methods would be most beneficial are those in congested environments in towns and cities, and accuracy is then especially needed to avoid damage to other buried plant.

Currently, BT does make use of trenchless methods, but only in particular circumstances. For the use to be extended, four criteria need to be satisfied:

(i) Much better cost-competitiveness compared to trenching;
(ii) Accurate plant location techniques for both metallic and non-metallic materials;
(iii) High steering accuracy, and
(iv) Solutions to providing two to, say, four ducts in close formation.

G. Cook (Head of Renovation Unit, Halcrow Water): It appears that most of the inspiration and investment in the development of renovation techniques has come from the contracting side of the profession. This has led to an array of possible methods which has confused the user and undermined the control of the designer. Mr Rees referred to consultants and contractors combining in the future. How does he see the possible future role for consultants, not only to give this field the impetus we all agree it deserves, but also to maintain the independence of engineering judgement which our clients expect?

R Tennant (Group Engineer (Highways), Westminster City Council): Donald Rees has called for an all-party agreement for the restoration of our underground assets, but this is only part of a local authority engineer's problems. The other part — sometimes the overriding part — is all-utility agreement on programming such restoration. Would the utilities accept some form of programme control by, say, the local authorities, and will the review of the Public Utilities Street Works Act (PUSWA) suggest such control?

Mr Clow mentioned a quick job — but need it be a quick and nasty one? At least such surface works are visible and can be dealt with; the mole leaves no surface mark, but has a potential for latent destruction which is difficult to trace. Developments in underground mapping are of prime importance in this respect.

Dr P M Zollman (Managing Director, ZED Instruments): Could Dr Tohyama enlarge on the phrases 'correction of direction wiggle' and 'target measuring techniques'? What is his opinion on the need for curved pipe-jacking if pipe lengths between shafts are to increase?

Dr Tohyama: In principle, sewers must be laid under the publicly owned land, mostly under roads. Therefore, the need for curved jacking still exists even if pipe lengths between shafts are to increase.

R. Tharby (Chairman, National Joint Utilities Group (NJUG)): I can assure Mr Tennant that the utilities are fully aware of the need to co-ordinate their activities. They have formed NJUG to speak with one voice, and are represented here today as an entity. NJUG would like highway authorities to get themselves together in a similar way. They are disunited at present; who knows what could be done together? We all await the PUSWA review report, and NJUG hopes it will recommend such action by the highway authorities.

Mr D F Rees: Mr Clow said that if the no-trench method were cheaper, contractors would use it. The trouble is that the whole of the NO-DIG field is in evolution, and pricing is most difficult.

If you use a trenching method, it is not enough to say 'divert the traffic'. Think of the cost of delays and the social costs, which can be enormous from traffic disruption. What are you diverting? Huge loads through streets not designed to take them — and thus further damage to such streets. One must wait a little more patiently for very shallow depth developments to come through evolution and to become economic through market scale.

Mr Cook asked my views on the consultants' role. I think that, as with manufacturers and contractors, there needs to be a far greater degree of specialization among consultants. They must see that their staffs receive specialized training so that they can concentrate on this type of work.

On the question of 'all-party agreement', I consider there is a need for a long-term policy and strategy to embrace the whole of the utilities. What is needed is a supermarket of services to draw from, such as could develop under healthy market conditions.

Mr J Thomson: In reply to Mr Heavens, there is a lot the pipe manufacturers can do. We cannot just use standard pipes with NO-DIG methods. Different types of pipes will be needed to match the new technologies.

I agree with Mr Clow that there are many places where it is not sensible to bore under, but I am sure you don't put ducts across motorways by means of shallow cutting! It is true that if statutory undertakers were charged, the taxpayer would pay. But he pays already. But it would focus attention on where the real charges come from. It is too easy for statutory undertakers to agree to disruption by trenches because it doesn't hurt or reflect on them.

On the question of 'how accurate?', the more accuracy we want, the more sophisticated and expensive it will be. But it is not always necessary to be so accurate.

I do not think it would be too difficult to install a multi-way duct by a NO-DIG method if you put in a bore using a heavy steel pipe. Then push in the multi-way duct to replace the steel pipe.

P A Banks (President, IPHE): I would like to ask Dr Tohyama if there is any legislation in Japan that requires utility undertakings to account for social costs.

Dr Tohyama: This is a matter that has not yet been precisely defined legally. Under the regulations trenches may not be dug in newly built roads. It is necessary, by contact with the local authority, to ensure that pipes and other services are laid while the road is being constructed. There is no limit to investment, but, of course, we have to compare various methods and their costs.

R N Craig (Sir William Halcrow & Partners): For innovation to be worth while there must be a good home market. This is essential if we are to have a healthy industry.

Do the authors consider there is a future for the use of multi-purpose tunnels?

Service connections to pipes are still a major difficulty, and this requires further research. There is also need for more research on surface mapping and site investigation. We now have the system of ranking of motorways during replacement — why could not this system be used for other roads during disruption by the installation of services? More account must be taken of social costs in comparing alternative methods of construction.

Dr Tohyama: Multi-purpose tunnels have been constructed to some extent in Japan. But I do not think the multi-purpose tunnel has a bright prospect. One problem is the difficulty in getting good co-ordination among the organizations involved from the viewpoint of timing of investment. Another difficulty is that the slope and the depth required for a sewer and a tunnel for other purposes are usually quite different from each other.

E W Flaxman (Chairman, Conference Organizing Committee): Lord Nugent, the President, Mr Rees and Mr Thomson have all asked for more Government responsiveness on infrastructure needs. I believe that Government is becoming steadily more aware of the needs, but I suggest there would be more impetus if there could be a serious sewer collapse — or, better still, a major sewage flood — in Downing Street!

G C Cox (Consultant, M McDowell Co-Partnership): I note that only fourteen of the conference delegates — about 3½ per cent of the total — are from district councils in the UK, yet they are the people at the sharp end, who need to know about the new technologies. There is no one here from the Department of the Environment, the Department of Industry or the Treasury — very little Government interest in something they should be involved in.

There is often great difficulty in obtaining statistics, but I suggest that the true figure for sewer collapses could be ten times the figure of 3000 a year quoted by Mr Thomson. In two neighbouring district councils, one reported 47 collapses and the other only three in the same period. The second district said that there were now not enough engineers left to report on collapses.

S R Kramer (Project Manager, Gas Research Institute, Chicago): We have an extremely complex infrastructure system, which must be protected. US gas utilities expend approximately $3.5 billion a year to install, maintain and replace the gas distribution piping system.

Mr Rees: I agree with Mr Cox that it is something of a reflection on the attitudes of Government ministries that they do not attend such an important conference. We require investment and, inevitably, risks have to be taken. There is not enough investment to reduce the risk factor — the whole thing is cyclic.

Abdullah bin Ali H Al-Harthy (Head of Public Health Engineering, Ministry of Defence, Sultanate of Oman): I would like to indicate the background to the problems which we face in the Gulf. From 1964 to 1970 I worked as a water engineer at Sharjah in the United Arab Emirates, when we were considering how to uprate water production in the Emirates. At that time nobody thought of doing anything except building roads and providing services. In 1970 the Gulf State Supreme Council and Abu Dhabi decided to construct sewage treatment plants, and until recently they were faced with many problems in crossing roads and other facilities.

In Oman the problem is more difficult. I was there immediately after the change of government in 1970, and our main preoccupations were improving the water system and the road facilities. There was no time to think of other things. Everybody wants water and roads.

One difficulty we face is that there is only one road that goes outside the country, connecting Sharjah with Aden. We have to use that road for all the traffic. The reason for my coming to this Conference is to acquire a knowledge of the latest engineering developments so that I can go back and discuss things properly with the many international consultants and contractors who are in the country.

Besides being an adviser to the Ministry of Defence, I am a director for the Ministry's public health service. The Ministry has 75 to 80 camps in the Sultanate, with from 500 to 25 000 people per camp, so we have big problems with water supply and other services. I want to ensure that any consultants or contractors who come to Oman are fully familiar with the new technology which has been introduced in this Conference. We will insist that modern systems will be introduced not only in Oman, but also in the Gulf States (I am also a member of the Gulf States Council of Local Engineers). We welcome all good ideas into Oman, and I shall be pleased to talk to anybody who is interested.

[To conclude the session Dr Tohyama showed a number of slides illustrating a membrane-wearing pipe-jacking method, being used successfully in Japan on pipes of 1100mm diameter. The membrane prevents groundwater ingress during and after jacking. Just behind the machine, grout is injected between ground and pipe.]

SESSION 2. Chairman: Priv.-Doz. Dr. Ing. D. Stein, Associate Professor, Ruhr-Universität Bochum, W. Germany

2.1 Trenchless pipelaying — yes please; a highway engineer's point of view

M F Hardy BSc, FEng, FICE, FIHT. *County Surveyor, Hertfordshire*

The paper examines the scale and cost of trench reinstatements and stresses the significance of other aspects including traffic delay costs, longer-term damage to carriageways and footpaths etc.

Together with the advantages and potential of trenchless pipelaying possible problems with the new systems are touched upon and some suggestions made for accelerating research and getting more positive feedback of information.

THE SCALE OF THE PROBLEM

Seemingly everyone knows about and comments upon the problems of trenches in highways, both carriageways and footways. The motorist and pedestrian have strong views which are reflected in correspondence and comment, regularly received by Highway Authorities, and the motoring organizations are never slow to remind those in charge of highway management of the problems both in the installation of services and during the subsequent maintenance.

But is this a subjective view and is the problem overstated? The facts indicate that in the United Kingdom it is likely that 1,800,000 reinstatements are carried out on carriageways and footpaths every year. Within the author's authority, with a population of nearly one million, the annual figure is in the order of 24,000. Put another way, in Hertfordshire, an average of six trenches are excavated per kilometre of road each year — a number corresponding to 100 openings per working day.

The other important aspect of the problem is, of course, the expenditure involved. It is highly likely that over £70 M is spent on road reinstatements each year, and this figure is likely to increase as development progresses and more services become available. Clearly the scale and costs of trenching are very substantial indeed and legitimately warrant attention from Statutory Undertakers and Highway Authorities.

The other equally important side of the equation is the growth in traffic. The current legislation governing trench reinstatements in U K came into operation in 1950. By 1982 the vehicle-kilometres travelled by cars and vans were 8.5 times the 1950 figures. For other goods vehicles the increase was over 3.5 times[1].

The combination of the increase in openings, coupled with the growth in traffic, clearly amplifies the importance of trying to get alternative means of providing services in highways without excavating and backfilling trenches.

OTHER IMPLICATIONS

Traffic costs

Reference has already been made to a general increase in traffic, but it is worth noting how that traffic is distributed among the various classes of road. The table below gives an indication of vehicle travel (billion vehicle-km) on each class of road in the U K based on 1983 statistics[2].

Motorway		31.45
Trunk road		44.94
Principal road (urban)		62.12
Principal road (rural)		41.63
Non-principal road		91.08
	Total	271.22

Legislation has prevented the use of motorways for statutory undertakers' plant, though it should not be forgotten that there are very substantial systems of surface water drainage both along the motorways and, at certain points, across.

Particularly on the Primary Routes (ie Trunk Roads and important Principal Roads) it is almost certain that any installation of new apparatus, or repair of existing equipment, will lead to the need for closing traffic lanes, and possibly closing the complete road altogether. Much thought has been given recently in the UK to the cost of traffic delays associated with any works which interfere with free traffic flow, and it is now possible to calculate the economic consequences of that disruption[3]. For example, taking a two-lane road with 15,000 vehicles per day and causing that flow to be interrupted by one-way working during 24 hours, delays are likely to cost in the order of £2,500 during that period.

In an exercise set up by the Transport and Road Research Laboratory simulating traffic disruption in the centre of the town of Reading it was shown that traffic delay costs were on average 1.25 times estimated engineering costs. In really difficult locations the comparative figure could be very much higher.

Subsequent damage to carriageways and footpaths

There is now little doubt that, with the best will in the world a reinstated trench will shorten the life of the adjoining carriageway or footpath in the longer term. After the severe winter of 1982/83 in the author's County a comprehensive survey was carried out to determine the extent of carriageway patching work arising from the impact of frost and snow. Of the total requirement for patching, approximately 30 per cent of cases were associated with trench reinstatements. Although statistically and inevitably some patching must occur on trenches, it is apparent that the reinstatements attract more than a proportionate share of the total.

Both short- and long-term problems with trenches can be associated with various technical problems:

- obtaining satisfactory compaction of all the layers and materials in the trench;
- excluding water which may reduce the load-bearing properties of the backfill;
- disturbing the structure of adjacent carriageways or footpaths during the trenching and reinstatement operations;
- having available suitable materials which can both accommodate the subsequent traffic loading, but at the same time, be suitable for handling, laying and compacting in trenches;
- achieving satisfactory joints and seals between the reinstatement and existing road to exclude water.

Despite attempts to adopt national standards each of these technical aspects continues to cause concern. There is no doubt that more research and tighter control of the whole sequence of trenching work will be required in future if we are to get better results — unless, of course, the need for the trench is avoided in the first place.

Administrative costs

The more frequent and difficult trenching becomes, particularly on heavily trafficked roads, the more it will be necessary to spend administrative money on supervision and inspection. In Hertfordshire there is an administrative charge to cover inspection and handling statutory notices etc. of 12½ per cent of the cost of reinstatement work. To this figure must be added supervision costs of the Statutory Undertakers and their contractors. The handling of

Notices and Charges for reinstatement can also be a tiresome process — particularly if you do not know who dug the trench in the first place! Many Authorities are finding it very difficult to produce the manpower to handle this type of work during a time of strict control of staffing and inspection manpower.

Other social costs

There are other often ignored detrimental aspects of trenching operations, especially in urban areas. Disruption to trading and pedestrian access to shops and other facilities; the creation of general noisy and messy conditions during construction; subsequent unsatisfactory appearance of carriageways and footways — all of considerable concern to the public and trader.

PUBLIC UTILITIES STREET WORKS ACT REVIEW

At the present time Professor Horne and his team are carrying out a review of the whole PUSWA situation in the United Kingdom for the Minister of Transport.

The current legislation was introduced in 1950, and, in effect, gave Undertakers the right to enter into highways for installing their services and repairing existing plant. In theory they pay for the cost of the whole works including the reinstatement, and any necessary subsequent repairs within a two-year period. However, as has already been pointed out, there are other costs involved which are now becoming increasingly relevant. Local Authorities are arguing that the legislation should be altered to take account of total costs so that any authority wishing to place services in roads should consider in the first place whether it is an economic proposition to do so, or whether some alternative should be found. Secondly, they should so arrange the operations in the highway to minimize total costs, including those associated with traffic delay, and longer-term deterioration of the carriageway.

ADVANTAGES AND POTENTIAL OF TRENCHLESS PIPELAYING

Clearly many of the factors discussed earlier in this paper indicate a substantial pay-off if the process of excavating, backfilling and reinstating the surface of trenches can be avoided, particularly on heavily trafficked routes. We shall be hearing during the conference about various alternative methods for avoiding trenches and, hopefully, something about the economics of the different systems which should then enable us to compare these new approaches with the traditional trenching methods. It will be interesting to hear the potential scope for using these methods at the present time with some indication of where they are most appropriate or, at the other end of the scale, where their limitations in use are likely to occur. It is worth remembering that we are talking not only about apparatus put in by Statutory Undertakers but indeed by Highway Authorities themselves who have direct responsibility for installing and maintaining very substantial lengths of highway drainage.

POSSIBLE PROBLEMS

The majority of services are located within urban areas, many of them in main streets which, additionally, are now carrying some of the heaviest flows of distributor traffic. Knowledge of the position and depth of existing services in roads such as these leaves a great deal to be desired. Indeed, one of the points made to Professor Horne's PUSWA inquiry has been the need to spend time and money on getting better locational information about existing services. However, it is unlikely that this work will be carried out very quickly, and it is an unfortunate fact that introducing any new equipment into the urban road subsurface environment is a risky business, even if you dig an open trench for the installation. Clearly there will be risks in using new technology boring systems to install services unless there is either:

— proper and reliable pre-determination of existing services, and/or
— some very positive means of ascertaining whether services are being interfered with during the boring process.

The author would be interested to know whether the new technology might mean that it will be economical to install services at lower levels than normal so that some of the shallower services which exist at the present time might be satisfactorily avoided.

Some of the advantages of the new technology will not materialize unless the whole of the operation can be kept clear of heavily trafficked carriageways. If it is necessary to introduce working pits in these carriageways to carry out the boring operation then, potentially, they could have the same traffic disruption effect as a complete trenching operation. A good deal of thought will have to be given to the siting of these pits — and indeed to the location of the whole system — if we are to optimize on the use of the new technology.

Hopefully, the conference will also discuss the important question of economy of scale. Unless the prices for trenchless pipelaying are competitive with the cost of alternatives, it will be hard to justify extensively employing these new techniques.

On the other hand if their potential and actual use proves to be substantial, costs will be reduced and their scope of application considerably widened.

THE FUTURE

Without doubt there must be widespread benefits if trenchless pipelaying can be made to work on an economic basis. Highway engineers would welcome the new system from many points of view. The author hopes that the conference will give some thought towards the pace of developing the new technology, to assess whether there is sufficient money and time being invested in research to develop the systems, and to get them into practical operation. Perhaps some conclusions could be drawn on this important aspect towards the end of the conference.

It is worth considering whether new legislation or Codes of Practice could be introduced to accelerate the use of these techniques by, for example, requiring service organizations to take account of the total costs of their operations.

As a final point, like so many other concepts, the idea of trenchless piping will not be used to full advantage unless its potential is understood by everyone likely to benefit from its introduction. Clearly this is one of the reasons for holding the conference in the first place.

However, it is not just a question of knowing about the systems. There is a need for positive assessment of the potential use and encouragement to apply the new systems followed by some positive and quick feedback on how the technology is progressing and whether the forecasted achievements are being obtained in practice. But perhaps that is the basis for another conference in two or three years time.

REFERENCES

[1]*Transport Statistics Great Britain 1972-82* (and earlier), HMSO.
[2]*Transport Statistics Great Britain 1973-83*, HMSO.
[3]Department of Transport, *QUADRO 2*, April 1982.

2.2 Trenchless mainlaying within British Gas

M Howe, *Assistant Division Manager*
British Gas Corporation, Engineering Research Station

P Hunter, *Principal Engineer*
British Gas Corporation, Engineering Research Station

Trenchless mainlaying using earth displacement and pipe bursting 'moles' is being increasingly used by BGC, the main aim being to avoid disruption to public use of highways etc. Additionally, there is a cost incentive to use trenchless methods if at all possible. One of the factors which has inhibited further use of these techniques is concerned with possible damage to road surfacings and adjacent plant.

This paper reviews the results of a theoretical analysis involving the application of the finite element method of stress analysis to the effect of an expanding cavity in soil. Physical modelling of the effects of moling using X-ray techniques have also been successfully carried out and the ground movements resulting from mains replacement by pipe bursting are being recorded at sites within BGC.

INTRODUCTION

British Gas have long realized the economic advantages of trenchless mainlaying. The technique of mains replacement by inserting smaller-diameter polyethylene (PE) pipe into existing gas mains has been used for many years. The technique of installing small-diameter PE mains in new ground by percussive moles has also become more generally applied as has an extension of this, the installation of PE pipe in dead mains using pipe bursting moles. On-line replacement in this way allows size-for-size replacement or greater to be carried out.

This paper outlines the wide range of studies being undertaken by British Gas at its Engineering Research Station (ERS) to support moling operations on site. The studies embrace theoretical and laboratory analysis of ground movement, development of ground movement measurement techniques and the development of a mole itself. The monitoring of numerous moling operations is also being carried out within the Regions of BGC.

THEORETICAL ANALYSIS

Preliminary theoretical calculations to determine the likely zone of influence around an earth displacement mole and its effect on buried pit cast-iron pipes, have been carried out by ERS.

The initial studies were undertaken on the assumption that the soil displacements occurred in an outward radial direction from the longitudinal axis of the mole and that the soil deformed at constant volume. However, the importance of volume compressibility of partially saturated soil was also briefly considered.

The effects of the passage of a 65mm mole close to mains of between 2 inch and 24 inch nominal bore have been determined. The mains were considered to be buried in a cohesive soil, classified as Group A according to the MCP for deep excavations[1]. The analysis enabled ground movements to be related to distance from the mole cavity and quantified in terms of a percentage mole diameter.

A typical result is shown in Figure 2.2.1. The constant volume theory indicates that the smaller pipes experience the highest total strain (bending plus axial components) but the strain level decays exponentially away from the mole. Larger-diameter mains ≥12 inch, are at minimal risk at all proximities apart from near impact. The method considered both homogeneous saturated and homogeneous unsaturated ground, but ignored the proximity effect of the ground surface and the effect of soil dilatancy. Therefore these simplified predictions are most likely to be applicable to cohesive soils at high depth/diameter ratios. It was also observed that as the percentage of air voids in soil was increased, the zone of ground displacement was reduced because of the compressibility of ground next to the mole cavity. This effect is beneficial to adjacent plant and should reduce the induced strain levels.

The influence of the ground surface on the distribution of ground displacement was investigated by applying the finite element method of stress analysis to an expanding cavity in soil. The analysis assumed a saturated cohesive soil, again deforming at constant volume. Six FE analyses were performed using two depths of cover and three cavity diameters. The results for a 65mm mole in a soil equivalent to MCP Group A with a depth of cover of 0.46 metres is given here as an example. The analysis allowed contour plotting of the distribution of predicted subsurface movement. Contours of total movement are shown in Figure 2.2.2. The decay of total movement in different radial directions out from the cavity edge are presented in Figure 2.2.3. This clearly indicates the surface effect on the distribution of ground movement since the movement above the cavity is greater than that below. There will always be a tendency for ground to move preferentially towards a free surface or void.

The ratio of upwards movement to downwards movement below the mole is approximately 4.5 to 1. For the 65mm hole at 0.46m depth the predicted upward movement at the surface is about 3.5mm. Figure 2.2.3 indicates that displacement in all directions around the mole will be minimal if the ratio of depth to mole diameter is greater than about five.

However, it must be remembered that in practice ground movement fields will be influenced by factors such as soil strength, local rigid zones, surface pavement structure and voids.

The theoretical studies are incomplete and other aspects of the problem are being considered. These include the influence of cracking and dilatancy of the soil on the distribution of ground movement, the effect of hard inclusions such as concrete on both the line and level of the mole and the soil displacement field. The type of nearby buried plant, its direction relative to the mole axis, and the influence of mole diameter also require consideration.

The zone of influence adjacent to the breaking out of an existing pipe using a pipe-bursting mole will also be investigated. However, the zone of influence due to radial expansion is likely to be less for the earth displacement mole, except in cases where joints and fittings have to be fractured and then expanded.

CONTROLLED EXPERIMENTS

In addition to the theoretical studies, a series of controlled experiments to determine influence zones caused by various moling techniques are being carried out. The investigation employs an X-ray technique to measure soil displacement. The technique has proved a simple, versatile and effective method of determining ground movements. A rectangular grid of lead on a 2cm pitch is created at the mid-plane within a special soil box. The box is then filled with soil and the grid is X-rayed to obtain its undisturbed dimensions (Figure 2.2.4). The mole head can then be passed through the soil at a controlled rate as shown in Figure 2.2.5 and the box X-rayed again to obtain the deformation of the grid. Figure 2.2.6 shows the radial deformation produced by the passage of a 65mm mole head. Figure 2.2.7 shows the soil displacement associated with the same mole head in the longitudinal plane. Here the ground movement promoted by the head is confined to within 3cm of the edge of the mole. The extent of movement observed in this experiment was limited, since the soil used was of low strength and contained a high percentage of air voids.

Further tests are planned using the technique to investigate the effects of pipe-bursting moles and to determine the trajectory of earth-displacing moles in the presence of differential soil strata and other plant.

FIELD EXPERIMENTS

To complement the theoretical studies, a programme of monitoring is being undertaken at selected sites within BGC Regions where mains replacement using pipe bursting is being carried out. Initially effort is being concentrated on monitoring the ground heave resulting from pipe bursting rather than earth-displacement moling.

Measurements are most simply obtained using a beam, 2.7m long graduated in 150mm intervals along its length and which is mounted on the ground surface above and at right angles to the proposed line of moling. Adjustable feet and an integral spirit level allow the beam to be set truly horizontal. The distance from the beam to the ground surface is noted at each graduation point before and after moling to obtain the movement profile. The photograph in Figure 2.2.8 shows the beam positioned over a tarmac footpath prior to moling.

To speed up the measurement process, ERS have also produced electrical surface profile equipment[2]. This consists of a guide beam and a moving head which incorporates a stylus connected to a potentiometer, to provide a continuous plot of the road surface on an XY plotter. A photograph of the equipment is shown in Figures 2.2.9.

The field measurement programme will involve observations at sites where 130mm, 170mm and 245mm moles are being used to break out 3 inch, 4 inch and 6 inch cast-iron mains, respectively.

COST COMPARISONS

The potential for trenchless methods within BGC

The replacement work currently being carried out in BGC can be categorized by location as follows; 58 per cent in the carriageway, 29 per cent in footpath and 13 per cent in unmade ground. Clearly with almost 60 per cent of the work being carried out in the carriageway the cost of reinstatement is a very significant factor[3].

If all this replacement work was to be carried out by open-cut trenching and subsequent reinstatement, the potential saving for a trenchless method is very significant. One of the BGC Regions estimates that if they carried out all their mains replacement by open-cut methods then they would spend at least £20 million over the next five years on permanent reinstatement alone[3]. Therefore if mains bursting could be applied extensively, the cost savings to the Region could be substantial.

Cost savings for the larger mains will be particularly high. This is possible since larger mains have fewer services than the smaller mains, necessitating excavations for service reconnection, also the bigger the main the greater will be the size of excavation and reinstatement required for open-cut work.

Further savings are possible if the amount of equipment needed on site to carry out a mains bursting operation can be reduced. Equipment required on site typically includes a large air compressor, a 5-ton capacity winch and lifting facilities. Therefore, ERS are developing a large diameter mains bursting tool[4] to overcome these disadvantages and further improve the savings over conventional trenching methods.

CONCLUDING REMARKS

BGC is continuing to monitor the performance of both earth-displacement and pipe-bursting moles. The results will be used to improve theoretical predictions of the effect on pavement structures and on other buried plant. These will be used in the development of British Gas specifications, codes of practice and joint industry consultative procedures.

The preparation and adoption of such procedures by all participating utilities is becoming more urgent since, as the economic benefits are becoming apparent, the techniques are becoming more widely used.

Much of the replacement work to date within BGC has concentrated on 3 and 4 inch mains; however, BGC mains replacement programmes indicate a demand for mains bursting equipment to replace 6 and 8 inch mains.

Several manufactures, as well as ERS, have responded to this demand and equipment has been developed to deal with these larger mains sizes.

ACKNOWLEDGEMENTS

The authors would like to thank their colleagues within BGC for their assistance in the preparation of this paper. The authors are also indebted to the British Gas Corporation for permission to publish.

REFERENCES

[1]Water Authorities Association/British Gas Corporation, *Model Consultative Procedure for pipeline construction involving deep excavation,* June 1984.

[2]Owen RC, Contribution to Symposium on highway maintenance and data collection. University of Nottingham, July 1983

[3]Dawson L, North Thames Gas. *Size for Size Replacement.* Paper presented at London and Southern Gas Association, December 1984.

[4]Smith B J, ERS unveils six cost-cutting mainlaying innovations. *Gas World,* August 1984.

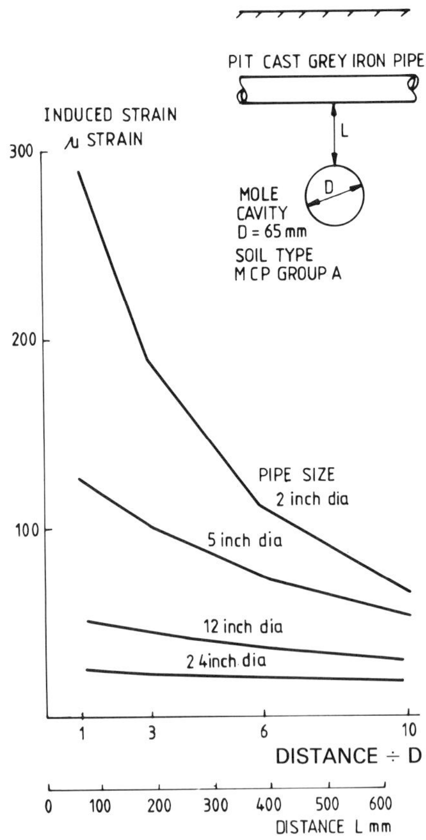

Figure 2.2.1. Strain induced in pipe v. distance L from mole cavity.

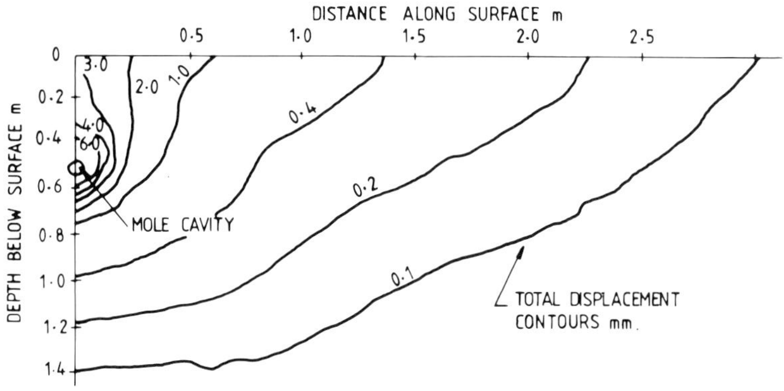

Figure 2.2.2 Distribution of surface movement around 65mm mole in MCP Group A soil.

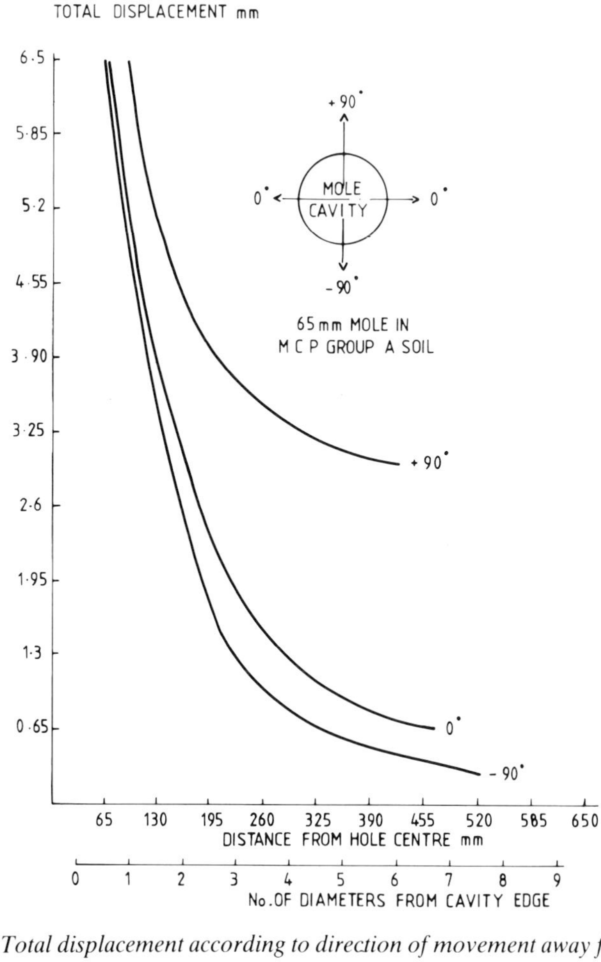

Figure 2.2.3. Total displacement according to direction of movement away from mole cavity.

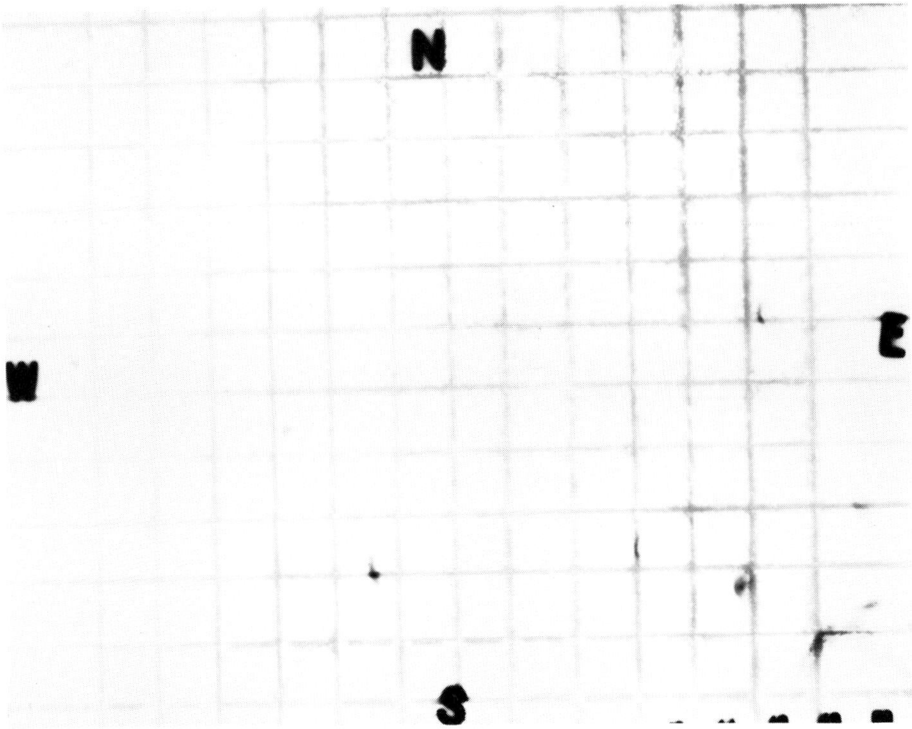

Figure 2.2.4. X-ray of 2cm grid prior to test.

Figure 2.2.5. 65mm mole about to enter soil box.

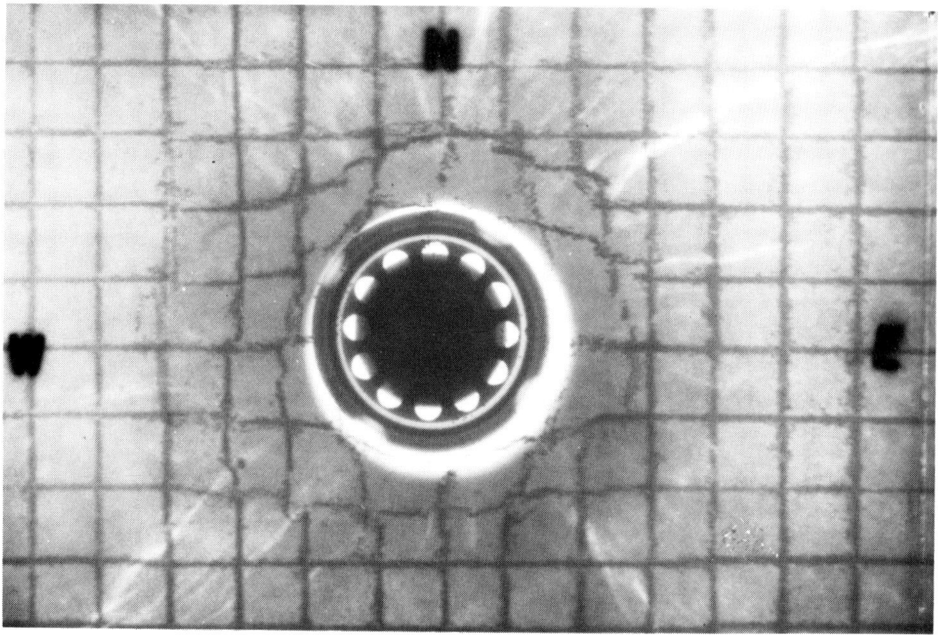

Figure 2.2.6. X-ray showing radial deformation after passage of 65mm mole head.

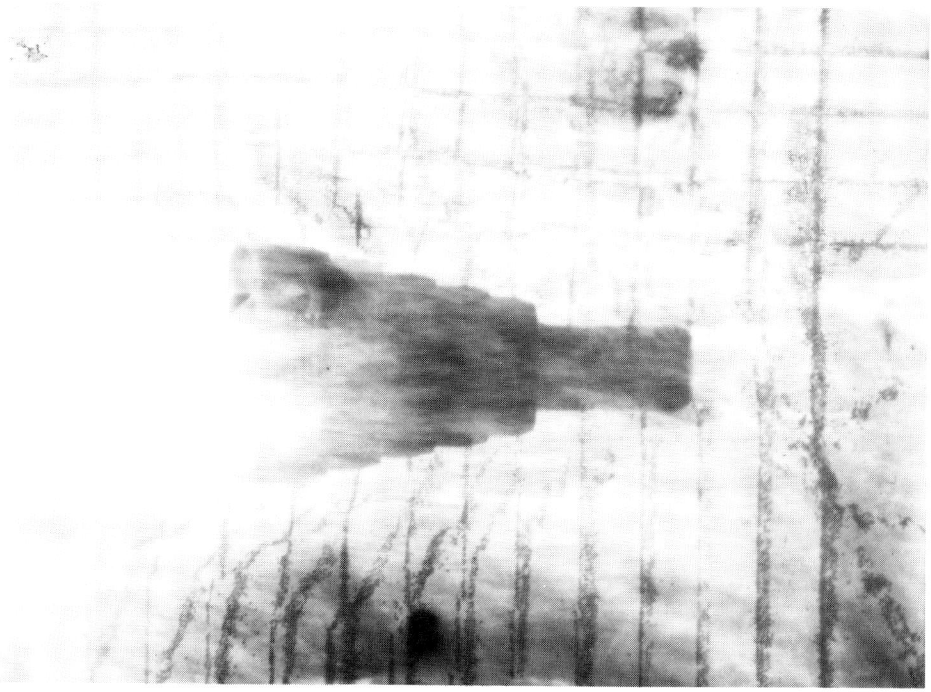

Figure 2.2.7. X-ray showing longitudinal deformation after passage of 65mm mole head.

Figure 2.2.8. ERS measurement beam in position prior to pipe bursting.

Figure 2.2.9. ERS electrical surface profile equipment.

2.3 Ground movements caused by trenchless construction

T D O'Rourke

Associate Professor,
School of Civil and Environmental Engineering,
Cornell University, Ithaca, NY

Trenchless pipelaying techniques may be classified according to convergent methods, in which the volume of excavated soil exceeds the volume of pipe installed, and expansive methods, in which the volume of pipe or construction equipment exceeds the volume of soil excavated. The ground movement patterns associated with convergent methods have been investigated extensively by means of field and model measurements. The results of these studies are reviewed, and the differential soil movements caused by convergent methods are related to bending strains in adjacent small-diameter pipelines. The ground displacement patterns associated with expansive methods in undrained clay are evaluated on a theoretical basis by modelling the construction technique as an expanding cylindrical cavity in a perfectly elastoplastic material. The radius of plastic soil deformation surrounding the expanding cylinder is related to the fractional increase of the initial radius of the cylinder. The bending strains imposed in adjacent small-diameter pipelines are shown to be a function of the radius of plastic deformation and the ratio of the undrained modulus to undrained shear strength of the soil.

INTRODUCTION

Trenchless pipelaying techniques can be divided into two broad categories, depending on how the ground is disturbed during construction. When the volume of excavated soil exceeds the volume of pipe installed, the surrounding ground will generally displace, or converge, toward the opening. Construction methods associated with a net volume loss in the *in situ* soil are referred to as convergent installation techniques. When the volume of the installed pipe or construction equipment exceeds the volume of the excavated soil, the surrounding ground will generally displace, or expand, from the opening. Construction methods associated with a net volume increase in the *in situ* soil are referred to as expansive installation techniques.

The potential for damage to adjacent pipelines, underground structures, and roads places an important constraint on trenchless construction methods. Accordingly, it is important to know how ground movement patterns are affected by soil type, excavation geometry, and construction procedure. In addition, an understanding of the soil displacement patterns can often help to evaluate the stresses and deformations which must be sustained by the installed pipeline.

This paper provides a review of ground movement patterns which can be caused by trenchless construction. The soil displacements associated with convergent installation techniques are reviewed by reference to previous field and laboratory studies of movements caused by soft ground tunnelling. Because there currently is little published information on expansive installation techniques, the soil movements associated with this category of construction are explored from a theoretical perspective.

CONVERGENT INSTALLATION TECHNIQUES

Convergent installation techniques include small-diameter (0.9 to 1.5m) tunnelling, microtunnelling (less than 0.9m diameter), and various types of thrust boring and jacking. Subsurface excavation performed with these methods can result in stress relief and ground deformation into the excavated cavity. The amount of soil displacement is related closely to the soil properties and ground water conditions. In addition, soil movement depends on the manner in which soil excavation is co-ordinated with temporary support from construction equipment and the subsequent installation of the pipeline.

To evaluate the influence of various trenchless methods on surrounding structures, it is necessary to understand the sources of soil movement, or lost ground, at the excavated cavity and relate this loss of volume at depth to the pattern of distributed movements above and surrounding the excavation. Detailed studies of ground loss and soil displacement patterns have been made for soft ground tunnelling (e.g., Peck, 1969; Cording, *et al.*, 1976; Attewell, 1977; Clough and Schmidt, 1977; O'Reilly and New, 1982). Although these studies have focused on tunnel diameters, which generally are larger than those typical of trenchless pipelaying, there is sufficient similarity in the procedures to use soft ground tunnelling studies as a basis for evaluating sources of lost ground and movement patterns.

Sources of lost ground

There are three principal sources of lost ground associated with soil tunnelling: 1) face and peripheral encroachment, 2) consolidation, and 3) local instability. Each of these is discussed under the headings which follow.

Face and peripheral encroachment. Cording, *et al.* (1976) have identified three primary sources of soil loss during shield tunnelling: 1) loss of ground into the face, 2) loss of ground over the shield, and 3) loss of ground during and subsequent to lining erection. These sources of soil movement are directly applicable to small-diameter tunnels, which usually are shield-driven. Moreover, there are similarities between the excavation and support procedures adopted in large-scale tunnels and certain types of microtunnelling. Some microtunnelling equipment, for example, will excavate an oversized cavity into which the pipe is installed by hydraulic jacking. Soil movement into the annular space between the excavated cavity and external pipe diameter is analogous to ground loss during and subsequent to lining erection.

The similarities with respect to face and peripheral ground loss between large-scale tunnelling and trenchless pipelaying mean that similar patterns of movement can be anticipated with the two general methods. Most forms of face and peripheral encroachment in medium to hard clays involve ground loss which is a relatively small percentage of the excavated volume. Small systematic losses of volume have been shown to result in repeatable patterns of soil movements for large-scale tunnels, which can be used as a basis for estimating the potential disturbance from trenchless pipelaying.

Consolidation. Tunnelling in normally to lightly overconsolidated clay can cause increased effective stresses as a result of long-term seepage into the tunnel. When these increased stresses exceed the *in situ* effective stresses, primary consolidation will occur. The consolidation leads to a volume loss which adds to the undrained soil displacement during excavation and installation of support. Cording, *et al.* (1976), Attewell (1977), and O'Reilly and New (1982) report on substantial increases in the volumes of surface settlement, which were observed over a relatively long time after tunnelling in soft clays and silts.

Long-term volume loss from consolidation will be related to the relative permeability between the lining and surrounding ground. Most segmental linings are relatively pervious with respect to silts and clays, even when they are well constructed and sealed. A continuous steel or plastic pipeline, however, is substantially less pervious than a segmental lining so that long-term consolidation will not be a problem unless sufficient longitudinal seepage paths exist to create a drainage boundary.

Consolidation of cohesionless loose sands may occur as a result of vibrations induced during construction. This could be a problem with percussive moles, which transmit a considerable amount of impact energy to the surrounding ground. Vibration-induced consolidation in very loose to loose sands could be a problem for both convergent and expansive pipelaying techniques.

Local instability. Heuer (1976) has pointed out that ground loss may be systematic, in which cas the volume of displacement is limited by the shield and lining geometry, or may be catastrophic, in which case there is a sudden collapse, run or flow of soil into the excavated cavity. Volume losses associated with local instability may be very large, sometimes exceeding the excavated volume of the tunnel where they occur. The principal sources of instability include rapid inflow of sands below the water table, large runs of dry cohesionless sands, and squeezing or very soft to soft clays. Local instability is most troublesome when it is not anticipated so that subsurface conditions must be evaluated for sudden changes in the groundwater level and soil properties

Pattern of soil settlement

Several researchers (e.g., Peck, 1969, Cording *et al.*, 1976, Attlewell, 1977, and O'Reilly and New, 1982) have shown that the distribution of settlement caused by tunnelling can be represented by a normal probability curve as illustrated in Figure 2.3.1. Studies performed with the aid of small-scale models (Atkinson and Potts, 1977) corroborate these findings and suggest that the normal probability curve gives a reliable estimate of the ground settlement pattern for excavation diameters consistent with the smallest microtunnels. It should be emphasized that the normal probability curve is a good representation of settlement resulting from face and peripheral encroachment, for which lost ground develops on a systematic basis as a relatively small percentage of the excavated volume. The normal probability curve generally will not give an accurate representation of distributed settlement for volume loss associated with the consolidation of clay or local instability. Consolidation of clay generally results in a broader pattern of settlement with less curvature, and local instability leads to an abrupt movement profile with more pronounced curvature.

O'Rourke and Trautmann (1982) analysed the response of jointed cast-iron pipelines to tunnelling ground movements in sand. Some of their results are illustrated in Figure 2.3.2, in which the settlement profiles of the ground and buried pipelines are shown. The analytical results pertain to an excavated tunnel diameter of 6.4m at a depth to tunnel centreline of 12.8m in dewatered sand. The nominal 100 and 300mm-diameter pipelines had wall thicknesses of 10.4 and 12.7mm, respectively, and were composed of 3.7m-long pipe segments joined at lead-caulked couplings, which were modelled as pinned connections. It was assumed that the pipelines were buried in loose sand at a depth of 0.9m from the street surface to top of pipe.

As O'Rourke and Trautman (1982) point out, the 100mm-diameter main behaves as a relatively flexible structure. Its deformed shape corresponds closely with the settlement pattern of the soil. The 300mm-diameter main, which is approximately ten times as stiff, behaves as a relatively rigid structure. Its principal model of deformation occurs as rotation of the joint closest to the tunnel centreline. Furthermore, it was shown that the maximum curvature imposed in the 100mm-diameter main was nearly equal to the maximum curvature of the soil settlement profile. These results were corroborated for similar soil and burial depths through analytical work by Harris and O'Rourke (1983), in which the maximum curvature of a 100mm-diameter cast iron main was shown to be approximately 90 per cent of the maximum curvature of the soil displacement pattern generated by adjacent deep trench construction.

Assuming that a normal probability curve will adequately represent curvature, the maximum curvature, K_m, from tunnelling settlements is given by (Attewell, 1977; O'Reilly and New, 1982):

$$K_m = \frac{S_m}{i^2} \tag{1}$$

in which S_m is the maximum centreline settlement and i is the distance to the inflection point of the settlement profile as shown in Figure 2.3.1.

The maximum settlement is related to the volume per unit distance of the settlement profile, V_s, by:

$$S_m = \frac{V_s}{\sqrt{2\pi}\, i} \tag{2}$$

Combining Equations (1) and (2) and introducting new terms for geometric parameters and volume loss gives:

$$K_m = \frac{\alpha}{4} \sqrt{\pi/2} \left(\frac{D^2}{i^3}\right) \tag{3}$$

in which α is the ratio of the volume of lost ground to the excavated volume and D is the diameter of the excavated cylindrical opening.

On the basis of measurements taken for field and model tunnels in clay (O'Reilly and New, 1982; Atkinson and Potts, 1977), it can be shown that $i = z/2$, where z is the depth to the tunnel centreline. This represents a good approximation, provided that z/D is greater than or equal to 1.5. Substituting for i in Equation (3) gives:

$$K_m = \alpha\sqrt{2\pi} \left(\frac{D^2}{z^3}\right) \tag{4}$$

from which the maximum bending strain, ε_b, in a small-diameter pipeline can be determined as:

$$\varepsilon_b = \alpha\sqrt{\frac{\pi}{2}} \left(\frac{D}{z}\right)^2 \left(\frac{d}{z}\right) \tag{5}$$

in which d is the outside diameter of the pipeline.

Equation (5) can be useful in predicting the maximum bending strain developed in small diameter pipelines, but requires an estimate of the ratio of the volume of lost ground to the excavated volume, α. This estimate must be made on the basis of judgement. It requires a careful evaluation of the ground conditions and tunnelling method in conjunction with previous construction experience. Field observations (e.g., Cording, *et al.*, 1976, Attewell, 1977; O'Reilly and New, 1982) indicate that α for clay with undrained shear strength exceeding 40 to 50 kN/m^2 has generally been between 0.01 and 0.02. Substituting these values into Equation (5), a range of bending strains can be estimated for a given pipe as a function of the tunnel depth and excavated diameter.

Figure 2.3.3 shows the maximum bending strain, expressed in microstrain, $\mu\varepsilon$, for a nominal 100mm-diameter cast iron main. Maximum bending strains are plotted as a function of excavated diameter for depths to tunnel centreline of 1.8 and 3.0m. It must be emphasized that the plot applies for excavated diameters typical of trenchless pipelaying in medium to hard clay, for which α is between 0.01 and 0.02. Furthermore, the plot also is based on the assumption that volume loss occurs as a result of face and peripheral encroachment.

For a given excavated diameter, the bending strain varies inversely as the cube of the tunnel depth. For example, the lower bound estimate of bending strain for a 900mm-diameter tunnel in Figure 2.3.3 decreases from 212 to 46 $\mu\varepsilon$ as the depth increases from 1.8 to 3.0m. The trends shown in the figure imply that trenchless pipelaying at depths exceeding 3.0m in medium to hard clay should not cause severe strains in overlying pipelines, provided there are no problems with local instability. Moreover, trenchless pipelaying under similar conditions at a depth of 1.8m would not generally lead to unacceptable strains, provided the excavated diameter is less than 900mm.

EXPANSIVE INSTALLATION TECHNIQUES

Expansive installation techniques include percussive moling, pipe driving, and on-line replacement. With on-line replacement an existing pipe is broken out by a percussive device and the opening is expanded to receive a new pipe, which is either pulled or jacked into place. The conditions which apply to on-line replacement are similar to those for an expanding cylindrical cavity. Several theoretical solutions exist for this problem, which can be helpful in visualizing how soil movements occur.

In this paper, it is assumed that the conditions of on-line replacement may be approximated by an expanding cylindrical cavity under plane strain in a perfectly elasto-plastic undrained clay, which is radially homogeneous and isotropic. Furthermore, it is assumed that the cylindrical cavity is buried at great depth with respect to the ground surface and that the *in situ* soil stresses are constant and equal in all directions. Solutions for this problem have been obtained in conjunction with the development of cavity expansion theory (Vesić, 1972) and soil pressuremeters (Gibson and Anderson, 1961; Palmer, 1972; Baguelin, *et al.*, 1972). There is evidence that this simplified characterization is a good general basis for judgement. In a review of soil pressuremeter testing, Wroth (1984) observed that an elastic-perfectly plastic model is satisfactory for test interpretation in undrained clays.

Problem definition

Figure 2.3.4 shows an expanding cylinder at depth to centreline, z, surrounded by a zone of plastic deformation with radius, R_p. It is assumed that the cylinder is sufficiently deep that the external pressure in the soil outside the plastic zone can be approximated as γz, where γ is the unit weight of the soil. The close-up view illustrates that the outside circumference of the plastic zone displaces a radial distance, u_R, as the initial radius of the cylinder, a, increases by a radial displacement, u_a. Because the elastic displacement at the initial radius, a, will be small compared with the plastic displacement, u_a is analysed only as a plastic deformation.

For the axisymmetric condition shown in Figure 2.3.4, solutions have been developed for the distribution of principal stresses within the plastic zone (e.g., Vesić, 1972, Baguelin, *et al.*, 1972). On the basis of this work, it can be shown that:

$$R_p = a_E \exp. 1/2 \left[\frac{p_i - \gamma z}{S_u} - 1\right] \tag{6}$$

in which a_E is the expanded radius, p_i is the internal pressure which expands the cylinder, and S_u is the undrained shear strength of the clay.

Recognizing that $a_E = a + u_a$, Equation (6) can be rewritten as:

$$\frac{R_p}{a} = \left(1 + \frac{u_a}{a}\right) \exp. \; 1/2 \left[\frac{p_i/\gamma z - 1}{S_u/\gamma z} - 1\right] \tag{7}$$

This modified form of the equation is expressed in terms of dimensionless parameters, which makes it easy to apply for a wide range of conditions.

By assuming that the clay in the elastic zone has Poisson's ratio, $\nu = 0.5$, so there is no volume change in the plastic and elastic zones, the equation of continuity for $R_p \gg u_R$ can be written as:

$$\pi(R_p + u_R)^2 - \pi R_p^2 = \pi(a + u_a)^2 - \pi a \quad (8)$$

in which u_R is the elastic radial displacement at R_p. Equation (8) reduces to:

$$2R_p u_R + u_R^2 = 2au_a + u_a^2 \quad (9)$$

It can be shown from elastic formulations that:

$$u_R = \frac{S_u(1 + \nu)R_p}{E} \quad (10)$$

in which E is the undrained Young's modulus of the clay.

By combining Equations (9) and (10), the following expression is obtained:

$$\frac{R_p}{a} = \left[\frac{2(u_a/a) + (u_a/a)^2}{\lambda}\right]^{1/2} \quad (11)$$

in which:

$$\lambda = 2\beta + \beta^2 \quad (12)$$

and:

$$\beta = \frac{S_u(1 + \nu)}{E} \quad (13)$$

Equation (11) is very useful for evaluating how different soil and geometric parameters affect the radius of the plastic zone. Figure 2.3.5 was developed on the basis of the relationships given by Equations (11), (12), and (13). It shows the radius of the plastic zone as a multiple of the initial radius of the existing pipe plotted as a function of the diametrical strain. The radius of the plastic zone is related to the increased radius of the existing pipe according to the undrained modulus, E, and undrained shear strength, S_u, of the clay. Clays often are characterized on the basis of E/S_u, which varies according to the overconsolidation ratio and plasticity of the deposit. Values of E/S_u for most clays fall between 1000 and 200 (Ladd, *et al.*, 1977), and the relationships in Figure 2.3.5 are plotted for E/S_u of 200, 600, and 1000.

The application of Figure 2.3.5 is limited by the assumptions underlying the theoretical analysis. As the radius of the plastic zone increases, the assumption of an axisymmetric stress condition related to γz becomes increasingly less consistent with the actual stress distribution. It is important, therefore, to restrict the use of the plastic expansion relationships to depths of burial which are large relative to the radius of the plastic zone. In this work it is assumed that Figure 2.3.5 will give a reliable indication of trends, provided that the depth to centreline of the pipe, z, is at least twice as large as the radius of the plastic zone. This constraint is illustrated in the figure by showing the ranges of application according to depth-to-diameter ratios for the pipe, z/D, of six and eight. For most clays, the range of application shown covers diametrical strains from 0.1 to 0.3. Limiting the range of diametrical strain also helps in controlling applications for cases which are consistent with the theoretical derivation. In the theoretical analysis, plastic strain rate and second order effects were neglected. These factors will become increasingly more important as the diametrical strain increases.

Once the radius of plastic deformation has been defined, it is necessary to relate this parameter to the pattern of deformation imposed in an overlying pipeline. Figure 2.3.6 shows the deformation which would occur along a horizontal line located at the boundary of the plastic zone around the expanded pipe. At any distance x with respect to the centreline of the pipe, the vertical displacement, δ_v, is given by:

$$\delta_v = \delta \frac{R_p}{(x^2 + R_p^2)^{1/2}} \qquad (14)$$

in which δ is the elastic radial displacement at $r \geqslant R_p$. From elastic formulations, δ is given by:

$$\delta = \frac{S_u(1 + \nu) R_p^2}{Er} \qquad (15)$$

Combining Equations (14) and (15) gives:

$$\delta_v = \frac{S_u(1 + \nu) R_p^3}{E(x^2 + R_p^2)} \qquad (16)$$

Equation (16) is an expression for the vertical displacement as a function of the horizontal distance. It can be shown that the maximum curvature of this distributed displacement occurs over the centreline of the expanded pipe at $x = 0$. Taking the second differential of Equation (16) at $x = 0$, we get an expression for the maximum curvature, K_m:

$$K_m = \frac{2S_u(1 + \nu)}{E R_p} \qquad (17)$$

If a small-diameter pipeline were located at the limite of the plastic zone as illustrated in Figure 2.3.6, then Equation (17) would give an estimate of the maximum curvature imposed in the pipe. For a pipeline with diameter, d, such that $R_p > 5d$, the maximum bending strain, ε_m, in the pipeline can be estimated by:

$$\varepsilon_m = \frac{(1 + \nu)S_u d}{E R_p} \qquad (18)$$

Equation (18) can be used with Figure 2.3.5 to evaluate the effects of on-line replacement. Consider, for example, that on-line replacement is planned for an existing 200mm-diameter sewer at a depth of 1.5m in undrained clay with $E/S_u = 300$ and $\nu = 0.5$. There is a 100mm-diameter cast-iron pipeline which crosses the sewer at a right angle at a depth of 1.0m. As a first estimate, it is assumed that the radius of plastic deformation should be confined to a distance equal to or less than the difference in elevation between the sewer and cast-iron main, such that $R_p/a = 5$. In Figure 2.3.5, it can be seen that this would correspond to a percentage increase in diameter of approximately 12.5 per cent. Hence, the expanded diameter would be limited to roughly 225mm. From Equation (18), the maximum bending strain in the 100mm-diameter main is predicted to be 500με. The predicted level of strain is likely to be an upper bound because it is assumed that the pipeline moves in one-to-one correspondence with the soil. In addition, any consolidation of loose or fissured soil around the sewer is not considered.

Because the use of Figure 2.3.5 is restricted to depths for on-line replacement which equal or exceed twice the radius of plastic deformation, there will be many conditions for shallow overlying pipelines, which are not valid for the theoretical considerations in this paper. Nevertheless, the theoretical derivations help to set upper bound limits on field performance and point out the critical soil properties and geometrical parameters which affect on-line replacement techniques.

As an additional consideration, the theoretical analysis can be extended to estimate the maximum internal pressure, p_i, required for expansion in various undrained clays at different depths. This value may be helpful in determining the amount of power or force required to operate the on-line equipment. It must be emphasized that this pressure does not include the force required to break the existing pipeline.

Referring to Equations (7) and (11), as u_a/a increases to infinity, it can be shown by combining and manipulating the equation that an upper bound on the pressure, p_i, is:

$$\lambda \exp.\ \left(\frac{p_i/\gamma z - 1}{S_u/\gamma z} - 1\right) = 1 \qquad (19)$$

from which the maximum value of the expansion pressure, p_m, can be derived as:

$$\frac{p_m}{\gamma z} = \frac{S_u}{\gamma z}\,[\ln(1/\lambda) + 1] + 1 \qquad (20)$$

Figure 2.3.7 shows the relationship between the maximum expansion pressure and the undrained shear strength for the range of E/S_u typical of most clays. The expansion pressure is related linearly to the undrained shear strength. Furthermore, the expansion pressure is most sensitive to the undrained shear strength, and is affected only modestly by E/S_u. This trend is in significant contrast with the relationship between the plastic radius and diametrical strain, in which the plastic radius is strongly influenced by E/S_u.

CONCLUDING REMARKS

Convergent installation techniques involve trenchless pipelaying methods for which the volume of excavated soil exceeds the volume of pipe installed. Small-diameter (0.9 to 1.5m) tunnelling, microtunnelling, (less than 0.9m diameter), and various types of thrust boring and jacking fall under this category. There are three principal sources of lost ground associated with soil tunnelling: 1) face and peripheral encroachment, 2) consolidation, and 3) local instability. Field and model studies have shown that volume loss from face and peripheral encroachment will result in a soil settlement pattern near the ground surface, which can be modelled by a normal probability curve. Analyses summarized in this paper imply that trenchless pipelaying at depths exceeding 3.0m in medium to hard clay should not cause severe strains in overlying pipelines, provided there are no problems with local instability. Moreover, trenchless pipelaying under similar conditions at a depth of 1.8m will not generally lead to unacceptable strains when the excavated diameter is less than 900mm.

Expansive installation techniques involve trenchless pipelaying methods for which the volume of the installed pipe or construction equipment exceeds the voume of the excavated soil. Percussive moling, pipe driving, and on-line replacement fall under this category. The conditions which apply to on-line replacement are similar to those for an expanding cylindrical cavity. Theoretical analysis of an expanding cylinder in a perfectly elasto-plastic undrained clay show some important trends. Expansion pressures are linearly related to the undrained shear strength and only modestly affected by the undrained Young's modulus. The relationship between the radius of the plastic zone of deformation and the percentage increase in pipe diameter depends closely on the ratio of the undrained modulus to undrained shear strength. The bending strain imposed in an overlying small-diameter pipeline is inversely related to both the radius of plastic deformation and the ratio of the undrained modulus to undrained shear strength.

The theoretical models developed in this paper help set an upper bound on deformations in the field and point out the critical soil properties and geometrical parameters affecting on-line replacement. It is recommended that additional work be pursued on the theoretical models as a basis for estimating performance, identifying important parameters, and planning field tests.

ACKNOWLEDGEMENTS

The assistance of M S Tawfik, who made helpful suggestions and reviewed the manuscript, is deeply appreciated. Special thanks are extended to L Crouse and A Avcisoy, who typed the manuscript and drafted the figures, respectively.

REFERENCES

[1]Atkinson, J H and Potts, D M, Subsidence above shallow tunnels in soft ground. *Journal of the Geotechnical Engineering Division, ASCE, 103,* GT4, Apr. 1977, 307-325.

[2]Attewell, P B, Ground movements caused by tunnelling in soil. *Large Ground Movements and Structures,* J D Geddes, Ed., Pentech Press, London, UK, 1978, 812-948.

[3]Baguelin, F, Jezequel, J F, LeMee, E, and LeMehaute, A, Expansion of cylindrical probes in cohesive soils. *Journal of the Soil Mechanics and Foundations Division, ASCE, 98,* SM11, Nov. 1972, 1129-1142.

[4]Clough, G W and Schmidt, B, *Design and performance of excavation and tunnels in soft clay: State-of-the-art report.* Presented at International Symposium on Soft Clay, Bangkok, 1977.

[5]Cording, E J, Hansmire, W H, MacPherson, H H, Lenzini, P A, and Vanderohe, A P, *Displacements around tunnels in soil.* Report No. DOT-TST-76T-22, U.S. Department of Transportation, Aug. 1976.

[6]Gibson, R E and Anderson, W F, *In situ* measurement of soil properties with the pressure-meter. *Civil Engineering and Public Works Review, 56,* 658, 1961, 615-618.

[7]Harris, C W and O'Rourke, T D, *Response of jointed cast-iron pipelines to parallel trench construction.* Geotechnical Engineering Report 83-5, Cornell University, Ithaca, NY, 1983.

[8]Heuer, R E, Catastrophic ground loss in soft ground tunnels. *Proceedings Rapid Excavation and Tunnelling Conference,* Las Vegas, NV, June, 1976, 278-295.

[9]Ladd, C C, Foott, R, Ishihara, K, Schlosser, F and Poulos, H G, Stress-deformation and strength characteristics. *Proceedings 9th International Conference on Soil Mechanics and Foundation Engineering, 2,* Tokyo, 1977, 421-494.

[10]O'Reilly, M P and New, B M, Settlements above tunnels in the United Kingdom — their magnitude and prediction. *Tunnelling '82,* Institution of Mining and Metallurgy, London, UK, 1982, 173-181.

[11]O'Rourke, T D and Trautmann, C H, *Buried pipeline response to tunnelling.* Paper presented to Europipe '82 Conference, Basle, Jan. 1982.

[12]Palmer, A C, Undrained plane strain expansion of a cylindrical cavity in clay: A simple interpretation of the pressuremeter test. *Geotechnique, 22,* 1972, 451-457.

[13]Peck, R B, Deep excavation and tunnelling in soft ground. *Proceedings 7th International Conference on Soil Mechanics and Foundation Engineering,* Mexico City, State-of-the-Art Volume, 1969, 225-290.

[14]Vesić, A, Expansion of cavities in an infinite soil mass. *Journal of the Soil Mechanics and Foundations Division, ASCE, 98,* SM3, Mar. 1972, 265-290.
[15]Wroth, C P, The interpretation of *in situ* soil tests. *Geotechnique, 34,* 4, Dec. 1984, 449-489.

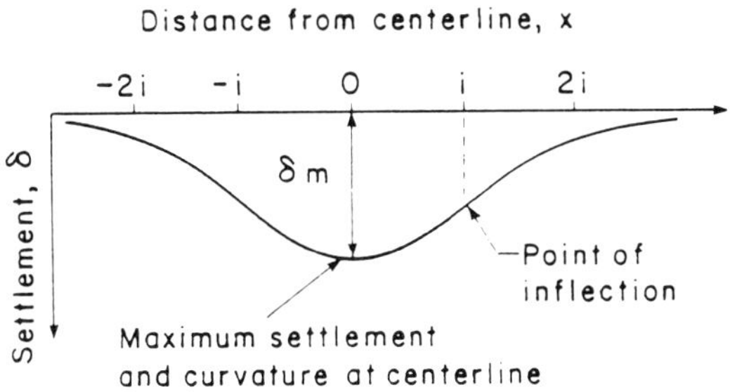

Figure 2.3.1. Normal probability distribution of settlement.

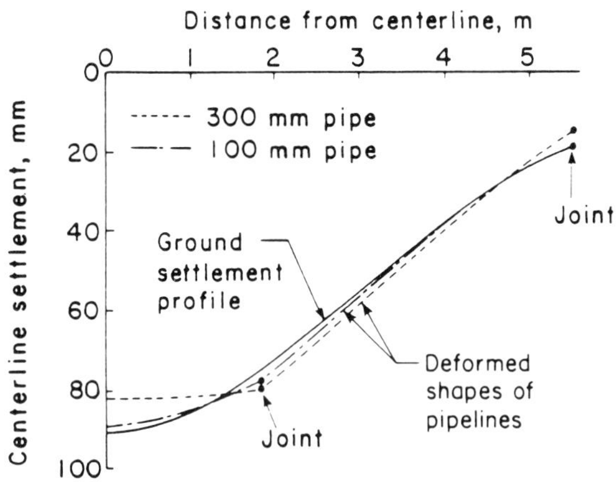

Figure 2.3.2. Comparison of ground settlements and pipeline deformation in loose sand (after O'Rourke and Trautmann, 1982).

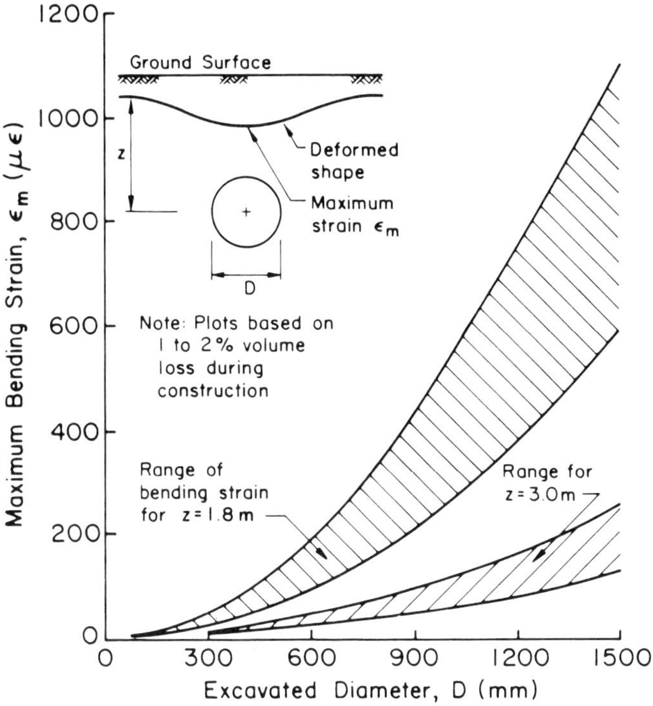

Figure 2.3.3. Maximum bending strain in a 100mm-diameter cast-iron main as a function of excavated tunnel diameter in medium to hard clay.

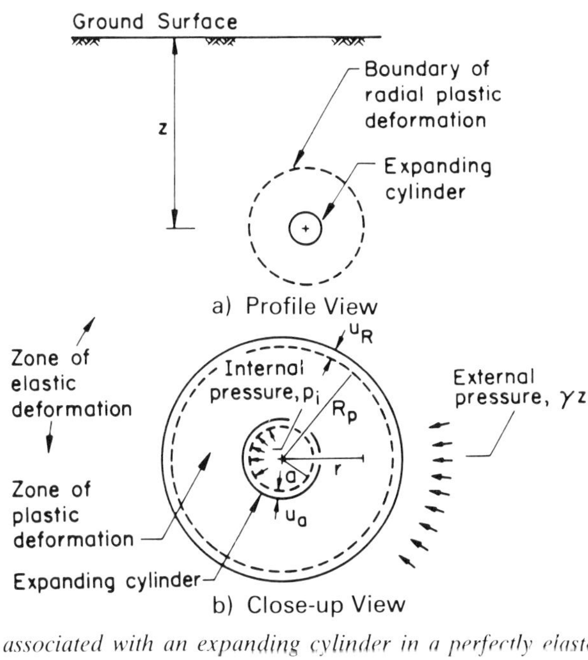

Figure 2.3.4. Geometry associated with an expanding cylinder in a perfectly elasto-plastic clay.

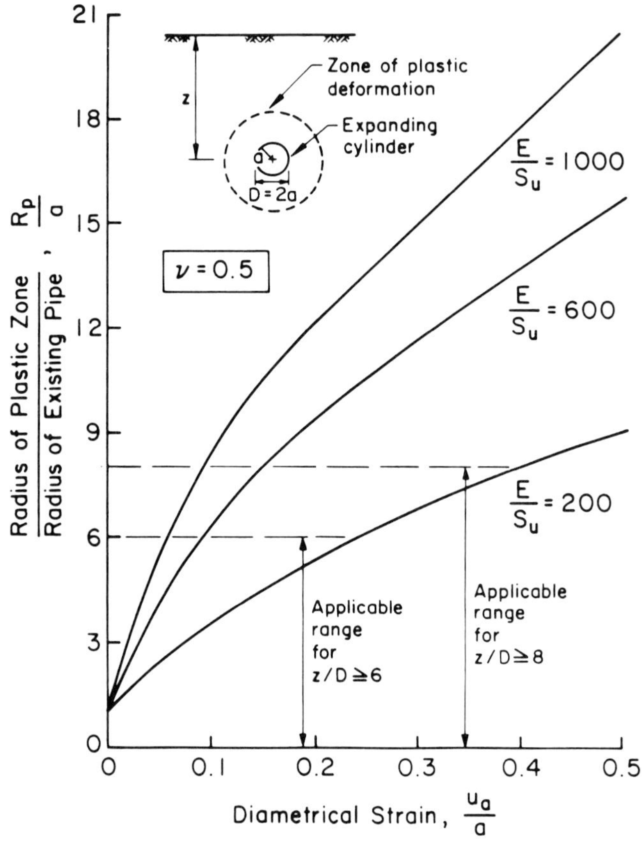

Figure 2.3.5. Radius of plastic deformation as a multiple of the existing pipe radius plotted as a function of the diametrical strain.

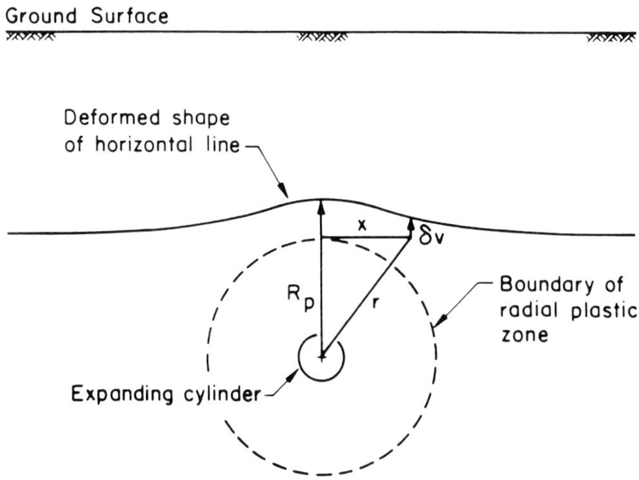

Figure 2.3.6. Profile view of distributed vertical displacements overlying an expanding cylinder.

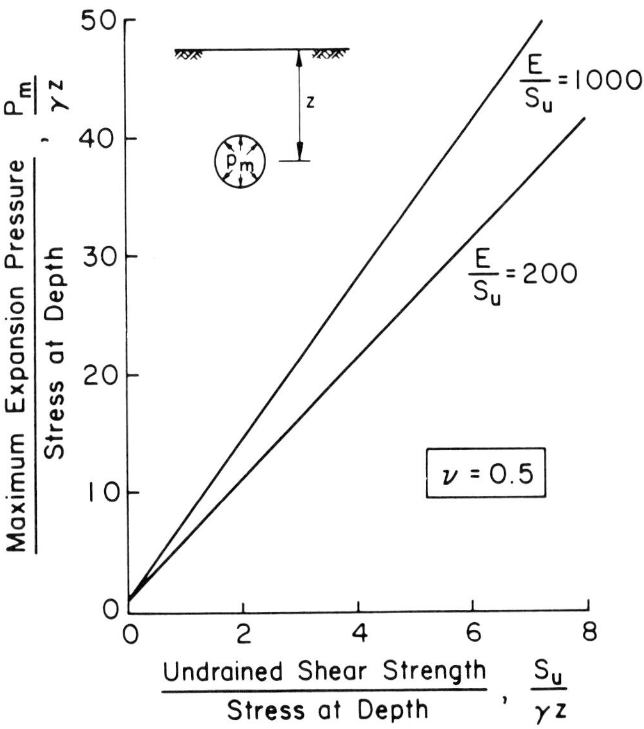

Figure 2.3.7. Maximum expansion pressure as a multiple of the stress at depth plotted as a function of the undrained shear strength divided by the stress at depth.

Authors' Introductions

Mr M F Hardy: Trenchless pipelaying — yes please: a highway engineer's point of view

I am here as a highway engineer, and in that capacity I am heavily involved in trying to achieve satisfactory management of traffic and sound economic maintenance of carriageways and footpaths, which means maintaining satisfactory standards at minimum cost. Trenching has a substantial impact on both traffic management and maintenance, particularly on main roads. I was interested in reading the papers that were presented this morning and listening to the tail-end of the discussion. Mr Thomson's figures that 30 per cent of new main construction and 58 per cent of replacement work is carried out in carriageways illustrate the significance of trenching to the highway engineer and, of course, the attraction of finding alternative means.

In the UK the rights of statutory undertakers and service organizations to lay services in highways are set out in legislation which was introduced in 1950. Those rights are the same, more or less, 35 years later, at a time when we have car and van traffic 8.5 times that which existed in 1950 and lorry traffic 3.5 times greater. Now Professor Horne's study which, as you know, is government-sponsored, is due to produce findings very shortly. To my mind the time is ripe to look at the impact of trenching in highways in the light of those 1985 flows and indeed of the conditions we are going to see in the next decade.

I have mentioned that some of my concern is about traffic on main highways. In my own county the traffic flow on many of the all-purpose roads, which form about 6 per cent of the total network, is 20,000 or more vehicles per day, and it is on those roads that we get into such trouble with trenching operations.

I would like to see thought now given to special arrangements to do all that is possible to avoid trenching on these main roads. All roads should not be treated the same in the legislation. I hope the Government is going to take this opportunity with Professor Horne's report to do something about legislation which is going to make it more attractive to this new technology.

We did an exercise in my own county a couple of years ago to look at patching requirements after a severe winter. We found that 30 per cent of the patching was associated with trench reinstatements. The impact of that expenditure is significant in the context of maintenance funds. About £90 million per year is spent on patching and £30 million probably is associated in one way or another with trenching operations. That is significant both nationally and locally.

Finally, I would like to make three general points:

1. Inevitably the vast majority of service work is associated with developed areas, where the streets normally already carry a wide variety of pipes and cables and other services, often shown on plans of very dubious authenticity. This lack of reliable information is a very real deterrent to the use of NO-DIG techniques in providing or repairing apparatus. We need fresh methods of locating and recording buried services, making use of modern computerized techniques.
2. The trenchless movement is unlikely to grow substantially unless it is pushed vigorously by all the parties involved, and unless the overall economics are taken into account. The economics would include the costs of traffic delays, reinstatement, maintenance and social costs. Why not set up some more positive trials — NO-DIG zones or NO-DIG routes — to enable us to assess the benefits and disadvantages?
3. Unless the new methods do get wide application, the costs will remain high and the use

limited. Hence the importance of this conference to me as a highway engineer and the follow-up spread of information and ideas. Everyone must be made aware of the potential of this developing technology and no one more so than highway engineers who have to deal with the results of trenches on the ground.

Mr P Hunter: Trenchless mainlaying within British Gas

The Engineering Research Station (ERS) at British Gas is carrying out a range of studies on the effects of earth-boring moles used to install the smaller diameter polyethylene mains and services in virgin ground and pipe-splitting moles used increasingly for the on-line replacement of cast-iron mains with polyethylene, size-for-size or greater. We are looking at the effects of the expanding cavity on both adjacent buried plant and the pavement structure. The work involves theoretical studies, laboratory experiments, controlled field experiments and the monitoring of the effects of operational moles. The aim is to prepare industry codes of practice for the proper use of trenchless methods and, as our services are frequently side by side with other utilities' services, to combine our findings with those of other investigators to produce joint industry codes.

In theoretical studies, the effects of the passage of a 65mm earth-boring mole close to mains of between 2-inch and 24-inch diameter have been predicted, and the soil/pipe interaction analysis applied to obtain the level of induced strain in the pipe. Figure 2.2.1 in the paper shows that the smaller pipes experience the higher strains since they are most likely to follow the profile of ground movement but the strain level decreases exponentially away from the mole. The larger mains are at minimal risk, apart from near impact, since they have high beam strength and resist the movement of the ground. The analysis enabled ground movement to be related to distance from the mole cavity, and it is quantified in terms of mole diameter.

The extent of ground movement around the expanding cavity is influenced by the degree of saturation of the soil. For example, with saturated soils having only one per cent of air voids, ground displacement decays to zero within about four cavity diameters from the cavity edge. With 12 per cent air voids the distance is reduced to within one diameter. These simplified predictions have ignored the effect of the ground surface, and assumed equal radial movement around the mole.

When the influence of the ground surface was taken into account using a finite element stress analysis, the preference for the expanding cavity to move towards the surface was demonstrated. Figure 2.2.3 shows the decay of movement in different directions from the expanded cavity. The ratio of upward movement above the mole to downward movement below is about 4.5:1. The movement is minimal if the depth/diameter ratio is greater than about 5. Further theoretical studies applying finite element stress analysis to a cavity expanding from 50mm to 200mm enables the pattern of ground movement to be further estimated. Figure 2.2.2 shows the contours of total displacement at a depth of cover of about 0.46 metres and shows the movement swinging upwards towards the free surface. The indication is that mains parallel are at less risk than mains crossing.

In the laboratory we have examined in more detail the likely zones of influence around a 65-mm mole by X-raying a soil box into which a 2-cm grid of lead shot had been introduced at mid-plane (Figures 2.2.4 to 2.2.7). The mole head was then passed through and the box X-rayed again. The zone of influence was limited to within about 3cm of the mole, though because of the high proportion of air voids, this was probably less than would occur in practice. Further experiments using this technique are planned for the near future.

Field experiments have been carried out to investigate the effect of an earth-boring mole passing beneath an established 100-mm cast-iron main. The results showed a build-up of strain in the pipe approaching about 200 microstrain as the mole passed the pipe, and a residual of only about 50 microstrain after the mole had passed, indicating some recovery of ground strain.

The vibrational effects from a 75-mm earth boring mole have also been measured. The result showed that the ground surface peak particle velocity decayed very rapidly away from the line of moling, from about 20mm per second 0.5 metres away to negligible values beyond about 2 metres.

The measurement of ground surface movement can be carried out using a simple beam (Figure 2.2.8), which is placed over the line of moling, levelled, and the distance between the road surface and the beam noted at 150mm intervals along its length before and after the passage of the mole. Alternatively we can use a semi-automatic road profile gauge as shown in Figure 2.2.9. This is a moving stylus attached to a potentiometer which gives a continuous plot of surface profile on an XY plotter.

On the operational side, it is planned to monitor the effect of the main-splitting moles which are used throughout British Gas for the replacement of cast-iron mains, mainly 3-inch and 4-inch.

The Engineering Research Station has also developed a main-splitting system, consisting of a towing section and a breaker unit. The breaker unit incorporates hydraulically operated fins which break the main into pieces and the conical shape of the body expands the bore of the main ahead of the protective sleeve. The breaker fins have a thrust capability of about 50 tonnes which enables it to burst virtually all types of joint and most types of fitting. It is presently undergoing field trials and a number of manufacturers have shown interest.

Professor T D O'Rourke: Ground movements caused by trenchless construction

In engineering we can make use of two approaches, empiricism, or practical experimentation and observation, and theory. I intend to talk about certain theoretical models which can be used to make sense out of trenchless excavation procedures. Theory is important and I think it comes at a ripe time now since there has been quite a lot of practical development which has shown that NO-DIG methods are viable. Theory is needed firstly, to provide a framework for interpreting field results and to guide empiricism; secondly to evaluate and to develop a predictive capability so that potential difficulties can be assessed systematically; and thirdly to provide a basis for design.

Trenchless pipelaying methods can be divided into two broad groups. The first group can be classed as convergent installations in which the volume of the excavation is greater than the volume of the pipe and equipment used to create it, so that the soil will tend to move towards the void. These installations include small diameter tunnels, microtunnels, jack- and thrust-boring methods. The other group can be referred to as expansive installation techniques, where the pipe and equipment volume is larger than the excavation volume, as in percussive moling, pipe driving and on-line replacement.

There has been much experience and study of convergent installation techniques which can be traced back to Terzaghi's work on the Chicago subway system in the 1940s. There is some very encouraging agreement between observations on ground movement in full-scale tunnelling and laboratory work at Cambridge University on a similar scale to the pipes installed by microtunnelling techniques. Thus there is a theoretical and empirical basis for making predictions, and some have been put forth in my paper.

The questions associated with expansive installation, however, are still awaiting a careful and considered theoretical treatment. Solutions already exist, based on cavity expansion theories or models which predict how much displacement will occur if new volume is introduced to the ground mass. It is assumed that the pipe is buried relatively deep with respect to its diameter so that soil is displaced radially from around the pipe. This will have limitations as the pipe approaches the ground surface.

However, the most important aspect to be dealt with in any modelling, any theoretical treatment of these problems, is the ground itself. Soil is very variable and the model should be sufficiently general, yet sufficiently simple, to take proper account of variations in soil properties. The paper gives a solution straight from plasticity theory. It is the only solution for which there is in fact a true closed-form solution, and I have tried to relate the volume of expansion in terms of a radius of plastic deformation and its influence on the displacements that would be imposed on a pipeline which crosses, let's say, an on-line replacement construction at right angles. It is assumed that there is no volume change, i.e. any volume introduced by virtue of the cavity expansion is transmitted in total to the surrounding ground. However, many soils are subject to compressibility. By starting with these kinds of models which predict or tend to over-predict the amount of bending strain imposed on pipelines above the expanded cavity, and then introducing suitable adjustments to account for compressibility, we have found that the amount of displacement is reduced substantially. In the example worked out in the paper, the maximum bending strain in the 100mm diameter main is predicted to be about 500 microstrain, but when allowance is made for only a 5 per cent change in air voids, that bending strain reduces to about 200 microstrain.

I hope the paper has shown the need for a balanced approach to the problems that we face, one that involves a healthy empiricism, and scepticism as to what theoretical results might say, but one which welcomes a certain amount of modelling input, so that together we can develop predictive capabilities for assessing the influence of on-line or trenchless replacement techniques on adjacent services, and for assessing the residual stresses that need to be carried by the pipeline.

Discussion on Papers 2.1, 2.2 and 2.3

R F V Aylott (Deputy City Engineer, City of London Corporation): Mr Hardy has given some statistics on the impact of services in one of the home counties. London has approximately a quarter of all the road openings in the UK — just under half a million per year — whereas it has just under four per cent of the total road mileage. That gives a very heavy concentration of openings. In the City of London there are about 3000 openings a year on 53 km of carriageway. Probably most activity is confined to some 30 key kilometres of the road network, so that we have a hundred or more openings per km per year in some of the more intensively used routes. This is in spite of the fact that we have 8km of pipe subway carrying 50km of shared trunk services. The scale of the problem, and hence the expenditure, is therefore quite horrendous.

In 1982/83 the highway authorities in the UK recovered just over £71 million in reinstatement charges from the public utilities — and this, in a sense, is for re-doing what has already been done. The total investment in trench reinstatement, including the original investment by the utilities, is probably nearly £300 million a year. The British Gas Corporation, with its intensive programme of mains replacement and improvement over the last 8 or 9 years, following the King report, have invested most heavily, but the Water Authorities are catching up. Gas and water together have contributed 74 per cent of the public utilities' activity in trench openings, and it is very encouraging that both industries are in the forefront of the new technology of trenchless pipelaying.

Some of the early attempts at pipe jacking were little short of disastrous, but, thanks to the pioneering efforts of James Thomson and others, we now have various techniques which are, to a very great extent, entirely trustworthy. We must now consider how to get the message across to those who have to decide which techniques to use.

I would like to hear the author's views on the question of alternative techniques being allowed within a contract. I am worried by the idea of the contractor nominating the method he will employ. I do not want to stifle innovation, but safeguarding the highway must be uppermost in our minds. At some point the views of the highway authority must be considered. I would accept that, in a town, there will be more public outcry if the services, rather than the highway, are disrupted, because you can go round the block to avoid a highway closure. But in a rural area a diversion may well add ten or fifteen miles to a journey.

In my view, the following are the key factors in the success of the NO-DIG concept. Where a new service is to be provided, the ability to locate accurately the existing buried services is of paramount importance. Where replacement is contemplated, the depth of the existing service is extremely relevant. There was an example last year of a gas board wanting to use the pipe-bursting technique actually within the structure of the highway, which I think is carrying enthusiasm for the NO-DIG concept a little too far. Other important factors are a knowledge of ground conditions, the relative costs of construction and proper supervision of the contract.

Accordingly, I would ask the following questions of the authors. What information should be assembled before a decision on the method to be adopted is taken? Who should take the decision on the method? Should the highway authority be consulted or not, and, if so, at what stage? Will the new techniques pave the way for more routeing away from highways in rural areas? Given the avoidance of surface disturbance, would this be more attractive to landowners, who would not have to suffer the anguish of possibly losing crops and dealing in compensation?

Is it possible to suggest some fairly generous rules of thumb on the minimum depth for using a new technique, having regard to the diameter of the service being installed? Perhaps Professor O'Rourke's work is moving in that direction. And can we be assured that we will not have too many unplanned excavations to recover moles which have become entrenched?

Mr M F Hardy: Mr Aylott's reference to the problems of road openings in the City of London supports my point that openings are concentrated in urban areas and, quite often, in specific locations in those areas. Very often that concentration of undertakers' apparatus coincides with a concentration of traffic, leading to traffic management and constructional problems.

On the question of alternative techniques, I would argue — and we put this argument to Professor Horne — that, certainly on major traffic routes, all options should be considered — the traditional method of trenching, the various NO-DIG techniques, and the possibility of using different locations altogether, eg, going on minor roads or even across private property. Each option should be costed.

It is true that, under current legislation, the highway authority, even if it has all this information, is not able to influence decisions except by agreement with the statutory undertakers and the contractors. This is where I would like to see some change in legislation.

Control of contracts is generally regarded as a difficult problem. But if the technique is well understood, we must get down to writing proper conditions of contract and specifications, detailing the method of carrying out all investigations and control of the work. I do not think this is an insoluble problem, bearing in mind the benefits to be derived from using NO-DIG techniques.

Mr M Howe: All of the utilities are aware of the advantages of using some of the methods we have heard described today. But it is perhaps worth pointing out to highway engineers that a considerable amount of mainlaying is going on by inserting new polyethylene main through existing mains. This requires only a short excavation at each end of the existing pipe, and small excavations at right angles to connect the new services.

It is pleasing for British Gas and the other utilities that the highway authorities recognize the considerable effort put in by the utilities, not only to develop new equipment, but also in the whole science of excavation and reinstatement of excavations. The specifications were not designed for the reinstatement of small narrow trenches, but for major road construction. The work that British Gas has done over the last six years on reinstatement materials, techniques and compaction equipment has been incorporated into a more practicable specification. I am pleased to say that highway engineers and the utilities have ongoing meetings to discuss how to derive the greatest benefit from the research undertaken by the utilities and continue to bring it into the operational sphere.

With regard to plant location, the National Joint Utilities have a test facility for evaluating plant location equipment. This consists of a representative selection of buried pipes and cables in various configurations to both standardize and simulate street locations. All existing commercially available equipment can be evaluated at this site and manufacturers are using it for development trials to improve effectiveness. In association with this site NJUG also have a performance specification for plant locators. Again the utilities are taking a strong lead to improve available technology.

Turning to the new technology presented at this conference, it is publicized by presenting papers at as many conferences as possible. In this way the whole subject is opened up for comment by users, highway engineers and others in the hope that general agreement and acceptance of the new technology can be achieved. Having used the term new technology it is worth recalling that installing pipes by moling is an idea some decades old now. The choice of method must always be based upon cost-effective considerations and conditions at site. Traffic flow is always an important consideration at the planning and operational stages. These principles are well established in all areas of street works on road maintenance, sewer renewal etc.

Professor O'Rourke: A lot of attention has been directed worldwide to the NO-DIG concept, and the good work done by British Gas and the Water Research Centre is widely recognized.

I would ask Peter Hunter and Malcolm Howe what happens in the pipe-bursting technique after the existing pipe has been fractured. This is often done with cast-iron pipes, leaving very sharp edges which could damage the new polythene pipe. What kind of point or line loads have they measured, and what precautions do they take to mitigate damage?

Mr Howe: After bursting, a pipe lining tube of 3mm wall thickness is pulled in behind the mole. The gas-carrying pipe is then inserted into the lining tube. The lining is inexpensive, and thick enough to protect the new pipe. No problems have occurred so far.

P. Lofthouse (Director of Operations, Southern Water Authority): Mr Hardy thinks the solution to the problem of excavations in highways and disruption of traffic lies in more legislation. I say emphatically No! to additional legislation. In my view the aim of this conference is to look for practical solutions. I believe they are here already.

There have been considerable developments in sewer renovation techniques, with tunnelling for large diameter sewers and pipe bursting for small diameters, and the costs of these methods are now proving to be economical. The experience of four pipe-bursting contracts in Southern Water shows that, although the first contract exceeded the cost of open excavation, the subsequent three contracts were progressively more viable financially.

To my mind, the answer is not to seek legislation but to push ahead with NO-DIG techniques and to ensure that they are sufficiently viable to avoid the need to consider open trenching. The best solution is the one which offers the minimum disruption at the lowest cost.

Mr Hardy: I am no lover of legislation, but I cannot see how the techniques we are talking about can be given added weight unless we have revised legislation and unless we get some means of putting all the costs into the equation. At the moment traffic costs and social costs are not taken properly into account. It is public opinion that will alter the situation, and the public are getting fed up with traffic delays on major routes, whether on the motorways on Bank Holidays, or in the middle of a large conurbation. They are putting pressure on elected councillors and politicians — and that is one reason why the Horne inquiry was set up. The highway authorities, statutory undertakers and others concerned with services must get together to take a lead in proposing the right sort of legislation; otherwise they may get something they do not want.

A. G. Munro (Engineer for Metropolitan Works, Melbourne and Metropolitan Board of Works): From the foregoing discussion it is my understanding that in the UK there is a considerable amount of conflict between utilities and road authorities. In Melbourne the relevant Act of Parliament gives the Melbourne and Metropolitan Board of Works the authority to construct services in private land, in public land, in roadways, etc. — virtually anywhere within the Metropolitan area. However, another Act prohibits the digging up of freeways (motorways).

Codes of Practice have been developed by agreement between the Board of Works and the Municipal Association of Victoria (a body representing local councils). These Codes cover such things as the location of services in roadways, responsibility for, and costs of, reinstatement, etc. Consultation occurs between Board and Council officers at the design and construction stages. I believe this works well and is beneficial to the community at large.

R H Morgan (Rickmansworth Water Company): The conflict between highway authorities and utilities will continue until the highway authorities recognize the fact that the streets serve more than one purpose. They are there not simply to carry vehicle traffic, but also to provide facilities for the utilities. This situation has been recognized in law for many years, and the sooner highway engineers accept it and promote the overall management of streets for their various purposes, the sooner the conflict will disappear.

D O Evans (Area Distribution Engineer, West Midlands Gas): As a founder member of the Midland Statutory Undertakers' Liaison Group (MSULG), which preceded the national group, NJUG, by a few years, I would like to emphasize that, through the activities of such groups, inter-utility wrangling is now much reduced. There is a commonality of purpose and response by the utilities to other parties as a consequence.

The MSULG co-operates in liaison committees with highway authorities, and we certainly advanced a long way in our dealings with the West Midlands County Council. Although we do not always see eye to eye, we at least have a chance to meet and discuss various issues, in contrast to those highway authorities who will not meet us to discuss any issue, or those who only want to meet us on their own terms. We also have some valued liaison committees with county authorities that go back much further than MSULG.

There are too many conflicting voices from highway authorities, and there is a real need for them to get their act together. It must not be overlooked that they also have apparatus beneath the surface — drains, and cables to street furniture and traffic controls. It would be unfortunate if real co-operation could only come about under the PUSWA Review, with Professor Horne or the Minister, Lynda Chalker, knocking our heads together.

D Wilson (Assistant County Surveyor, Berkshire County Council, and representing the Institution of Highways and Transportation): In making a plea for highway engineers to recognize that the highways exist for more than one purpose, and that therefore co-operation between highway authorities and public utilities is absolutely essential, both Mr Morgan and Mr Evans expressed a desirable objective.

However, it must be stated that the *primary* function of a highway is for the passage and repassage of persons and goods. Historically this has come to mean that the utility "delivering" his "goods" has a right to use the highway (or, more correctly, the soil under the highway) as a receptacle for his apparatus. It must nevertheless be recognized that this is a secondary, and fortuitous, function.

The plea for co-ordination is agreed, provided it is for the purpose of progress, not for finding scapegoats; for the overall public benefit, not for the sake of any one sectional interest. It is to be hoped that Professor Horne's report, and subsequent action by the Government, will create a climate in which such "public" co-ordination is facilitated. Then engineers will be able to get back to being the engineers they are supposed to be, seeking the best-managed solutions to the many problems facing us all with an ageing infrastructure.

G C Cox (Consultant, M McDowell Co-Partnership): Mr Hardy quotes in his paper a figure of 1.25 times the estimated engineering costs for the centre of Reading as being the social disruption cost. How does he reconcile this with the figure of between 3.0 and 10.0 for the same area reported by Dr Denis Peters of the Water Research Centre? Are they using the same model?

Could the authors of papers 2.2 and 2.3 compare the measurements of ground movement — I refer to Mr Hunter's excellent displacement vector diagram — and microstrain resulting from pipe expansion with those of typical — and I mean typical, not good — normal trenching procedures?

The expansion method has been criticized because of damage to adjacent services. But surely traditional alternatives are in most cases more damaging. I suggest that future trials should include normal trenching as one of the standard procedures for comparison. This would give more practical guidance than figures of microstrain in isolation.

J Thomson (Consultant): Would the authors of papers 2.2 and 2.3 tell us something of the influence of ground conditions on impact moling? How do varying ground conditions affect the operation?

R B Rosbrook (Development Engineer, Southern Water Authority): I have studied the costs of sewer pipe bursting and find that it is becoming quite competitive with other methods, purely on civil engineering costs, without considering social costs. I doubt whether legislation is necessary, and I think the paper by Glennie and Reed this afternoon will support that view. Clearly, as more work is done, these costs will come down still further.

Mr Howe: Mr Cox queried the use of microstrain measurements. Just as in structural and other branches of engineering, safe working levels of stress and strain are established for various materials, in this area of study, too, microstrain is measured so that it can be compared with the maximum working strain permissible in the material concerned. In the reported work, the measured values of microstrain were below the accepted maximum working values, and therefore the techniques were acceptable.

Mr P Hunter: In answer to Mr Thomson, in general, the stiffer the ground the higher the strain level will be in the pipe, because of the improved mechanical transfer of soil displacement to pipe deformation. Conversely, in soil with air voids, the ground movement will be reduced because it is compressible. The road structure itself can be important. With a main expanded under a thick road structure, little ground movement can be transmitted to the surface, because the localized expansion will be absorbed in the adjacent soil. We are investigating the roles of the road structure and the ground below on the profile of ground movement, both by controlled experiments and by monitoring operational moles.

Mr Howe: It is worth pointing out that gas mains are customarily laid about 2ft deep under footways and 2½ft under roads. At these depths the soil is usually found to have high air voids and these conditions tend to favour moling.

Professor O'Rourke: Soil conditions are extremely important. We have spoken of trenchless excavation being divided into expansive and convergent methods. A percussive mole, for example, would be an expansive method, since it is creating extra volume. But if the level of vibration were suitable, and the material into which the mole was being inserted was a loose cohesionless sand, it is likely that appreciable consolidation, or volume loss, would occur. Volume loss could exceed the amount of expansion caused by inserting the device.

During the construction of the Washington DC Metro, an air rammer was used at one location to drive 3-inch diameter holes to receive tie-backs. It created a significant amount of vibration. The soil was a loose, cohesionless material, typically with blow counts of 5 to 10. The settlement at the road surface as a result of the anchors being installed was several inches.

We have found that the compressibility of the soil can have a dramatic effect on the degree of disturbance which may be conveyed to adjacent utilities. From preliminary modelling work we can see that, for changes in the degree of saturation of 5 to 10 per cent at constant water content, the zone of significant displacement may be confined to perhaps three radii from the centre of the pipe being replaced. This is encouraging. However, we can also encounter soils which are incompressible. In fact, there are some conditions where dilatancy can occur and there will actually be a volume increase.

As far as trenches are concerned, we need to differentiate between shallow and deep trenches. By shallow trench I mean one which is 4 to 5 feet deep, does not require support, involves relatively low stress changes, and is generally open for not more than 48 hours before backfilling. Under these conditions, the displacements will be more or less elastic. Unless the main is very close to the edge of the excavation and heavy loads are imposed, the distortion or displacement of an adjacent main will not be excessive.

On the other hand, with deep trenches, say 10 to 16 feet deep, with some form of sheeting and timber support, open for long periods, and subject to variations in workmanship, very large displacements of adjacent pipes can be generated. In these circumstances there will be great advantages in using a trenchless on-line replacement method, where, if the soil compressibility is suitable, the zone of influence may be confined within a few diameters of the pipe being replaced.

Dr Ing D Stein (Session Chairman): In Hamburg, using the Telemole system, in sandy ground, and with overburden of 2 to 2.5m, we have measured ground movement, which we found to be only about 1mm. If you use the best equipment and work carefully, I think ground movement need not be a problem.

Mr Hardy: I cannot answer Mr Cox's question about the relationship between traffic delay costs and construction costs directly, but will look into this and speak to him later.

Mr Munro is correct in identifying a utilities versus roads conflict in this country, and I hope we can take the opportunity of the PUSWA inquiry to get future arrangements on a better footing. The Melbourne system of Codes of Practice is an interesting approach. We have had for many years in Britain what is called a "model agreement", but it has not been adopted entirely either by the statutory undertakers or by the highway authorities. Both sides may to some extent be criticized for that. However, the "model agreement" is outdated, too detailed and too rigid. We need something that offers more flexibility.

E W Flaxman (Chairman, Conference Organizing Committee): Several speakers today have referred to the current review of the Public Utilities Street Works Act, under Professor M R Horne. We did invite Professor Horne to this Conference, but he is too busy completing his report to the Transport Minister, Mrs Chalker, to be able to attend. He hopes to submit his report by about the middle of May. The Institution has undertaken to send him a short summary of points arising from the Conference which are relevant to his inquiry.

2.4 The comparative costs of some methods of trenchless construction

D J Irvine

Technical Services Manager
Central Engineering Department, Tarmac Construction Ltd

The results of a study to compare the costs of installing and renovating pipes are discussed. Various methods of construction are considered including work in trench and tunnel and a cost model is defined. Different site conditions are taken into account and include good and bad ground under both roads and fields. The estimated Gross Unit Costs of constructing new pipelines are given excluding maintenance, running and social costs. The costs of renovating existing pipes are based on average cost equations based on a number of completed contracts. Graphs of Gross Unit Cost v. effective pipe diameter are presented and some conclusions drawn.

INTRODUCTION

During recent years there has been increasing concern at the total cost of installing services in trenches. It has been suspected that in addition to the direct cost of the works on site there are significant additional social costs, particularly in urban areas. The development of alternative trenchless construction has been stimulated by, among other things, the prospect of reducing the social costs while trying to achieve direct costs which compare favourably with those of trenched construction.

To provide better information on which to judge the benefits of currently available techniques, the Water Research Centre commissioned a series of studies of the technical and economic aspects of trenchless construction. This paper describes work undertaken by Tarmac Construction Limited to compare the direct costs of various methods of laying and renovating sewer pipes. The study was intended to form one part of an overall review and hence indirect and social costs were excluded.

THE OBJECTIVES

Economic studies are concerned with making comparisons between a number of alternative ways of investing resources in order to select the one which gives the optimum future return for the investment.

The difficulties of economic comparisons are well known. There is a problem in ascertaining the true initial cost of a project since this can be greatly influenced by the position of the appraiser within the economic system. Judgements vary on degrees of responsibility for risk and damage and hence on what constitutes an element of direct cost. For instance, individual promoters of projects frequently exclude consideration of the wider social costs and limit themselves to insuring their immediate third party risks. Nevertheless the risk of long-term damage to adjacent roads, buildings and services must be increased whenever ground is disturbed by construction and the cost of this damage will be ultimately paid by the owners of the damaged roads, buildings etc. unless the promoter or his contractor have been negligent.

Therefore, to simplify the study of direct cost it was limited to a comparative assessment of the contractor's costs plus the cost of engineering and supervision for constructing sewers up

to 1500mm diameter by the following methods:
(a) Traditional pipelaying in trench;
(b) Trenchless pipelaying systems;
(c) On-line replacement of pipes;
(d) Renovation of existing pipes.

Thus the resulting information could be combined with separate studies which concentrated on the social and other costs associated with the various methods.

THE COST MODEL AND DEFINITIONS

The cost model required a 600 metre length of pipe to be installed using gravel bedding and with manholes at specified distances. Each pipeline was to be constructed as a single contract and the pipe diameter, site conditions and construction method remained the same over the whole length.

Four separate sites were envisaged:
(i) Greenfield in good ground (dry stiff clay);
(ii) Main road in good ground (dry stiff clay);
(iii) Greenfield in bad ground (soft wet conditions);
(iv) Main road in bad ground (soft wet conditions).

Definitions of cost

The various costs were defined as:
(i) Nett Rates - The Contractor's unit costs before adding overheads and profit;
(ii) Tender Unit Cost — The cost per metre of construction including reinstatement, ie the Contractor's selling price;
(iii) Gross Unit Cost — The cost per metre of the total scheme including the cost of the Engineer's design and supervision.

An allowance was made in every case for the costs of design and engineering supervision of the works by adding a percentage uplift to the tender unit cost. The Gross Unit Cost of each method was calculated as at May 1984.

Definitions of technical terms

It was necessary at the outset to clarify and define the meaning of some technical terms, particularly those used for describing recent innovations or affected by them. These definitions were co-ordinated with similar terms used in the technical studies being carried out under the auspices of the Water Research Centre:

(i) *Construction in trench* — for the purpose of the study this term was used for installing new pipelines on new lines although it was recognized that on-line replacement of existing pipelines may also be done in trenches.
(ii) *Trenchless construction for new pipelines* — the installation of a new pipe in the ground without the need for continuous open cut. The new pipe will have adequate structural and hydraulic qualities without incorporating any prior existing pipework.
(iii) *Trenchless construction for on-line replacement* — methods providing a new pipe on the line of an existing pipe. Typically a percussive mole is used to expand (and consequently destroy) the existing pipe to provide a clearance hole for the new pipe.
(iv) *Microtunnelling* — a method of constructing small diameter tunnels using remotely operated equipment where no man-access into the tunnel is possible (ie generally where the internal diameter of the tunnel is less than 900mm)
(iv) *Sewer renovation* — methods which improve the structural and/or the hydraulic performance of the original sewer fabric. The methods generally fall into two groups suitable for either man-entry pipes or non-man-entry pipes, although some techniques cover the whole range.

THE METHODS CONSIDERED

1. New construction

Trenches
Structural calculations were done to ascertain the pipe classes required for the different site conditions. In some cases Class B bedding was found to be insufficient and Class A and Class E beddings were provided. For convenience in pricing, vitrified clay pipes were used for the small diameters and concrete pipes used in the larger diameters.

The excavation was done using a hydraulic backhoe excavator and the trench support was traditional sheeting and strutting varied to suit the ground conditions. Some allowance was made for obstructions including crossing services etc.

The costs of reinstatement to the surface were included and allowed for stripping and replacing topsoil and re-seeding in fields, and both temporary and permanent reinstatement in roadways.

In all cases the total cost of building the manholes was calculated and then distributed as a cost per metre of pipeline. Seven manholes were allowed in the 100mm to 750mm diameter pipe range while the larger diameters had four manholes.

Timbered headings
These costs were calculated for working in good ground only. Hand excavation was allowed with mechanical handling of the spoil up the shaft. On completion of the pipe the heading was backfilled with concrete leaving the timbering in place. The working shafts were used for constructing the manholes in all types of tunnelling.

Segmental tunnels (dry firm ground)
These tunnels were constructed using boltless smooth bore segments. Where the hydraulic design of the system required large diameters of pipe the tunnel was used running full. For smaller service diameters the tunnel was lined either with an *in situ* concrete channel or concrete pipes laid with a concrete bed and haunch (Figure 2.4.1). The lining methods were based on specifications used by some water authorities.

Segmental tunnels (soft wet ground)
In this case bolted concrete segments were used. An *in situ* concrete lining was provided where necessary for hydraulic reasons. As with the tunnels in good ground, the smaller diameter pipe capacities were provided by constructing either an *in situ* concrete channel or a concrete pipeline within the tunnel (Figure 2.4.1).

An allowance was made for using low-pressure compressed air to assist in stabilizing the tunnel face and excluding water.

Specialist tunnelling systems
In these cases subcontract prices were obtained from the various specialists and their assistance is gratefully acknowledged. The costs of manhole construction and any working shafts and access pits to be provided by the main contractor were then added to the subcontractor's price for the tunnelling.

As before, tunnels were lined where necessary to provide smaller diameter service pipes.

Pipe jacking
The subcontract prices for this method were provided by the Pipe Jacking Association. Again the costs of manholes and working pits built by the main contractor were added. An additional allowance was made in bad ground for dewatering using wellpoints.

2. On-line replacement using trenchless construction

Tunnelling and pipe jacking
The on-line replacement of pipes can be done using several techniques of trenchless

construction which are also used for installing new pipes.

Segmental and minitunnelling can be used to follow the line of the old pipe and to construct a new structure in the ground. It is necessary to prevent flows through the original pipe and this may entail overpumping. Live lateral connections can be remade fairly quickly by working within the new tunnel and installing temporary drainage arrangements.

Pipe jacking can also be used in the same way although reconnection of laterals presents a slightly greater problem since the whole of the pipe lining is moved and the laterals cannot be reconnected until the main line is complete.

The costs of these methods of on-line replacement are considerably influenced by the number of lateral connections and the need for overpumping and diverting flows.

Moling

The use of impact moling has been undergoing considerable development particularly in the gas industry.

Size-for-size replacement of pipes involves some enlargement of the existing bore prior to installing the new pipe and this may disturb the surrounding ground and any lateral connections. At the present time the laterals are usually remade by local excavation in pits but there are several remote-controlled devices in the course of development.

The costs of the methods tend to vary considerably depending on the number of lateral connections and whether access and receiving pits are required in addition to the existing manholes.

3. **Renovation techniques**

These methods included sliplining, precast gunite and other types of preformed lining. The costs were calculated using equations derived by WRC from the records of a wide variety of contracts in various parts of the UK. The original cost data had a wide scatter and the 95 per cent probability upper and lower limits to the costs had been derived on a statistical basis.

Since details of the contracts providing the original cost information were not available it was not possible to determine how closely the parameters of the equations reflected the conditions assumed in the cost model.

THE COMPARISON OF GROSS UNIT COSTS

The estimate of the contractor's costs was done in a formal manner. Quantities were calculated, a programme drawn up and the necessary construction resources allocated. As previously stated, the costs of designing the permanent works and the engineering supervision on site were allowed by adding a percentage to the Tender Unit Cost.

The Gross Unit Costs of the various methods considered are shown on graphs in Figures 2.4.2, 2.4.3 and 2.4.4. The original graphs were drawn to the same scale and printed on clear film to allow convenient comparison.

Since the costs are based on a single hypothetical model, the curves reflect general relativities and trends rather than absolute costs which can be applied directly to a particular case. The more the case differs from the cost model the less valid is the comparison.

It should also be recognized that some of the methods are highly capitalized, which may reflect against them in the present depressed state of the market. The high initial cost of plant means that high utilization is required to reduce the unit cost. It is a matter of conjecture how the price relationships might change if such methods were assured of higher utilization than it is customary to assume at the present low volume of total demand.

Another point to be noted is that some of the methods are more sensitive to changed physical conditions than others. The Gross Unit Costs allow for the Contractor's risk but under many standard forms of contract the Promoter carries the risk of the additional costs arising from unexpected physical conditions or artificial obstructions. Thus, a proper assessment of total direct project costs should include some allowance for such potential liabilities.

The graphs show that if social and disruption costs are ignored then in the cases considered:

(a) traditional pipelaying in trench is less expensive than any other method of installing new pipes;
(b) the cost of renovating man-entry pipes is competitive at the deeper depths;
(c) non man-entry tunnels are cheaper than pipe construction in either heading or man-entry tunnels;
(d) in good ground timbered headings are cheaper than man-entry tunnels for constructing small-diameter pipes but are more expensive for larger diameters;
(e) in bad ground the costs of forming man-entry holes by pipe jacking, segmental and minitunnelling are all very similar;
(f) Man-entry tunnel and pipe jack systems become less competitive for the small sizes of service pipeline because the excavation and installation of the main lining are in effect temporary works allowing the construction of the main pipeline. However, if the unused part of the tunnel space could be used for installing other services then the unit costs would be reduced considerably. The provision of a tunnel is an asset which may be under-used in economic terms if only one service runs through it, although the provision of easy access may reduce the maintenance costs.
(g) The increase in cost of pipes in tunnels at diameters below 900mm is due to the additional cost of lining the tunnel for hydraulic reasons together with an increase in the number of manholes provided.

CONCLUSIONS

For the cost model considered the direct construction costs of pipes installed using trenchless construction appear to be higher than the direct costs of equivalent construction in trenches.

Nevertheless the direct cost of a project to the promoter is only one element of the total cost borne by the community at large. There is a need to establish for each method the overall cost including the social and other indirect costs. This will provide a less parochial basis for comparison and assessing the best total cost-benefit.

The problems involved in implementing the results of such comparisons, which could be to the disadvantage of individual promoters in the short term, must be a matter for conjecture.

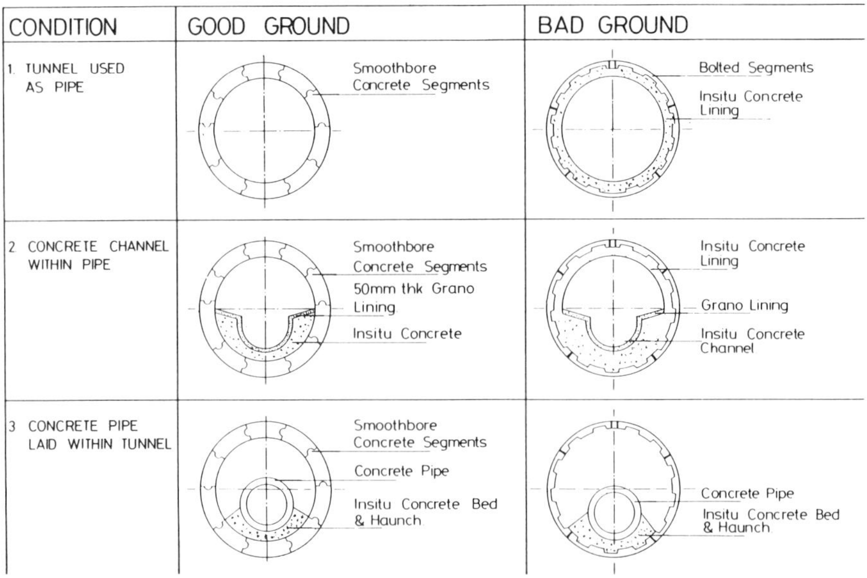

Figure 2.4.1. Pipes in segmentally-lined tunnels: methods of forming nominal service pipe diameter.

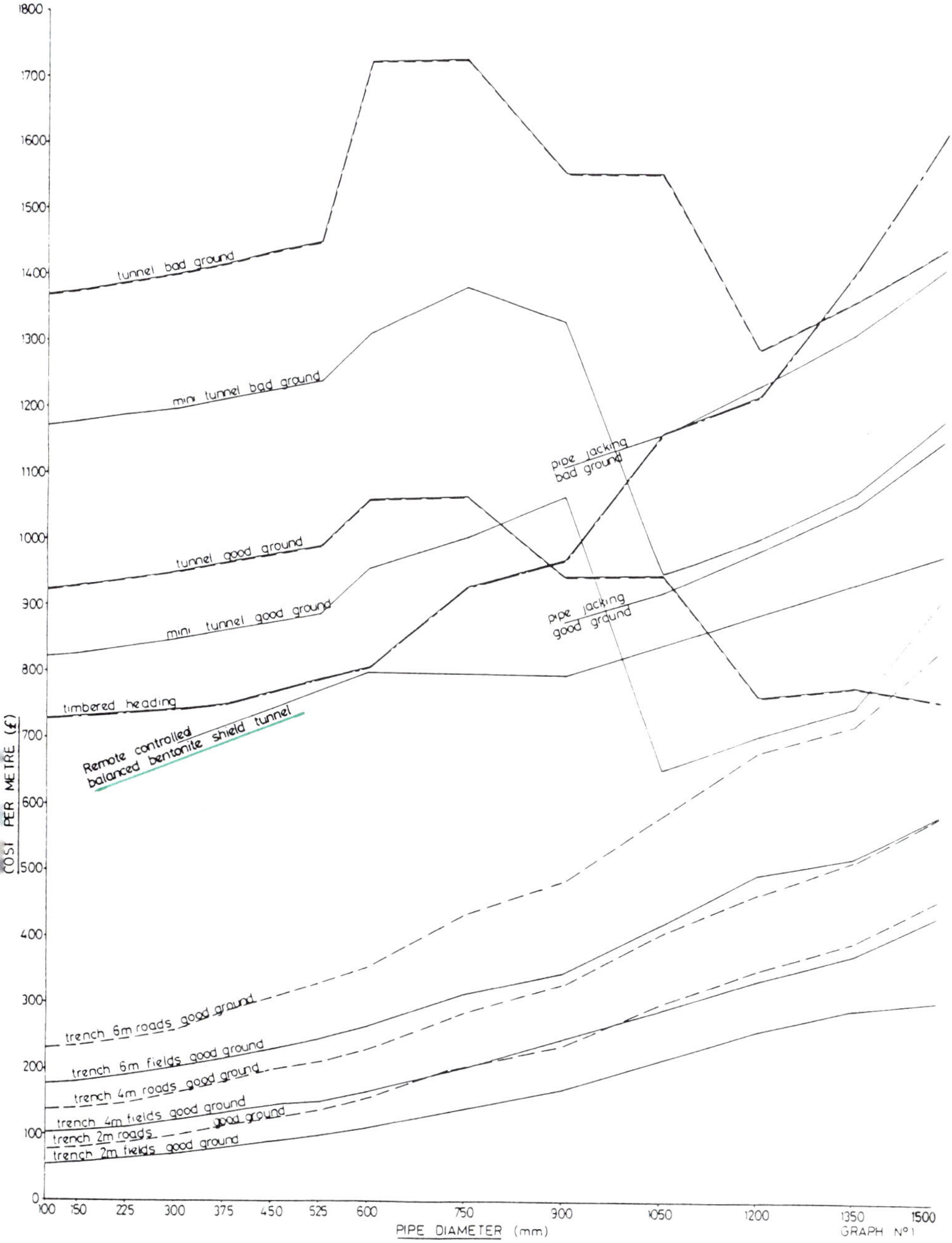

Figure 2.4.2. Comparison of gross unit costs: new pipes in trenches and tunnels.

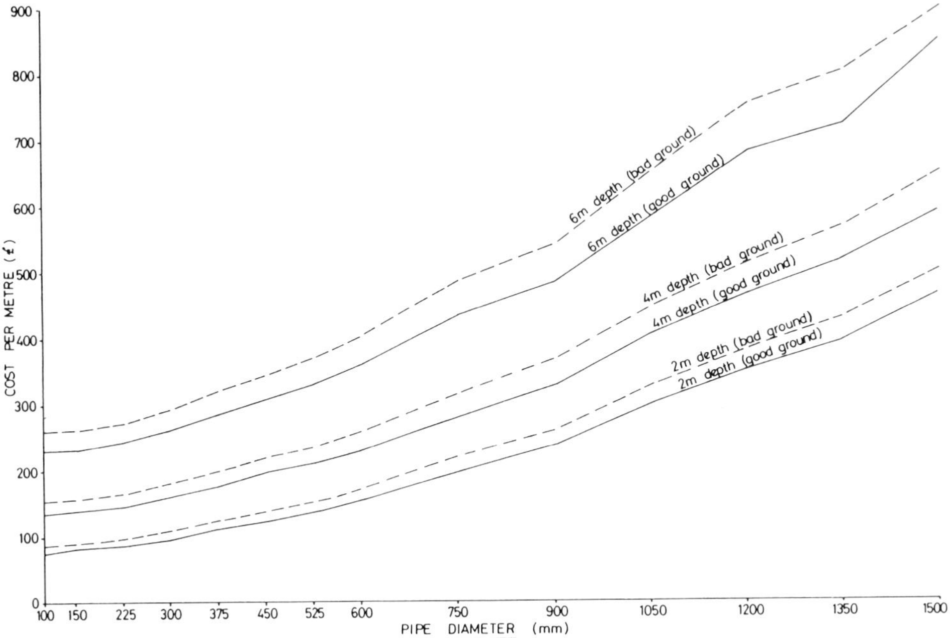

Figure 2.4.3. Comparison of gross unit costs: pipes in trenches: good and bad ground.

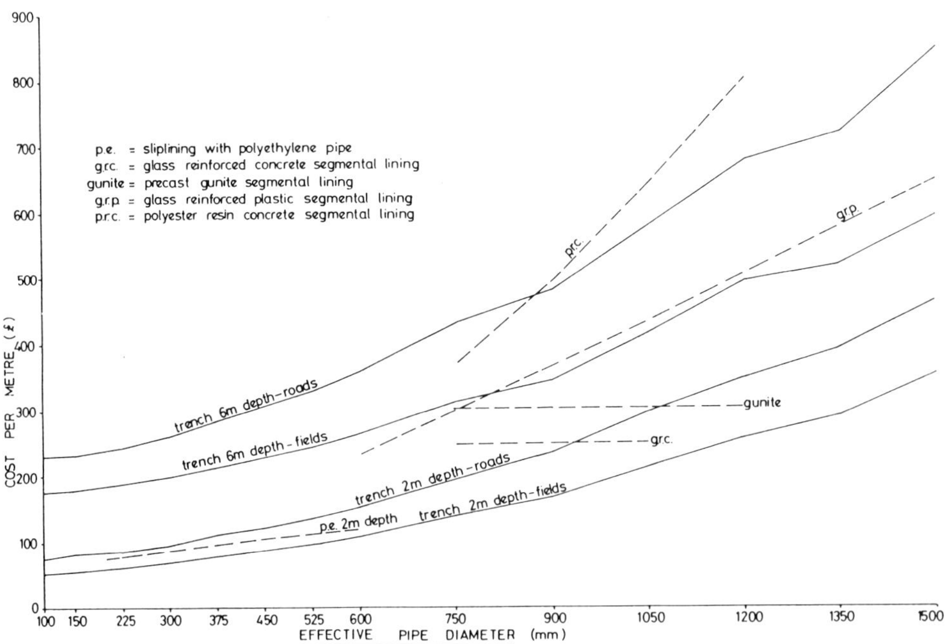

Figure 2.4.4. Comparison of gross unit costs: new pipes in trenches: good ground; and pipe renovation.

2.5 Social costs: trenchless v trenching

E B Glennie and K Reed *Water Research Centre Engineering*

The costs of trenchless pipelaying are compared with trenching, with the emphasis on costs other than the direct contract costs. This is done using two case studies, of sewer and water main replacement. In both cases on-line replacement by moling was the technique used. Traffic disruption and business losses were the principal costs in a "critical" location. Highway reinstatement and gas main replacement were also important.

The potential for trenchless pipelaying in the water industry is considered, in the light of current approaches to sewer and water main replacement. No market predictions are made, in view of the lack of reliable statistics.

INTRODUCTION

The degree to which trenchless pipelaying will replace trenching depends on:

(i) technical factors — what trenchless methods are able to do;
(ii) the engineering costs, and
(iii) the effect on others — for example motorists, businesses, utilities and highway authorities.

The technical factors are discussed in other papers at this conference. The engineering costs are considered by Mr Irvine in the paper that precedes this one. Here we use two case studies to compare the effect on others of trenchless pipelaying and open-cut trenching. We restrict our attention to quantifying the effects as far as possible, and do not consider questions of legal liability.

Then we summarize the approaches to rehabilitating sewers and water mains that are being developed by WRC and the water industry, and consider what they mean for trenchless pipelaying.

Reliable figures are very hard to obtain, and we have not tried to make the bricks of the market predictions without the straw of current spending details. The best data available have indicated that in England and Wales slightly less than 1,000km of sewers are renewed/replaced each year, and a rather shorter length of water mains. In addition approximately 2,000km of new water mains are laid each year by Water Authorities. Our approach is to assume the overall laying rate will remain constant, and to guess at the proportion that might be laid by trenchless methods.

CASE STUDY A — ST. CROSS ROAD, WINCHESTER

The technical aspects of this work are described in the ALH/Southern Water paper on the "Replacement of small-diameter pipes by pipe bursting". Here the costs are considered, both to the Water Authority and to others (Table 2.5.1).

The costs to others are often difficult to quantify, and business losses in particular are confidential to Southern Water and the businesses concerned. However, sufficient information is available to provide a general comparison of costs between the trenchless technique that was actually used and open-cut trenching.

.A 1,100m section of 9-inch sewer under St. Cross Road was rehabilitated (Figure 2.5.1): most of this by moling, some by sliplining a PE pipe and a short length by trenching. St. Cross Road is the main access to Winchester from the south, and carries 10,000 vehicles per day. It is narrow for a main road, 5m wide in places. The sewer runs down the centre of the road for much of its length.

Traffic delays

The road was officially closed on Monday 8 October 1984 for a period of nine weeks. However it remained open to local traffic and continued to be used by some through traffic. The main diversion route was along Badger Farm Road and Romsey Road, which normally carry about 11,000 vehicles per day. For the first two days there were considerable delays on this route. After this, traffic settled down and journey times during peak hours were five minutes longer than before the closure.

Two thousand vehicles per day transferred from St. Cross Road to Badger Farm Road and Romsey Road. The remainder used other routes, including the bypass to the east of the City. The cost of the delays and extra distance travelled has been estimated by WRC to be at least £27,000 per week based on a Department of Transport formula. Over nine weeks this comes to nearly twice the cost of the engineering work. Several bus routes were diverted.

The weekly cost of delays would have been more, if trenching rather than moling had been used. This is because St. Cross Road continued to be used by some through traffic, even though officially open for local traffic only. Also, the works would have lasted more than three times as long, and therefore the total cost of delays would have been more than trebled.

Before the road closure, explanatory leaflets were distributed in the area affected, and special diversion signs were made. The total cost of these preparations, which were felt to be well worthwhile, was £8,000.

Business losses

When work was started, considerable anxiety was expressed by the businesses along St. Cross Road over the probable loss of trade.

Southern Water's attitude is that it is prepared to consider making *ex gratia* payments to those affected. Claims for loss of profits should be based on audited accounts. Clearly these claims are confidential to the businesses concerned and Southern Water, and (at the time of writing) are not yet settled. We will therefore restrict ourselves to general remarks.

The businesses that may have been affected were these:

(i) *At the southern end of St. Cross Road.*
A parade of shops including a newsagent.
Two public houses serving lunch-time snacks.
No work was done in front of these shops, but traffic was diverted and the passing trade reduced.

(ii) *In the central/northern part of St. Cross Road.*
A garage that serves petrol and sells and services cars.
A big hotel with a restaurant for non-residents.
A snooker club.
A car tyre and exhaust fitting centre.
A hotel with restaurant and bar.
A sauna.

(iii) *In Southgate Street at the northern end of St. Cross Road.*
A restaurant.
A garage, with a car showroom but selling no petrol.
A public house.
A hotel.
Several shops and offices.

Since local access to vehicles and pedestrians was maintained, the businesses likely to be affected are those that depend on passing motorists for sales, e.g. of petrol and newspapers. Losses to all businesses would have been much greater had it not been possible to maintain local access.

Claims were expected to be substantial, "thousands of pounds per week". Whether the effect on businesses was as great as at first feared remains to be seen.

The work was carried out during October and November to avoid the summer holiday and the pre-Christmas seasons, which are particularly important to businesses in the City.

Economic theory indicates that if business lost in one place is transferred elsewhere, then the nett loss is zero and should not be included in an economic appraisal. This isn't much comfort to those who lose! Perhaps the theory should not be followed here.

To summarize, the fear of business losses was a significant factor, even with a trenchless technique being used. Losses would have been much greater if trenching had been used because:

(i) local access to traffic would have been impossible, and
(ii) the works would have lasted more than three times as long.

Other factors

Noise and dirt created by the works was small, much less than would have been caused by trench excavation and backfilling.

The use of granular (DoT Type 2) backfill at the limited number of excavations allowed immediate permanent reinstatement. Trenching would have required permanent reinstatement at a later date, costing an estimated £27,500 (WRC estimate) and causing further disruption.

The use of NO-DIG techniques avoided the need to replace a gas main at an estimated cost of £80,000.

These last two are of course direct engineering costs, but extra to the sewer renewal contract.

CASE STUDY B: NEWCASTLE STREET, BURSLEM

The previously described case history for the installation of the sewer at Winchester illustrated a lot of problems in relation to the social impact and associated costs. The case study to be described for on-line replacement by moling of a 150mm cast-iron water main did not present such extensive problems.

The water main was approximately 660m of 150mm diameter cast-iron pipe situated in the carriageway of Newcastle Street which is the main road into the town centre of Burslem, Staffordshire. The pipe was some 80 years old and was in a carriageway whose surface consisted of 200mm of hot rolled asphalt on top of 200mm of concrete. The main was subject to structural failure and 6 bursts had been experienced in the year prior to replacement. Traffic flow was maintained during installation work with the use of temporary traffic lights.

The installation of a sleeved 180mm outside diameter MDPE pipe with replacement of all fittings and renewal of 46 service connections was carried out by Severn Trent Water. A development contract was let with a specialist moling contractor being assisted by a general civil engineering contractor who carried out preparatory work for the moling contractor. It was anticipated that the use of the on-line replacement technique would be carried out in one-third of the time required by conventional trenching methods, leading to considerably less disruption to both public and traffic with some significant saving in cost.

The installation of new service connections was carried out before the mains replacement to allow speedy reconnection afterwards. The actual period for replacement of the main and reconnection of services was three weeks. No undue difficulties were encountered in the moling procedure, although there were stoppages caused by equipment malfunctions. Excavation had to be carried out to recover the mole where steel couplings, ductile iron pipe or repair collars had been installed. Most of these locations had been identified in advance.

Moling rates varied between 5.1 metres/hour and 20 metres/hour, and were higher at the end of the contract than at the beginning.

Costs

The costs are shown in Table 2.5.2, which shows that the main saving was in permanent reinstatement. However, the contract costs were high because:

(i) this was a development contract: costs can be expected to decrease as experience is gained, and

(ii) two contractors were used: it was felt that savings would have been made if a single contractor had been employed.

A total of 63 excavations were made, 12 entry/reception holes, 37 holes for service/side street connections, 8 excavations at collars, couplings, etc., and 6 unplanned excavations. In addition the highway surface was disturbed in a few places.

The reduced length of time for which traffic was disrupted was also felt to be a significant factor.

The water supply had to be cut off from consumers connected to the section of main actually being replaced during any one day. No consumer was cut off for more than a day.

SEWER RENEWAL: THE NO-DIG POTENTIAL

In case study A, the social costs caused by the sewerage works were significant, even with a trenchless technique being used. It is regarded by the Water Authority as being unusual in this respect. How widespread are business losses, traffic disruption and other costs caused by sewerage work? A general discussion has been given by Peters[1] and will not be repeated here.

We used a national newspaper cuttings service to obtain articles on sewer works from local newspapers throughout the country. In the six months from July to December 1984, 33 cases were reported. Of these 28 mentioned traffic disruption, 8 losses to businesses and 2 damage to buildings. In all these cases, the sewers were laid in trenches. There were presumably other cases that did not reach the local press.

Some quotations (not from case studies A or B) give the flavour of the reports:

"Angry traders ... claim workmen's mess is turning shoppers away and fear the Christmas boom will pass them by". (Similar quotes could have been taken from two other cases).

"Sewer works which will cause the digging up of — promenade during the peak summer season has been criticized by members of the town council".

"Traffic problems in the town were one of the reasons — Ltd. was delayed in going ahead with a project to double its plant in —".

Clearly the timing of works is important, to avoid the Christmas shopping boom or the peak holiday season.

Other precautions can be taken to minimize the effect of sewer works on the public.[2]

In 1981/1982 £175 million was spent by Water Authorities in England and Wales on renovating or replacing old sewers and building new ones. This was made up of over 1,000 schemes, dividing in approximately equal numbers into cost bands less than £50,000, £50,000 to £200,000, and more than £200,000.

A significant proportion of this work was carried out on "critical" sewers which are:

- big (more than 900mm diameter);
- made from brick;
- deep (more than 5m);
- under a busy road, a main shopping street, access to an industrial site;
- or in any other situation where failure would cause disruption.

The proportion of work that is on "critical" sewers can be expected to rise as the "sewerage rehabilitation strategy"[3] is put into practice. It is in these critical locations that trenching is most disruptive and trenchless techniques will show the greatest saving for sewer replacement.

At present it is only in the larger diameters (more than 1,500mm diameter) that a trenchless technique (tunnelling) is used in more than a small proportion of cases. With the smaller diameters, most work is still done by trenching.

The recent growth of the sewer renovation market is of interest when considering the future of trenchless sewer replacement (Figure 2.5.2). Spending on renovation has grown from less than £2 million in 1979/1980 to nearly £10 million in 1984/1985 but is still a fairly small proportion of total spending. It is cheaper to renovate than to lay a new sewer, but renovation can only be done where:

(i) the old sewer is still in a reasonable structural condition, and
(ii) some reduction in sewer cross-section can be accepted.

Moling and tunnelling techniques are not so limited, and the potential market may be greater. How far the technical potential is realized depends on:

(i) the engineering costs — will they come down as the techniques become established?
(ii) the attitude to traffic disruption and business houses — how should they be taken into account when choosing the technique for sewer renewal?

WATER MAINS RENEWAL: THE NO-DIG POTENTIAL

A strategy for the rehabilitation, including lining, renewal or replacement, of water mains is currently being developed by a WAA working group and WRC, and will be presented in a Manual this year.

Historically water mains have been viewed as long life assets able to perform their required functions with minimal maintenance and no additional investment. The general approach has been to deal with problems on an operational basis as they occur.

This reactive approach has been questioned in recent years: firstly where operational costs are so high that investment in new pipework or renovation of existing pipework may be more economic. Secondly, methods of maintenance may not be adequate to ensure that mains fulfil their function. This may result where:

— restrictions on manpower and other resources prevent adequate maintenance;
— dereliction is too far advanced for normal maintenance work to be effective;
— there is a demand for better water quality, more water or a more reliable supply.

As a result many Water Authorities and Companies have embarked on mains renovation and renewal programmes with investment of large sums of money and the possibility of greater sums being required in the future.

The principal reasons given for embarking on mains renovation and renewal are:

(a) deterioration of water quality (especially discoloration);
(b) poor flow and pressure and lack of carrying capacity;
(c) high burst frequency.

The former two of these reasons are by far the most common and are caused by internal corrosion. Obviously the degree and rate of corrosion is a function of the type of water carried, with upland soft waters presenting the greatest problem. Normal procedure for rehabilitation has been to clean the pipeline and line with cement mortar or epoxy spray coating.

In 1983, 800km of main were relined, mostly by lining with cement mortar. Lining is often considerably cheaper than renewal (Table 2.5.3) particularly in carriageways where reinstatement charges may be high. The true cost of lining may be increased if pipes are in a poor structural condition and have to be repaired to deal with bursts arising from the application of the lining process including cleaning and subsequent pressurization of the system.

Pipes may be judged to be in a condition beyond which relining can be applied and in this instance the renewal option would be chosen.

It has been estimated that in 1982/1983 676km of water main was renewed, for structural or other reasons, at a cost of £40 million[4]. In addition, three times this length of new main is laid each year to supply new consumers or meet increased demands. Almost all of this work was done by open-cut trenching. The extent to which trenchless techniques will penetrate this market is uncertain. Case study B is encouraging, but here there was only one service connection every 15 metres and therefore comparatively few excavations were needed.

In the gas industry, moling is widely used for mains replacement, and is considered to save money, particularly on reinstatement. An important factor in the cost of moling is the number of service connections that have to be remade, since a hole has to be dug at each. Are water service pipes more frequent than gas ones?

CONCLUDING REMARKS

The potential for trenchless pipelaying depends on the engineering costs relative to trenching, technical factors and the costs to others.

(i) The engineering costs include permanent reinstatement of the highway. The degree to which this is avoided by trenchless techniques is crucial to its potential for water main replacement. Hence the frequency of service connections will be important.

(ii) Traffic disruption and business losses may be very large when replacing sewers and water mains in critical locations by trenching. Trenchless techniques reduce them considerably but do not eliminate them.

(iii) The question remains of how costs to others should be taken into account in the choice between trenching and a trenchless technique.

Some have become direct costs, for example the replacement of gas mains affected by sewer trenches.

If extra engineering costs are incurred in order to avoid social costs, this should be recognized so that other projects in a tight capital programme are not jeopardized.

(iv) It has proved to be impossible to make precise statements about current or past spending: and this has made looking into the future even less possible than usual. It would help greatly if each Authority were to record centrally the length of sewer or water main laid and the method used to lay it, at the end of each contract. We understand that one Water Authority already does this.

ACKNOWLEDGEMENTS

The authors have relied heavily on information freely provided by others: in particularly by the staff of Southern and Severn Trent Water Authorities and the Winchester City Engineer's department. We thank these Authorities for permission to publish this information.

We would like to thank Mrs J Whitaker for doing the typing.

REFERENCES

[1]Peters, D C, *The social costs of sewer rehabilitation.* BHRA conference on the planning, construction, maintenance and operation of sewerage systems, Reading, September 1984.

[2]Read, G F, Sewer renovation and reconstruction in Manchester. *Municipal Engineer, 1,* 27-28, June 1984.

[3]The WRC/WAA *Sewerage Rehabilitation Manual,* available from the WRC Engineering Centre.

[4]Lackington, D W, *Survey of the renovation of water mains.* IWES Symposium on the deterioration of underground assets, London, December 1983.

TABLE 2.5.1. Case Study A: Costs to the Water Authority and others

Item	Trenching (estimated)	Moling (actual)
Sewer rehabilitation contract	£132,000	£149,000 (£95,000 for moling)
Duration	30 weeks	9 weeks
Gas main replacement	£80,000	Nil
Permanent reinstatement	Permanent reinstatement along whole length 1,100m @, say, £25/metre = £27,500.	Immediate permanent reinstatement, except for 65m open cut @, say, £25/metre = £1,600.
Disruption to traffic	Closed to through and local traffic. Several bus routes diverted.	Closed to through traffic — £27 000/week. Open to local traffic. Several bus routes diverted.
Business losses	More than with NO-DIG because of restricted local access. Longer duration, and need to close road again for permanent reinstatement.	Thousands of pounds per week? 3 hotels, 2 garages (1 selling petrol), several shops.
Noise, dirt and smell	Considerable.	Small. Excavation at manholes and side connections only.
Adjacent buildings	None.	None.
Damage to roads, etc. to which traffic was diverted.	Not important.	Not important.
Other costs.	As for NO-DIG.	Traffic routing £8,000. Public relations.

TABLE 2.5.2. Cast Study B: Costs to the Water Authority and others

Item	Trenching (estimated)	Moling (actual)
Main replacement contracts. (Service renewal work is excluded. It would have been the same for both trenching and moling).	£22,400	£11,800 (Moling) £13,300 (General)
Materials (including temporary reinstatement)	£12,000	£10,000
Permanent reinstatement	£17,000	£8,000
Duration	12 weeks	3 weeks
Disruption to traffic	Control by lights on main road.	Control by lights on main road.

TABLE 2.5.3. Water mains renovation and renewal

Lining (cement mortar or epoxy)	Trenching	Moling
To deal with internal corrosion.	Where pipe structure has deteriorated, or extra capacity is needed.	Where pipe structure has deteriorated, or extra capacity is needed.
£20/metre for 150mm main.	£80/metre for 150mm main (Case Study B).	£65/metre for 150mm main on development contract Case Study B).

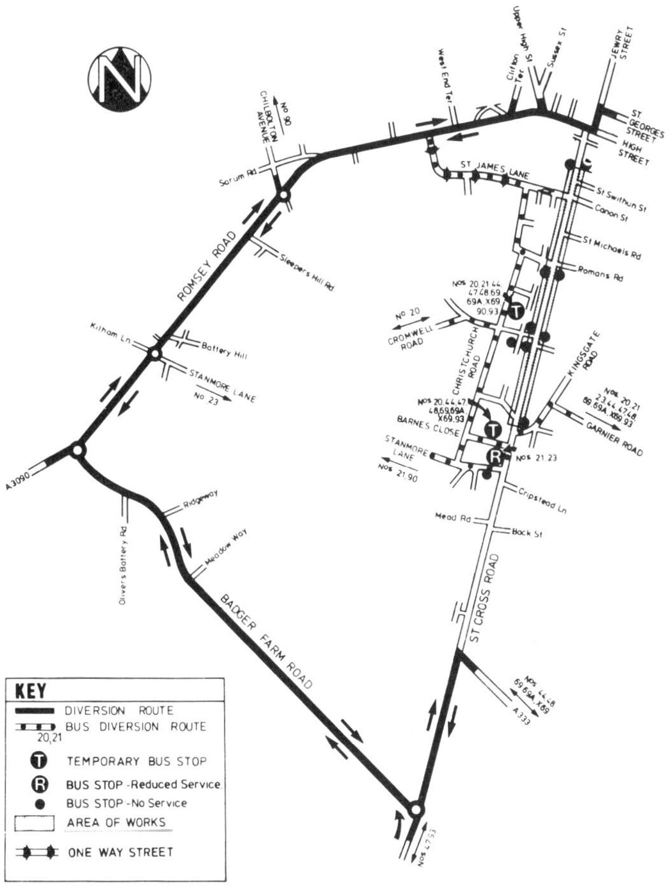

Figure 2.5.1. Case study A: Southgate Street/St. Cross Road diversion.

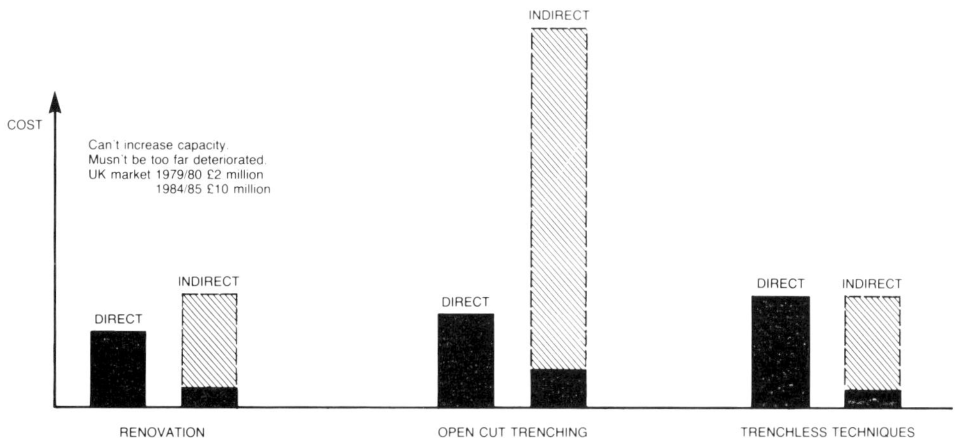

Figure 2.5.2. Sewer rehabilitation methods: costs and capability.

Authors' Introductions

Mr D J Irvine: The comparative costs of some methods of trenchless construction

The paper describes the objectives and results of a comparative study of the costs of different methods of installing pipes in the ground, both in trenches and using trenchless construction.

Because the opportunities for permutating costs are enormous it was necessary to define a few limits. Therefore a cost model was set up in which a 600-metre length of pipe would be installed. The object of the comparison was to work out the gross costs for various methods of constructing the pipe, though there are gaps in the study. The construction methods studied provide three groups of structures. First, pipes which support the ground and act as the conduit, secondly, a construction process forming a hole in the ground which is then lined down by some method, and thirdly various methods of renovating pipes *in situ*.

The costs for pipe jacking stop at a somewhat large diameter, not because it is impossible to jack pipes of a smaller diameter, but merely because at the time of the study there were no data on the cost of smaller diameters. It should be noted that large diameter pipe jacks can be lined down using just the same methods as with segmental-lined tunnels, for which costs are also given.

A variety of ground conditions were considered to get a feel for the cost sensitivity of the various construction methods. Bad ground was taken to be wet silt, and good ground was represented by a firm dry clay. Also the cost effects of going under green field sites and through roadways were examined.

Figure 2.4.1 gives an indication of the methods that were considered for lining down large diameter tunnels in order to achieve the necessary hydraulic characteristics for the pipe line. It was necessary to make a judgement here based on work Tarmac had carried out elsewhere, since what can be achieved in lining down depends in part on the physical size of the men doing the work, and the outputs they can achieve.

The paper points out that when large diameter tunnels are lined down, in fact two structures are built. Firstly, a hole in the ground is made and then a hydraulic structure is built. Where a tunnel is used running full, down to say 900mm diameter, the total cost is roughly proportional to the cost of the tunnel and reduces with the diameter. The difference between the cost of the tunnel lining and the total cost is made up of the cost of access shafts, manholes and the like and on-costs. These costs decrease only very gradually with the diminishing diameter of tunnel. There is an increase in cost at effective pipe diameters below 900mm when a hydraulic channel is constructed within the tunnel since a larger primary tunnel is required to give sufficient working room. At about 600mm effective diameter a pipe is a convenient carrier instead of the channel.

The general conclusions of the study concerning work in trench in good ground were that the direct cost of trench construction is cheaper than trenchless construction for the cost model considered. It was difficult to be precise on the cost of renovation because there was an enormous spread in the available cost information from actual contracts. However, it appeared that in bad ground renovation would become a gradually cheaper option to laying new pipes as depth increased and ground conditions worsened. Again, only direct costs are considered in Figure 2.4.3.

There are some points that I would like to make about Figure 2.4.1 because it strikes me that we are doing two jobs here and getting only one benefit. We are putting water or sewage through a very expensive tunnel leaving a lot of wasted space which could be used for other services. The use of multi-purpose tunnels was referred to earlier today and I suggest that when we look at the initial costs of construction we are looking at only one aspect. In building a structure that is going to be used for a number of years we really need to consider its total cost including maintenance and social costs, and compare that with the total benefit obtained during its life.

So I suggest that in future we must consider the whole life cost of a pipe installation. We should not assume that one type of pipe in the ground will give the same service as another, particularly where we are constructing what are, or could be, multi-purpose tunnels. In addition, as many speakers have said previously, we must consider the total cost to society at large. The cost to the promoter alone should not be the sole criterion.

Mr K Reed and Mr E B Glennie: Social costs: trenchless v. trenching

Mr Reed: In reviewing our paper Edward Glennie and myself would like to mention those aspects which may influence the choice of the method of construction to be adopted in the installation of a new pipeline or the replacement of an existing pipeline. These were noted in the paper as being the technical factors — what are trenchless methods able to do and what is the best technique for the job in hand; the engineering costs, the bottom line of any contract — how much will it cost, and will trenchless construction costs decrease in the future; the effects on others — these include both quantifiable effects which in themselves can be identified and costed, and intangible effects which may be difficult or even impossible to quantify and cost. I shall leave the finer details of these effects for further comment by Mr Glennie. I will deal in a little more detail with the two former aspects.

During this conference experiences are going to be shared from all over the world and in this context the techniques and systems that are currently available would seem to be numerous. We have the means to perform work with a trenchless construction technique, but the inevitable question is, at what cost? Mr Irvine has just presented his paper showing that the direct costs for trenchless construction for small diameter pipelines are significantly greater than the open-cut alternative. Details in previous work have already shown that the consideration of social costs can make the trenchless option less expensive than the open-cut alternative, but social costs are not necessarily the responsibility of the client, and at the present time there is no means whereby they can be considered in project appraisal. I feel that until this becomes the case, open cut will probably remain less expensive and indeed may appear more attractive to a prospective client. So one asks the question, are there any means whereby the direct costs of trenchless pipelaying can be reduced? In several papers we will hear or have already heard of the market potentials that have been identified in specific countries. James Thomson's overview looked at a wide spectrum of markets and the statistics in relation to the total available market are staggering. If this market can be tapped by the trenchless construction techniques then inevitably costs will decrease, particularly where capital-intensive machinery and equipment are being used, and must be maintained and used in order to recover development costs and allow even further development to be carried out.

The majority of development in the microtunnelling aspects of trenchless construction has been taking place in Japan and Germany and there the scope is that much greater with the installation of first-time sewerage on a much greater scale. The potential, albeit on a much more limited basis, does exist in this country but remains to be exploited. As part of WRC's work in the tunnelling aspects of trenchless construction we have already commissioned two preliminary reports on the subject, the first, a "state-of-the-art" review looking at available methods, and the second a study of comparative costs. WRC will continue to support work in this area and in this respect will be funding part of work which has recently been commissioned by the Construction Industry Research and Information Association

(CIRIA), who have appointed consulting engineers Binnie & Partners to carry out a study which will be entitled *The Investigation into Trenchless Construction Methods for Installation of Underground Services*. This study will be investigating trenchless construction techniques as applicable to all utility services and has the co-operation of all utilities.

Mr Glennie: We have heard about the construction costs of trenchless techniques compared to trenching. What about the costs to other people? My talk is based on two case studies that are described in our paper and on previous work by Dennis Peters of WRC. For the case studies we owe a big debt of gratitude to staff of Severn-Trent Water, Southern Water and Winchester City Council. Both case studies are of the replacement of old pipes, not the laying of new ones, and they are both concerned with pipe-breaking techniques applied to diameters up to 300mm.

The first point that I want to make is that social costs are important. In certain critical locations they can be several times bigger than the construction costs, and these locations tend to be in town centres and on main roads where a substantial amount of rehabilitation of sewers and water mains is being done. Well; what are these social costs? A complete list is given in our paper and I will comment on some of them. Firstly, disruption to traffic. Works on or under busy roads result in many thousands of vehicles being delayed or diverted, and the costs estimated from the extra time needed and the distance travelled, using a Department of Transport formula, can be many thousands of pounds per week. This was so in our first case study where journey times into Winchester were only five minutes longer than usual.

Three factors contributed to keeping down the traffic disruption costs. Firstly, careful planning of the diversion routes. Secondly, keeping the road open while the work was going on, officially for access but in practice for some through traffic also. This was made possible by the trenchless technique that was used. Thirdly, the works took 9 weeks compared to an estimated 30 weeks for trenching.

Several factors determine the cost of traffic disruption, including the volume of traffic using the road, the road width, the position of excavations, the length and capacity of alternative routes if they are necessary, and finally the length of time for which the road is restricted or closed.

Business losses are generally caused by access to the business being restricted or cut off — access for goods vehicles or the public. Shops that depend on passing trade are particularly vulnerable, as are garages selling petrol. In our first case study it was possible to maintain access for vehicles to all businesses. This would not have been possible had trenching been used. Again, the second factor in keeping down the business losses was the short duration of the works.

Now two more factors, damage to the highway and to adjacent pipes, ducts and cables. These are both, to some extent at least, direct costs paid by the statutory authority. Here I would like to distinguish between relatively shallow pipes, perhaps at a depth of one metre, and deep ones at a depth of two metres or more. Consider first the renewal of fairly deep sewer pipes of which Case Study A is an example. Trenching can cause ground movement and damage to rigid pipes nearby whereas with pipe breaking ground movement is much smaller and more localized and damage to mains and services is likely to be minimized. In Case Study A lead-in trenches had to be excavated, six manholes were constructed and fifty laterals reconnected, this in a length of over 1000 metres. There was a very clear saving in highway reinstatement costs over trenching. If it is correct to assume that deep sewers tend to have relatively few connections, then we can conclude that damage both to the highway and to other services will be minimized by trenchless techniques.

Turning now to the relatively shallow pipes, at say one metre depth. In this case trenching does not necessarily cause significant ground movement. In urban situations there are likely to be services crossing or adjacent and these may be damaged if a trench is dug. They may also be damaged when the pipe-breaking technique is used if they are very close to the pipe. Lateral connections may be fairly frequent. The one every 15 metres in our second case study was perhaps unusually few. An excavation has to be made at each lateral, so there must come a point when it is equally easy and equally cheap to dig a trench along the whole length. It is therefore less clear that in the case of pipes of up to 1 metre deep trenchless techniques cause less damage than trenching. However, British Gas, as we have heard, use moling techniques widely for gas main replacement and believe that significant savings can be made.

In conclusion, some general remarks. Firstly, social or indirect costs are important, especially in town centres and under urban main roads where a substantial amount of rehabilitation work is done. Trenchless techniques reduce these costs significantly but do not eliminate them. Secondly, I believe that procedures for appraising and financing work should encourage engineers to take these indirect social costs into account. This may mean higher direct engineering costs, but the experience of British Gas and of our two case studies suggests that at least for the smaller diameters the engineering costs need be no higher.

Discussion on Papers 2.4 and 2.5

M P O'Reilly (Head of Ground Engineering Division, Transport and Road Research Laboratory): Mr Irvine points out, in conclusion (f) which he draws from his graphs, that the economics picture would be completely changed if the unused part of a sewer tunnel could be used by other services. It has always seemed to me to be rather odd to dig a large hole in the ground and then fill all but the space required immediately with concrete. This makes it impossible to use this space, often created at considerable expense, in the future when increased capacity is needed. A few years ago I saw in Liverpool a sewer built 100 years ago to serve 1000 people. It is now serving 100 000 people, because the hole was big enough and the extra space can now be used. Should we not consider back-filling with polystyrene or similar material, which could be fairly easily removed when the extra space was required?

The data given in Mr Irvine's paper are average values, and actual costs would be spread around these averages, as is shown by the case studies of Glennie and Reed, where all factors were included. In their Case Study A, moling proved to be cheaper than trenching because the latter method would have necessitated the replacement of a gas main.

The accompanying figure shows the idealized results of an examination made by the Northumbrian Water Authority for TRRL of contract work done for them over the last fifteen years, valued at £40 million. The diagram shows both average and 90 per cent confidence limits of the equal cost depth to invert for trenching and tunnelling for sewers with internal diameters up to 2100mm. As can be seen there is considerable spread about the average. Traffic costs were also considered independently by Alastair Dick and Associates, and the results are summarized in the accompanying table. The results of this study are to be published in due course.

Given the scatter in the data and the volatility in the contracting industry, both from financial constraints and the multiplicity of new and innovative developments, I think many more schemes should be put to alternative design and tender. The recent experiences of design-and-build for major bridgeworks shows that designers do not always identify the most economical scheme[1,2]. If the client is to get best value for money, we must harness the ingenuity of the contractor as well, and the single engineer/consultant design approach certainly does not do this.

[1]Clements, L, The Kessock Bridge design-and-build contract and proposals for managing similar contracts. *Proc. Instn civ. Engrs*, Part 1, *76*, 23-34, February 1984.

[2]Prescott, T A N, Stevenson, W M C and Nissen, J, Foyle Bridge: its history, and the strategy of the design-and-build concept. *Proc. Instn civ. Engrs*, Part 1, *76*, 351-361, May 1984.

Estimated costs due to traffic delay

Traffic demand reference flow	User costs due to delay (£ per week in 1979 prices)					
	Shuttle working			Carriageway width restrictions		
	Length of works (m)			wide single	single	dual
	20	100	300	2-lane	4-lane	2-lane
Peak flow = 0.67 × capacity	<500	1000	3000	<500	2000	3000
Peak flow = capacity	2000	4000	9000	2000	11 000	17 000
Peak flow = 1.33 × capacity	7000	10 000	19 000	6000	22 000	32 000

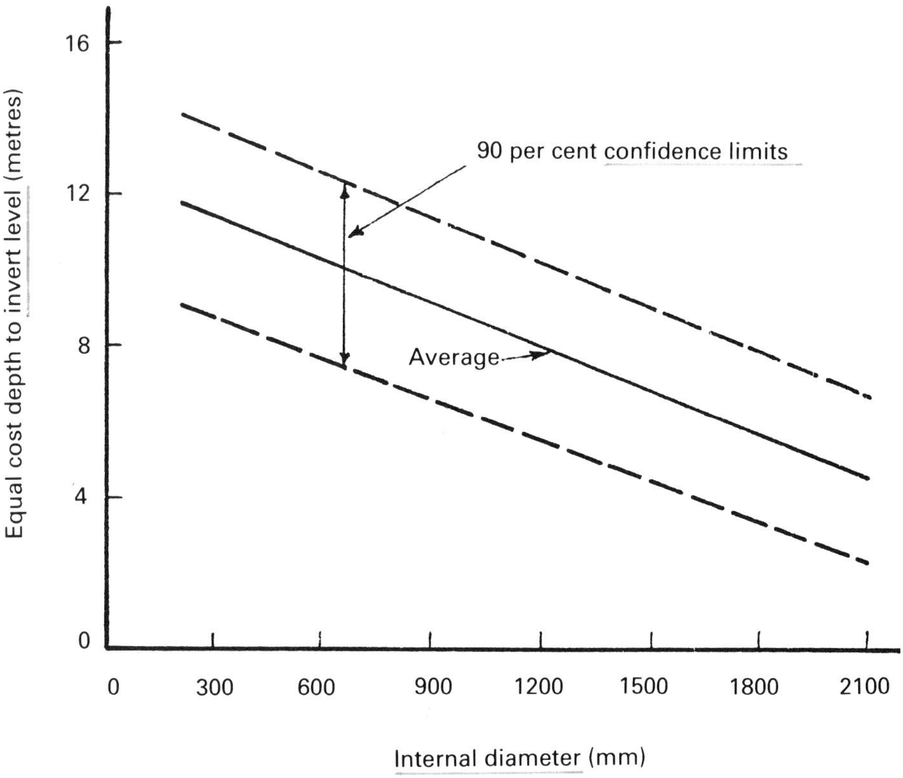

Idealized relation between equal cost depth and internal diameter for pipes and tunnels

Mr D J Irvine: I agree with several of the points made by Mr O'Reilly. My suggestion of multi-use tunnels is, perhaps, hoping for Utopia, but, if all the authorities involved did negotiate as they do in Melbourne, it might be possible. Mr O'Reilly mentioned a very old sewer in Liverpool, which is handling more than its designer intended. The additional space must also give better opportunities for maintenance in dry weather, and is available to carry surcharge in peak conditions. That is worth money to someone, and it ought to be taken into account by some means. The problem is that our accounting methods are very rudimentary. The accountants ought to start looking at what is to be built and how it is to be operated and put a value on it, and contrast that with the capital cost of achieving it.

Concerning the offer of alternative designs, as a contractor, I feel that it would be a very laudable thing if market forces were allowed to dictate, but we are up against an arbitrary ceiling: money is committed, not according to need, but according to other judgements. In such a situation I doubt whether market forces are the total answer. Contractors will put forward alternatives where they think they have a cost advantage that will be reflected in the price, but then someone will have to decide whether what is being offered corresponds to what is required. Once again, we are back to value considerations.

A Rooke (Director, Iseki Poly-Tech., Inc.): The gross unit costs given for bentonite slurry shields in Figure 2.4.2 of Mr Irvine's paper appear to be quite accurate for pipes down to 900mm internal diameter. However, the recent introduction of the Unclemole machine (to be discussed in paper 4.4 tomorrow) will reduce the cost for pipes of 250 to 500mm i.d. considerably.

Also on Figure 2.4.2, the cost of 900mm pipe-jacking in good ground seems to be excessive at about £870 per metre. Would Mr Irvine comment on this?

On Figure 2.4.3 the cost of trenching in bad ground does not seem sufficient to cover the costs of interlocked sheet piling and dewatering often needed in these conditions. Can we have a breakdown of the costs included for these items?

B Wilkinson (Principal Engineer (Sewerage), Yorkshire Water Authority): Both papers indicate that the engineering costs of trenchless methods are greater than for trenching, and Tables 2.5.1 and 2.5.2 of the paper by Glennie and Reed show the engineering costs per week of contract period for moling work to be twice those for trenching methods.

Speakers in the first session this morning bemoaned the fact that the market for new methods was too small. One is led to wonder whether these innovative contractors are trying to recover development costs too quickly, or are expecting the client to bear too great a proportion of the commercial risk. Would the authors like to comment on the relationship of the cost charged to the client to that incurred by the contractor?

Mr K Reed: It is very difficult to give an indication of the relationship between costs charged and costs incurred, particularly in view of the capital-intensive nature of the equipment being developed. I think it is accepted that some risk is involved on both sides, and we would hope that, perhaps by direct negotiation, some equitable arrangement can be arrived at so that there is some remuneration to a contractor to cover an element of his development costs. But, certainly, a client should not be expected to bear the major part of development costs in what can only be considered as early trials.

Mr E B Glennie: Following Mr Wilkinson's comment, I think it should be pointed out that, although the cost per week of moling is agreed as being twice the cost of trenching, the total cost for doing the job was approximately the same in our case studies.

Mr Irvine: I am pleased to hear that Mr Rooke has some new machines entering the market, and all contractors will look forward to getting information about them.

It is possible that the cost of pipe-jacking in good ground is overstated on Figure 2.4.2, and there are various reasons. More jacking and receiving pits have been included than might be considered necessary by some operators. Accuracy is a function of the length of pipe jack and the size of the jacks used, and a judgement was made. Also there was some difficulty in getting information on some prices, so again judgement was used. In addition, an element was allowed for write-off of capital plant. So, overall, the pipe-jacking costs in Figure 2.4.2 may be overstated by 10 per cent, or even, possibly, as much as 20 per cent for some conditions.

Mr Rooke queried the difference between trenching costs in good and bad ground. The reason for the rather small differences is that in good ground it is still necessary to put in shoring to protect the operatives. The difference is generally the extra over for perhaps driving interlocking sheets a bit deeper when working in wet ground.

SESSION 3. Chairman: Professor H. P. S. van Lohuizen, Technical University, Delft, Netherlands

3.1 Development of impact moling techniques in the UK: design, use and cost considerations

P T Etherton *Merstan Impact Moling Ltd, UK*

British Gas have been instrumental in promoting the use of impact moling equipment throughout the industry. Other utilities could benefit from their experience; investigations suggest that impact moling techniques still appear to present something of a mystery to many Distribution Engineers. The scope of this paper is to provide a historical background, and illustrate the benefits of using impact moling equipment generally.

INTRODUCTION

Three major factors have influenced the current rapid adoption of impact moling systems. They can be summarized as follows: (i) there is a constant requirement in modern society to extend and renew the service infrastructure within constantly restricted financial budgets; (ii) permanent reinstatement costs of paved surfaces have risen alarmingly over the past few years; (iii) while the model agreement and narrow trenching techniques have improved reinstatement procedures greatly, control of the quality of materials and labour involved still represents a problem area.

HISTORY

Air-driven impact moles were first invented by the USSR Mining Institute in Novosibirsk in 1960 approximately 24 years ago. The first machines to be manufactured by Odessa Plant were non-reversible and to a large extent unstable. They were, however, simple in design and rugged in manufacture and met with acceptance in Comecon countries.

Introduced to Germany in 1966, the UK in 1968, and the USA in 1971, the equipment initially met with an enthusiastic reception throughout the West. In these countries a totally different growth pattern of acceptance of the moling technique emerged. An examination of the methods adopted by distributors to introduce the equipment, together with other factors which influenced the use of impact moles may prove informative.

The USA experience

The major impact moling company in the USA, whose sale of the equipment amounts to approximately 90 per cent of the current market, manufacture the machinery under licence from the USSR. Current annual sales of the equipment amount to approximately 2,300 units annually. All utilities, cable television installation etc. use impact moles as standard equipment. The fairly consistent sales figures over the past four years indicate a replacement market now being serviced as opposed to the supply of equipment to the first-time customer. Aggressive marketing by a large American company with an extensive existing customer base within the construction industry obviously was instrumental in introducing the equipment successfully.

Due to the vastness of the areas involved, sales outlets, however, are developed through hardware houses and then subsequently to small distribution companies, perhaps only one or two men strong, who are responsible for a small area. The hardware houses are trained by the manufacturer in the use of the equipment. The distributors are therefore able to develop techniques and train clients on a virtual individual basis, being co-ordinated by the hardware house, and ultimately the manufacturer. This marketing strategy fulfils one of the basic requirements for the successful introduction of moling equipment, that is, individual training to first-time users which enables operators to obtain satisfactory results in using impact moles. This in turn provides the operator with enough confidence to overcome initial reservations when faced with the prospect of changing from traditional open-cut procedures to a mechanized system.

Europe

Firm Essig, a West German family company, successfully modified imported USSR equipment to ensure reliability and directional stability. Initial involvement of a qualified design engineer ensured ancillary equipment was produced in response to the requirement of the industries being serviced. The fragmented nature of the utilities who in the main operate on a town or regional basis enabled individual engineers to adopt the new techniques for purely cost saving reasons, without requiring approval from a national co-ordinating body. The development of an alternative West German-manufactured product followed quickly behind Essig's success. The demand for reliability, however, has resulted in the preference for USSR design equipment capturing approximately 70 per cent of the European market of around 2,000 units per year. As in the USA experience, once again, the "hands on" management style of a family business resulted in close attention to training and development which guaranteed successful introduction of the equipment.

United Kingdom

Between 1968 and 1970 some 200 non-reversible USSR-produced machines were initially imported and sold via a UK trading house without the benefit of any back-up engineering services, or operator training. Numerous are the events of impact moles appearing unexpectedly in the centre of roads or railway lines, and the majority of the machines were lost. The net result was the discrediting of moling techniques, and reluctance among distribution engineers to use impact moles generally, their use being reserved for specialist applications only.

The subsequent successful introduction of a West German-designed machine with good directional stability went some way to re-introducing impact moles in the UK. Numerous authorities purchased machinery and engineers continued to appreciate the potential of such equipment. With notable exceptions within the gas industry moling techniques, however, were not adopted as first choice for methods of pipe installation, but still viewed as a specialist art for occasional use. With the USA and European experiences to draw on there must have been some factors which have restricted for so long the use of impact moles within the UK on a large scale. I feel these can be summarized as follows: (i) equipment was required to enable operators to simply detect other services prior to impact moling; (ii) the need to change existing planning systems to encompass the use of moling equipment; (iii) insufficient financial pressure to reduce unit installation costs; (iv) traditional reluctance of operators to adopt new techniques; (v) requirement of reliable low maintenance equipment with good directional stability. Whilst all the above factors are relative, the present financial climate ensures that items (i), (ii), (iv) and (v) will automatically be accommodated to follow (iii).

COST EFFECTIVENESS

The following data have been extracted from the ERS Report E363 Issue 2 March 1984 prepared by Mr D J Smith of ERS, and Mr A Garnham of Northern Gas and serves as an illustration as to the total monies expended by BTG on traditional trenching methods. From these data it can be appreciated that any percentage reduction in these costs can reflect substantial savings nationally (Tables 3.1.1 and 3.1.2).

The average cost of laying 180mm polyethylene pipe in a carriageway can be estimated at approx. £54 a metre, footpath £29 a metre, and in open ground £22 per metre. These figures indicate that the higher savings to be achieved by using trenchless techniques are to be found in congested urban areas.

Impact moling techniques eliminate permanent reinstatement costs which can be illustrated as follows: 1982 Kent County Council published permanent reinstatement costs, four courses as per model agreement, type 1 road £68 per m^2, approx. £34 per lin.m. average trench width 300mm. Medium size contracting company carrying out own permanent reinstatement costs within British Telecom contracts, approx. £18 per metre. The advantage of using a system which eliminates these costs totally is obvious. The growth of trenchless techniques in British Gas, which include insertion and moling methods, can be illustrated by Figure 3.1.1 and Table 3.1.4.

A comparison between traditional costs and installation of the same pipe by impact moling techniques can be seen in Figure 3.1.2 and Table 3.1.3.

It is appreciated that there are major differences between the water and gas industries which will limit the scope for using impact moles for installing water services. These can be briefly summarized as follows: (i) safety requirements resulted in legislation being adopted which provides for the ownership and maintenance of the house water service being the consumer's responsibility; (ii) there is a limited planned replacement of existing services to house owners, and no major mains replacement policies to change from metallic to polyethylene or plastic alternatives are envisaged in the immediate future; (iii) with a low cost of the end product costing £1 per cubic metre a loss of 25 per cent of clean water nationally does not at present pose a problem; (iv) as the product supply is not a limited source, politics would tend to dictate capital expenditure allocated as opposed to engineering requirements; (v) cleaning and cement-lining of existing mains can be easily achieved at a reasonable cost of approximately £15-20 per metre, and structural problems probably only exist in 10 per cent of installed water mains.

To summarize this section, it can be concluded that using impact moling equipment is always cost-effective. The recovery of the capital expenditure required to purchase machinery will inevitably be a function of the length which can be achieved within a given timespan. Within British Gas this recovery time can be as low as five weeks. But in some areas of low usage it could be considerably longer.

Financial appraisal as to viability of purchasing equipment will inevitably be the responsibility of the distribution engineers who must examine workloads, interpret the amount of work which could be carried out by impact moling, and subsequently prove that the savings achieved in reinstatement costs justify the capital expenditure required.

DESIGN AND PRINCIPLE OF OPERATION

Impact moles are available in both fixed and moving-head configuration. Both types of design exhibit different forms of characteristics through varying ground conditions. It is generally felt that moving-head equipment is marginally more stable, but many thousands of fixed-head machines are forming impacted bores throughout the world with a degree of accuracy sufficient for the purpose.

With a view to impartiality and being a manufacturer of fixed-head equipment the position on this subject is fairly clear, that is, both types of equipment put holes in the ground, and it is the end user who must decide which type of equipment he prefers. For the purpose of the remainder of the paper I shall limit its contents to fixed-head equipment of the type I manufacture and distribute.

The design and principle of operation of the reversible pneumatic undergrond piercing tool is explained in the following description, and accompanying Figure 3.1.3. The machine body consists of two parts: cylinder 1 and anvil 2 connected rigidly between each other. Striker 3 with front and rear bearing surfaces is positioned inside the body. The front bearing surface has machined flats allowing for passage of compressed air. Air distributor 4 enters the striker tail part and rests upon shock absorber 5, which provides the decrease of inertia forces arising in the slide-valve parts. The shock absorber is pressed in nut 6. Elastic valve 7 protects the machine inner chamber against penetration of foreign particles. Compressed air is supplied through hose 8. Nut 9 together with air distributor 4 forms a screw pair and allows for the axial displacement of the air distributor relative to the machine's body. The inner surface of the body and the external surface of the striker form chamber A. The striker's inner chamber and air distributor form chamber B, which is connected with a source of compressed air through the distributor's axial channel and air supply hose.

When the striker is in the position shown in Figure 3.1.3, compressed air flows from chamber B to chamber A through ports C. Due to the difference of striker working areas at the ends of chambers A and B, the striker begins to move to the right. As the striker approaches the extreme right position, air escapes from chamber A into atmosphere through striker ports C and the shock absorber exhaust holes. The air pressure in chamber B first stops the striker and then moves it to the left, the striker delivering a blow against the anvil. The cycle is then repeated. Under the action of blows the machine is driven into the ground. The body back motion is prevented by frictional forces arising between the body and the ground. The underground piercing tool holds the given direction due to its axial symmetry and length, and the use of auxiliary cutting heads and flutes.

The air distribution system provides reliable operation and faultless starting of the tool in any position.

Reversibility is accomplished by changing the axial displacement of the air distributor which can take two operating positions: front position for direct run of the tool, and rear position for reverse run. In Figure 3.1.3 the front position is shown by continuous lines, and the rear position is shown by dotted lines. The valve is displaced from one position to another by turning or pulling the hose. When the distributor is in its present position, the tool operates as previously described. When the distributor is in the rear position, compressed air is let into the front chamber earlier than in the case of direct run, and the striker stops before it reaches the anvil. On the contrary, the return air is exhausted later, so the striker delivers a blow against nut 6. Under the action of these blows the underground piercing tool moves through the hole in reverse direction.

Ground effects

The type of ground and soil conditions in which the punchers are to operate are an important factor in the overall performance and accuracy of the unit. The factors to be taken into account are as follows: (i) the type of soil and its degree of homogeneity; (ii) the degree of compactness of the soil; (iii) the moisture content of the soil and its ability to be deformed and displaced; (iv) the critical depth defined as the minimum point of entry and line of travel below the surface; (v) the contour of the ground above the line of travel; (vi) the presence of obstructions, such as large rocks.

A brief discussion of the above factors and their effects upon the performance of the impact mole follows.

The type of soil and its degree of homogeneity is a major factor in determining the course the device will tend to follow. As a general rule, homogeneous type soils of wet and/or hard clays will cause the device to rise upward toward the surface; homogeneous sandy soils will cause the device to run a fairly level course; and homogeneous sand and gravel type soils will cause the device to run downward. The degree of compactness and moisture content of the soil are factors that govern the rate of penetration of the device. As the tool penetrates into the ground it displaces the ground forward and to the side. It can be obviously concluded that in hard well-compacted soils, this displacement will be slower than in lesser compacted soils.

Therefore the rate of penetration is slower in hard compacted soils and increases in soils of lesser compaction. Similarly, wet soils will flow and deform more readily than dry soils and thus the rate of penetration is greater in wet soil conditions. The highest speeds were obtained in moist sandy soils of 12 to 17 per cent moisture while the least speed was obtained in clay and loam type soils of 4 to 6 per cent moisture. Another effect of soil moisture content on performance is best understood by analysing the principle of operation of the underground piercing tool, that is, a frictional force between the tool body and soil is required in order to produce motion. As the moisture content increases, more slippage is evident.

Special considerations

Difficulties may arise when the underground piercing tool is used for driving holes in running, moist, made-up, peaty grounds, or in grounds with cavities and water pockets i.e. when cohesion between the ground and the tool is insufficient to provide its motion, and/or when the driven hole collapses behind the tool as it is penetrating.

In this case it is advisable to drive rigid pipes or conduits into the ground which can either be driven out and replaced by plastic ducting or the casings themselves utilized for laying the required services etc. The impact mole is connected to the pipe or conduit via steel wedges or optional attachments, the impact mole providing the driving force to punch the steel tube into the ground. Connections between the piercing tool and the pipe attachments provide a rigid transmission of the blows necesssary for penetration when driving the pipe into the ground.

Numerous types of ancillary equipment for attaching to, or to be used in conjunction with impact moles have been developed by manufacturers. The increasing use of the equipment within British Gas has resulted in specialist equipment being developed for this particular industry. It would be anticipated that the use of impact moling techniques within the water and other industries will result in similar development patterns emerging.

Impact moles are basically divided into two types:

(i) Service, and
(ii) Mains laying

A brief description of the type of equipment used and the method of operation is as follows:

M40/55

Procedure for forming impacted bore:-

(a) Excavate entry and exit pits at correct depth to sizes specified;
(b) Check for existing services; if in doubt excavate pilot holes;
(c) Place launcher in entry pit and insert impact mole;
(d) Use optisights or boning rods to align mole and launcher to exit pit;
(e) Connect air supply via a lubricator and the mole will form an impacted bore as the machine travels through to the exit pit;
(f) The machine can then be disconnected, and removed from the exit pit. The air supply hose can then be used to draw in the service. If sufficient clearance is available the pipe can be installed after formation of the bore;
(g) In a blind bore situation, the machine can be placed in reverse, and used to tow in the service while travelling in the reverse mode.

M71, 95 & 130 — Larger services

Scope of use:-

When installing polyethylene or plastic pipes and ducts all machines can be used to tow in pipes or ducts whilst simuntaneously forming an impacted bore. This is achieved by connecting the pipe or duct to the rear of the machine.

Larger bores are formed by using the 95mm and 130mm equipment with associated expander housings.

The procedure for operation is as follows:

1. Entry and exit pits should be constructed at a depth of cover between 6 and 10 times the diameter of the machine, or machine plus expander.
2. It is always advisable on bores in excess of 15m to form a pilot bore for using the larger equipment.
3. To assist speed of penetration when expander housings are being used a winch cable can be threaded through the pilot bore, and the equipment subsequently winch assisted.
4. For ducts or pipes less than 95mm or 130mm respectively the machine can then be fired along the pilot bore, and the service installed. If the bore requires increasing an expanded housing is attached to the mole, and a third winch-assisted pass can then be made. Expander housings can be used to increase the bore by approx. 90mm maximum

depending on the prevalent ground conditions.

5. The pipe can then be simultaneously installed during the formation of the bore by the use of tensioning cables, and frames.

PIPE PUSHING USING THE M155 OR M250

The system of pipe pushing for installing steel ducts is a cost-effective alternative to traditional thrust boring systems. The main advantage is the limited working space required, and the elimination of the requirement for a reactive wall. All fixed-head machines can be adapted for this purpose, and the procedure is to attach the impact mole to the pipe by means of metal wedges. The stretching action of the piston drives the open-ended tube through the ground, and resultant spoil inside the pipe is removed by the injection of water under pressure.

PIPE EJECTING

The smaller machine can be used to eject existing steel pipes, the adaptor being connected to the front of the machine and to the pipe to be ejected. The pipe is then hammered out of the ground whilst simultaneously towing in the replacement service.

PIPE BURSTING

Details of the various systems available will be the subject of a separate paper, and therefore it is not within the scope of this paper to elaborate on the technical aspects of the various systems available.

It may serve, however, to clarify the difference between cast iron cracking systems as initially developed by D J Ryan and British Gas, and sewer replacement systems as initially conceived and developed by myself, Mr John Marsh (Engineer) and the Water Research Centre.

With cast-iron systems used to date, mainly within British Gas, an impact mole with cutting blades fractures the pipe whilst simultaneously towing in a plastic, or low density polyethylene sleeve.

This sleeve protects the new service. Fracturing clayware pipes does not require cutting blades, and a new pipe is installed without a sleeve. The sizes and weight of the pipe required to be installed, together with the aggressive conditions encountered on this type of work, require, however, a totally different engineering approach to that of fracturing cast-iron pipes at shallow depths.

CONCLUSION

There is without doubt scope within all utilities to use impact moling equipment. The degree and viability of using the systems described will undoubtedly vary from utility to utility. Responsible manufacturers would always be willing to indicate the possibilities as well as the limitations of the technique discussed. Engineers, planners and operators alike, however, should always approach the installation of the service by selecting as first choice any system which will avoid traditional open-cut techniques. Social implications should also not be ignored totally. The recent installation of cable television around Swindon illustrated that it may be marginally cheaper to hand-excavate across the lawn to install a service at a shallow depth, but would you like it?

TABLE 3.1.1 Excavation and reinstatement expenditure by surface category (estimate) 1982/83

Surface category	New mains £k	Mains replacement £k	New service £k	Replacement service £k	Small hole work £k
Unmade ground	6,304	6,080	12,676	9,690	8,355
Footpath	3,711	17,418	4,833	26,384	10,951
Carriageway	12,383	85,251	10,582	26,800	34,641
	22,398	108,749	28.091	62,874	53,947

GRAND TOTAL = £276,059,000

TABLE 3.1.2

Soil displacement mole application	Average saving estimated by districts for the techniques
Mains replacement	38%
New mains	21%
Service replacement	44%
New services	30%

TABLE 3.1.3 Cost Comparison (£)

	Open cut				Impacted bore			
	Excavate, lay backfill & temp. reinstatement per metre	Permanent reinstatement per metre	Cost per metre incl. perm. reinstatement	Total cost as illus. incl. perm. reinstatement	Excavate & backfill entry & exit pits & temp. reinstatement	Form inpacted bore & lay per metre	Reinstatement of entry & exit pits	Total cost as illus. incl. permanent reinstatement
(A) (6m)	30.30	4.32	34.62	207.72	42.87 (3m @ 14.29)	16.00	10.08 (3 × 3.36)	148.95
(B) (10m)	14.29	3.36	17.65	176.50	42.87	9.50 (only 7m)	10.08	119.45
(C) (1m)	11.42	3.36	14.78	14.78	21.43 (one only)	—	5.04	26.47
(D)	7.09	—	7.09	70.90	21.43 (one only)	4.96	—	71.03
				469.90				365.90

TABLE 3.1.4 Estimated usage of each technique in British Gas (%)

	Trenchless methods			Narrow trenching			Conventional open cut		
	77/8	82/3	87/8	77/8	82/3	87/8	77/8	82/3	87/8
Main replacement	5	33	50	—	21	25	95	46	25
Service replacement	20	39	50	—	—	—	80	61	50

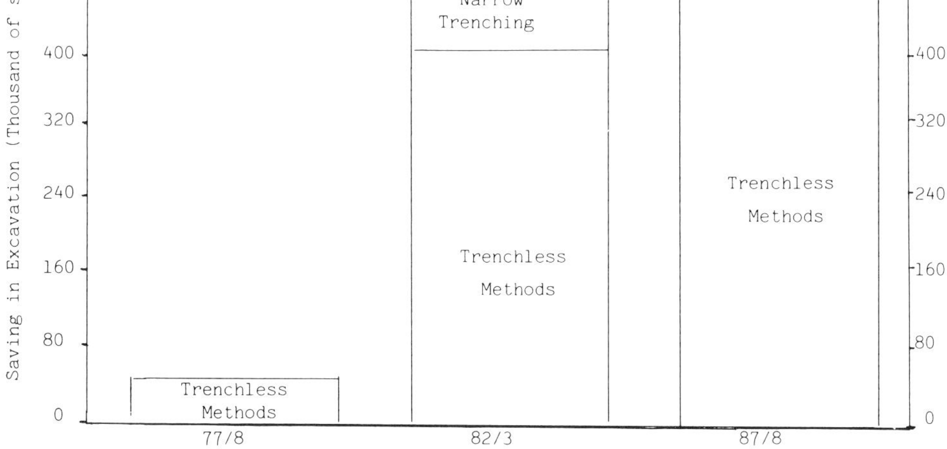

Figure 3.1.1. Impact of "new" techniques.

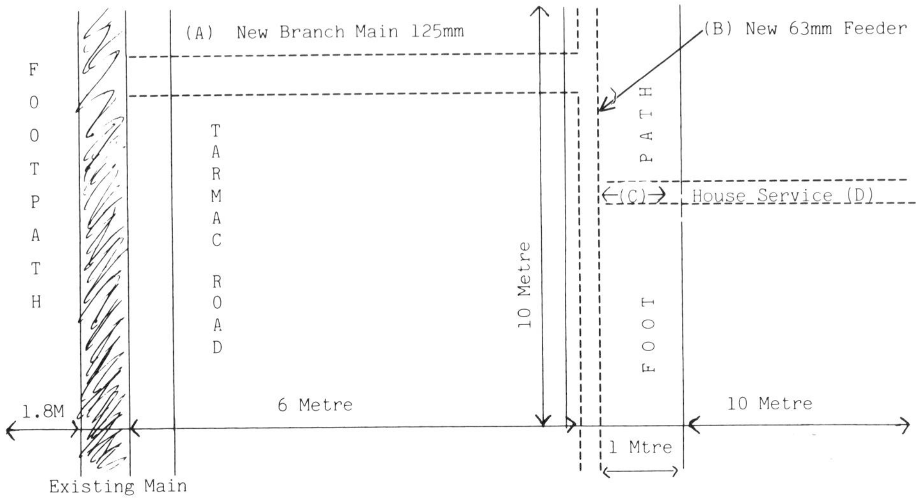

Figure 3.1.2. Extension of new polyethylene gas pipework from existing cast-iron main.

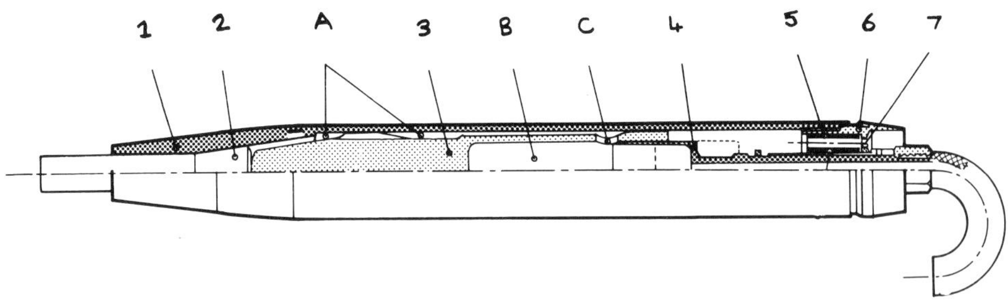

Figure 3.1.3. Reversible pneumatic underground piercing tool.

3.2 Auger boring for pipeline construction

A G Munro, MIE Aust, *Acting Engineer for Metropolitan Works, Melbourne and Metropolitan Board of Works*

W D McMurdie, MIE Aust, *Retired: Previously Area Construction Engineer, Melbourne and Metropolitan Board of Works*

The Melbourne and Metropolitan Board of Works has developed an auger boring system which is used for the construction of new pipelines and the replacement of old pipelines.

The system which has been developed over the past 27 years has played a key role in the programme of sewering the Melbourne Metropolitan area.

This paper traces the development of this system, the general state of the technology, special purpose methods and its application including the benefits, limitations and economics.

INTRODUCTION

Melbourne is the capital city of Victoria and is the second largest city in Australia. The Melbourne and Metropolitan Board of Works (MMBW) is the hydraulic servicing authority for the greater Melbourne area which covers approximately 2000 km^2 and serves a population of 2.5 million.

Since 1973 the Victorian Government has required land subdividers to provide all public services prior to sale of the land. This includes stormwater drainage, sewerage, water supply, electricity, gas and telephone. Prior to this however, services were not required to be provided by the subdivider but had to be constructed by the relevant service authorities after development had taken place. Hence the MMBW has been involved in construction of sewers and water mains in these backlog areas over several decades and anticipates completing this programme by 1987.

Residential subdivisions, which form a large majority of the backlog area, commonly consist of privately owned and fenced blocks of approximately 16 × 35m. A typical block is occupied by a detached house of 140m^2, a garage of 37m^2, a landscaped garden often including a swimming pool and a paved outdoor living area. Sewers are mostly located in the rear of properties, and where the topography permits, the most economical layout is to have the pipeline serve two rows of properties.

The use of machine open cut is sometimes not possible or economic due to resultant extensive property restoration. For this reason there was a need for a method of construction of on-grade pipelines with a minimal amount of earth removal and surface disruption. Auger boring met this need.

Another requirement was the ability to construct pipelines across road reserves with minimal disruption of vehicular traffic, to other services and to the pavement. By excavation of small shafts in the street, or in adjacent properties, the disturbance can be minimized.

Auger boring is a technique for excavating a borehole for the construction of a pipeline to a specified grade. A boring machine is set to a particular alignment and gradient in one shaft and a hole is bored to adjacent shafts. Boreholes can be enlarged by fitting a larger cutting head (reamer) to the auger train and re-boring as the augers are drawn back toward the machine. This is called back reaming.

The geology of the Melbourne area is diverse and includes silts, sands, clays, mudstones, siltstones and basalt. Auger boring is most applicable to mudstones and siltstones (locally known as reef) and to clay.

HISTORY AND DEVELOPMENT

The MMBW commenced the development of a boring system in 1958 by the purchase of a 4.25kW, air-driven lightweight boring machine for 150mm pipeline construction. As directional accuracy was to be achieved by the beam strength of the augers, core boring equipment, consisting of a cutter head and auger tubes fitted with external helical strips, was first used.

It was soon realized that the machine needed to be more rugged and powerful and to have a larger helical clearance on the auger tubes. Hence a second heavier 7.2kW air-driven machine was purchased. The clearance of the augers was increased by replacing 11mm helical strips with 19mm strips. Difficulty was still experienced with core blockages, so the length of bores was limited to 7.6m. Special plain barrelled, butt-jointed concrete sewer pipes complete with rubber rings and metal clips were purchased for installing in the boreholes.

In 1960 it was decided to develop a hydraulically driven boring machine, the first one being five times more powerful than the latter air machine. Associated with this was the development of full-face augering in lieu of core boring. Full-face augering was carried out with a three-legged tungsten carbide-tipped cutting head, whereby the spoil was transported along the auger flights instead of through the centre of the tube. This necessitated augers with small tubes to provide adequate clearance.

By 1961 the hydraulic borer had proved to be superior to the air-driven models. Gear-type hydraulic pumps and motors were used operating at pressures of up to 7600kPa. Initially the feed cylinder was parallel to, but off the line of, the augers. "Machine chatter" occurred with this system but was overcome by placing the feed cylinder co-axial with the augers. This latter development required a short bore (back-bore) at the rear of the machine to accommodate the protruding cylinder.

The size of hole that could be bored was increased by the development of a 380mm diameter reamer. This permitted the laying of a conventional 225mm diameter rubber ring jointed pipe in lieu of the specially designed plain barrelled pipe.

By 1962 the system was extended to permit boring, not just from one shaft to another, but from the machine shaft through one or more intermediate shafts. With this came the development of a steady bearing for use in intermediate shafts to guide or correct the direction of the augers. Instead of using augers for the full length of the bore, extension rods were developed to transmit the rotation and thrust to the head. Bearings were incorporated in the extension rod system to maintain alignment and minimize torque loss.

At this stage it was realized that the beam strength of the augers was not as important as had been previously thought. Also with continuing spoil movement problems, particularly in clay, augers were now manufactured on smaller diameter tubing (150mm).

A further advance in the thrust mechanism was the development of a scissor principle system in 1968. This system which used a shorter thrust cylinder and a lever arrangement, enabled the machine to fit wholly within the shaft, thereby eliminating the need for the back-bore and reducing set-up time (Figure 3.2.1).

In 1970 the auger system was further upgraded to improve the transmission of spoil and to reduce friction. Spoil transmission was improved by changing the pitch of the flights from approximately $0.7d$ to $1.2d$ (where d = the external diameter of the auger). To reduce friction and minimize gouging of the bore, a 38mm wide hard facing strip was fixed to the edge of the flight. This also increased the life of the augers.

By 1971 the MMBW had a complement of six machines. At this time there was no known commercially available boring machine or contracting company able to accurately bore hard Silurian mudstone from a shaft as short as 2.4m.

1978 was the MMBW's peak production year for auger boring, achieving an output of 21km from 14 machines. 1979 saw a consolidation of the equipment, achieving an average output of 2km per machine per year. It also saw the application and development of other types of machines for shorter bores. This included a commercially available lightweight petrol-driven borer and a MMBW-manufactured air-driven machine.

Head and reamer techniques progressed from using removable bits held in place by set screws and disposable push-in bits to a more satisfactory fixed tungsten carbide bit. The size of back reaming was increased concurrently with the development of the auger boring system.

In recent times contracting firms have established auger boring systems for similar applications and the installation of other small diameter conduits.

TECHNICAL DETAILS OF CURRENT EQUIPMENT

Power packs

The boring machines are operated by hydraulic oil power provided from an above-ground unit. The most recent power packs are trailer-mounted silenced units weighing 2.85t. They contain two gear-type hydraulic pumps. The larger one, a tandem pump, consists of a 44mm diameter cell and a 19mm diameter cell which when combined provide three differing flow rates. This pump supplies the power for the rotation of the boring machine and provides speeds of 58, 29 and 87rpm respectively. The smaller pump provides the power for the thrust ram on the boring unit. Both pumps are powered by a 6-cylinder diesel engine and have a noise level of only 70dBA. Other features include intercoolers, thermo cut-outs and full-flow oil filters on both the feed and return lines.

Hoses 18m long are desirable; however, up to 45m lengths can be used but with considerable pressure loss.

Boring units

The boring unit which is set up in the bottom of the shaft consists of the boring machine and a base (Figure 3.2.1). The boring machine consists of a hydraulic motor to provide the rotational power to the augers. Piston hydraulic motors have been found to be the most suitable type for this application. The base supports the machine and provides the forward and reverse motion by means of a thrust mechanism (Figure 3.2.1). It is designed so that a saddle bearing can be attached to either end of the base, hence boring can be carried out in both directions from one set-up. Currently, most bases are fitted with adjustable cast-iron slides (or wear strips) which are not lubricated.

There are two basic types of boring units in use, the co-axial thrust ram type and the scissor ram type. Both types have their particular advantages; however, the current trend is toward the scissor ram design. The thrust is normally 5.6t but by increasing the pressure, a thrust of up to 7.8t can be achieved.

Heads and reamers

Basically heads are used as the cutter for forward boring, while reamers enlarge the hole when pulled back toward the machine by the augers.

Heads are available for clay, soft reef, hard reef and basalt in sizes 190mm and 300mm. An important aspect of cutter head design is to ensure the ready flow of cuttings back past itself. Flights are added immediately behind the cutter head tips both to transport cuttings and to assist accuracy, giving lateral support to the head. Heads are usually fitted with drag-type cutters as fluid flushing is not required and the thrust can be lower.

Clay heads are designed to slice through the clay and clear it from the cutters to prevent clogging. This type of head is also suitable for boring in soft mudstone. Reef heads are used mainly in soft to fairly hard mudstone, but due to their geometry, tend to clog when used in clay. A 190mm hard rock head has been made by modifying a commercially available 140mm drag bit. It has been found to be suitable for the hardest mudstone and performs reasonably well in vesicular basalt. A 300mm hard rock head has been built, but has been found to be more suitable for full-face boring in hard mudstone.

The main feature of a reamer is also its spiral configuration to ensure continual removal of spoil from the cutter teeth to avoid clogging. Additional tail spirals are sometimes used to assist with the spoil transmission and to ensure a clean hole. The MMBW has nine sizes of reamers ranging from 250mm to 815mm which cope with most ground conditions from clay to hard mudstone.

Augers and rods
Augers and extension rods are used to transmit the thrust and rotation to the head or reamer. Augers are also used for transmission of spoil from the head or reamer to the nearest shaft.

It is believed that bore alignment is affected more by the gauge of the cutting head in relation to that of the lead augers, rather than by the stiffness of the augers. Hence, as previously mentioned, the original concept of ultra-stiff auger tubes has been abandoned in favour of a semi-stiff section of 6 to 10m in length. This has allowed the use of small tube diameters of 100mm and 125mm, which together with an increased flight pitch (greater than 1.2 *d*), has greatly increased the efficiency of transmission of spoil.

Three sizes of auger exist, 190mm, 300mm and 530mm, and are available as lead or spoil augers. Lead augers are threaded to achieve a degree of rigidity whereas pin-jointed spoil augers are cheaper to manufacture and quicker to fit. To avoid jamming, double start square threads are used to join the lead augers. To improve clearances spoil augers are 5mm smaller in diameter than lead augers. Alternatively, a clearance auger fitted with drag cutters to enlarge the bore is used as the last lead auger. All flights are now fitted with a 19mm-wide hard facing strip.

Extension rods are jointed, using spring clips to retain the pins in position. Some rods are fitted with bearings on wheels to support them in the bore hole and to reduce friction. Augers and rods are generally 900mm long with some 450mm lengths. A series of adaptors join various thread types and sizes, and threads to pin joints.

BORING PROCEDURE
Shafts are excavated along the line to be bored at approximately 11m centres. This distance can be varied to suit ground conditions, accessibility of the site, and the economic desirability of placing shafts where they can also be used for the construction of property branch connection points. The machine shafts are required to be 2.4 × 0.9m inside the support system and a depth of 300mm below invert level of the finished pipeline. Over-excavation should be avoided as it can create problems of instability of the boring machine if the augers jam. Intermediate shafts, either 1.8 × 0.8m or 1.4m diameter are sunk to 150mm below invert level. Survey reference points are established at each shaft for line, level and distance, or alternatively a laser beam is used as an above-ground reference.

The boring unit and base are carried to the machine shaft on four-wheeled hand trolleys and lowered into the shaft from a four-legged gantry. The base is set up to level and grade on three levelling jacks and aligned and fixed in position by side screw jacks.

Boring commences (Figures 3.2.2 and 3.2.3) with screw-joined lead augers, the first three being supported by the saddle bearing fixed to the boring base. After the advance of approximately eight lead augers, pin-joined spoil augers are used. When an intermediate shaft is reached, the position of the bore is checked and minor corrections are made using a steady bearing installed in that shaft. Large variations can be corrected by backreaming the hole for a distance up to 4m to permit greater deflection of the auger. Gross variations can be overcome by using the short bed borer in the intermediate shaft and boring back to the previous shaft. The steady bearing remains in position until the head has progressed approximately three auger lengths. Once the number of augers fitted is equal to the greatest distance between shafts, extension rods are used. Every third rod is a bearing or wheeled rod to support the system and minimize friction losses. Spoil removal is from the last shaft that the head has passed through.

When a bore is completed in one direction it can be either reamed out to a larger diameter as necessary, or the rods and augers withdrawn to the machine shaft and removed. The machine is then turned around on the bed and the boring commences in the other direction in a similar manner.

Normal reaming is done by fitting the required size reamers and pulling the augers back toward the machine. In basalt however, reaming to 300mm diameter is usually carried out in a forward direction following a 190mm pilot hole. Spoil transmission is toward the machine and removal is at the shaft closest to the reamer as with boring.

When boring and reaming are completed in both directions, the shaft is filled to just below the invert level of the pipe. Rubber ring-joint pipes are jointed in the shaft and pushed along the line by the boring unit or by a hand-operated jack. Pipes 300mm diameter and larger are laid on skids to avoid the build-up of clay in the sockets and to make jointing easier (as the pipe spigot is at the correct height). The annulus around the pipe is filled by a grout mix.

APPLICATION AND CAPABILITIES OF THE SYSTEM

Auger boring is most applicable to the following situations:

- steeply sloping ground;
- highly developed and treed areas;
- under or close to buildings, retaining walls, and other structures;
- under roads, rail tracks or other areas where the surface disturbance is required to be minimized.

In soft rock ground it has significant benefits in the reduction of complaints of property damage due to the elimination of blasting in tunnels.

In basic terms the auger boring system has the capability of full-face boring holes of 190mm and 300mm in self-supporting ground up to hard mudstone. Holes can be reamed to 250, 300, 400, 460, 530, 635, 685, 740 and 815mm diameters. Total distances of approximately 60m in each direction can be achieved with intermediate shafts spaced at 10 to 13m apart. The maximum distance which has been bored between shafts is 36m and this was limited by spoil transportation and auger friction.

Auger boring is predominantly used in self-supporting clay, Silurian and Devonian reef and to a lesser extent in slightly weathered basalt. The reef consists of mudstone and siltstone which is normally in a high to medium state of weathering. The equipment can readily bore rock with a seismic velocity of 1070m per second.

The main limitations in the use of the system are some ground conditions, flatness of grade and obstructions. Significant differences in hardness of the ground often result in deflection of a bore whilst boring across the interface, hence this should be avoided if possible. Preferably, lines should be designed at 1 in 80 or steeper so that practical variations in boring may result in lines of 1 in 100 or at the worst 1 in 200. Obstructions including underground services may limit the placement of the shafts, and may also require complete exposure prior to boring to ensure safe conditions.

It is very difficult to generalize on production rates, and even more difficult to give cost comparisons as no two lines are identical. Under favourable conditions penetration rates of 4m per hour can be achieved, but average production rates of 2m per working hour or 1700m per machine per year are more realistic. Auger boring is not a low-cost method of construction; however, in the correct application it is an economical method. In a typical sewerage reticulation area the cost of bored lines may be 1.25 times the cost of open-cut lines. Boring cannot compete in clay where open cut is practicable. It can, however, compete in reef open cut, particularly when there are restrictions on blasting. Boring is usually considerably cheaper than shaft and short tunnel excavation.

SPECIAL PURPOSE METHODS

In addition to the general methods which have been described several special purpose techniques have been devised. Briefly they include:

Basalt boring

Basalt boring has been carried out using tricone, point attack and drag bits. All methods require a continuous water supply and a high thrust, particularly with the tricone head. Although the MMBW has not carried out an extensive amount of basalt boring it has achieved bores of up to 400mm diameter and 50m in length.

Sleeve boring
A technique has been developed for pipeline construction under rail tracks where the hole must be permanently supported. The hole is supported by the augers as it is bored and, by the use of a swivel coupling, the pipe or a sleeve is drawn into the hole immediately behind the backreamer. This method is not applicable to running ground.

Short bed boring
A short bed borer using 450mm-long augers has been developed to bore from shafts only 1.4m in length. This is used particularly where obstructions restrict the size of shaft which can be sunk.

Slope boring
A boring bed has been mounted on an adjustable frame to bore holes at angles between 25° and 41° to the horizontal. It has many uses particularly for construction of property connections where the sewer is in a street which is in a cutting.

Alternative borers
The MMBW has purchased and developed lightweight petrol- and air-driven borers for use on short and less accurate boring. They are used under trees, sheds, concrete driveways etc., and are quicker and easier to set up.

SEWER REHABILITATION
The bore replacement technique is a recent development by the MMBW for same-size sewer replacement. The method was developed to be used in lieu of open cut for on-line or off-line replacement where the same sized pipe is required. By using the existing sewer pipe cavity, economic advantages were foreseen in carrying out minimum excavation where excavation and refill costs associated with conventional trenching would be very high. Improvements, refinements and user experience have now produced a technique that is continually being utilized with a cost saving in the order of 20 to 50 per cent compared with conventional replacement.

The auger boring equipment has been adapted for this work where the old concrete pipes are reamed out and new pipes simultaneously drawn in.

The boring machine is installed in a shaft excavated adjacent to a manhole or at a house branch connection point. Augers of a smaller diameter than the existing sewer are pushed through the sewer to the next shaft and a backreamer and shield assembly are attached to the end of the augers. The boring machine rotates the augers and the reamer whilst the shield, due to the presence of a bearing, remains steady. A length of pipe is attached to the shield and the boring machine draws the assembly towards it while reaming out the old pipe (Figure 3.2.4).

The reamed-out pipe material is drawn forward to the machine shaft by the augers from where it is hand-loaded into a bucket and hoisted to the surface. As the reamer progresses, additional lengths of pipe are added at the rear shaft and equivalent lengths of augers removed from the machine shaft. The pipes are towed in by clamping the rear of the pipe to a chain attached to the shield. This is necessary to avoid stressing the solvent-cement joint in the uPVC pipe. uPVC has generally been used as the pipe replacement material; however polyethylene has also been used successfully.

This technique yields progress rates of up to 8m per operating hour. Sewage flows are bypass-pumped, often on a 24-hour basis, so that work can proceed for several days without the need to complete a length each working day. As the new pipe is installed between shafts, the house branches are reconnected and sewer flows resumed downstream of the work.

To date, 3.3km of 150mm, 225mm and 300mm diameter sewer has been successfully replaced with uPVC and polyethylene pipe using this technique.

Now that this technique of same-size bore replacement has been proven, the next phase of development proposed is to upsize the pipe by reaming out a larger diameter than the original pipe to allow larger pipes to be drawn in. Reasonably self-supporting ground will be required for this to succeed; however the shield does provide some measure of support.

Should upsize bore replacement prove to be successful, it will exhibit considerable benefits compared with conventional open-cut, on-line, or off-line replacement. Bentonite slurries can and have been used to provide additional ground support and reduce pipe friction when using this method.

CONCLUSION

Trenching as a method of installing conduits is becoming less acceptable, particularly in metropolitan areas, due to its disruptive effect on everyday life. As reinstatement costs rise alternative trenchless methods are becoming more economical.

The auger boring system is a practical construction method which can bore holes of 190 to 815mm in diameter in excess of 60 metres in length at specified grades as flat as 1 in 100 in ground conditions varying from clay to basalt.

The advantage of the system is its ability to bore holes accurately on line and grade from a small shaft. Thus it has been extensively used in areas restricted by other utility services, buildings and trees.

ACKNOWLEDGEMENTS

The authors are grateful to the management of the Melbourne and Metropolitan Board of Works for permission to publish this paper. They acknowledge the assistance of many officers of the MMBW for their contribution and assistance.

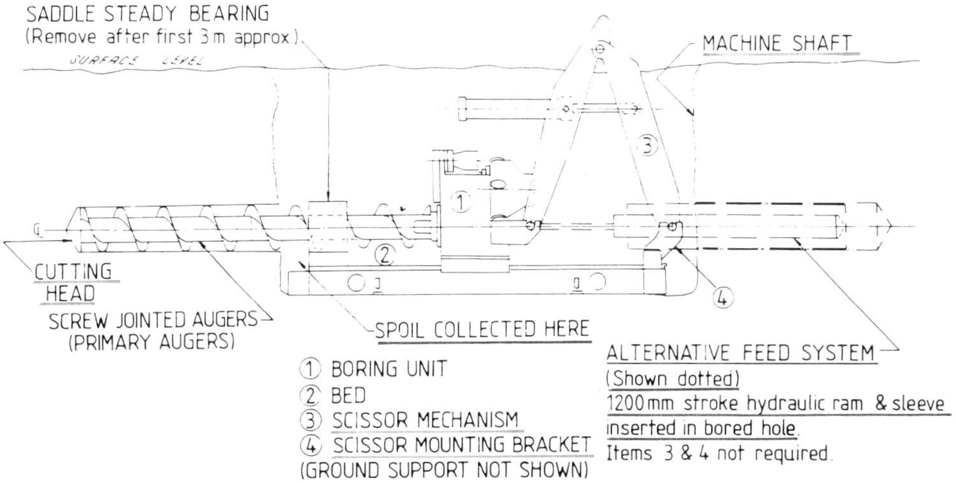

Figure 3.2.1. General arrangement in machine shaft.

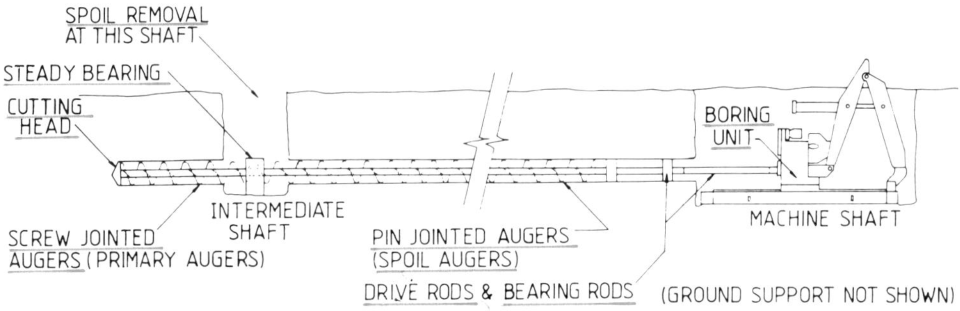

Figure 3.2.2. General arrangement: machine shaft and intermediate shaft.

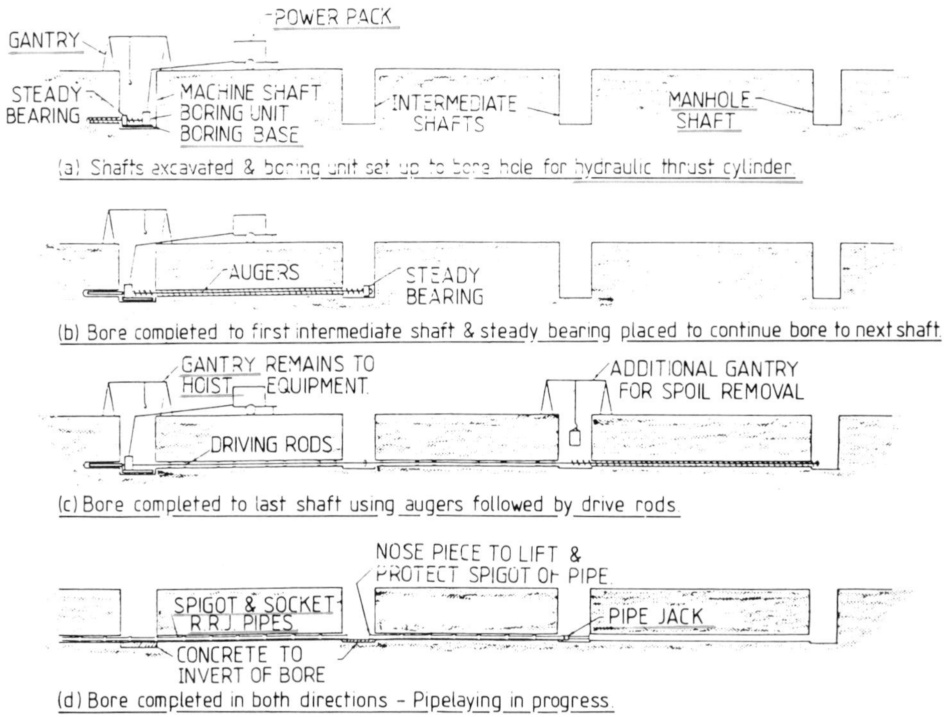

Figure 3.2.3. Auger boring sequence.

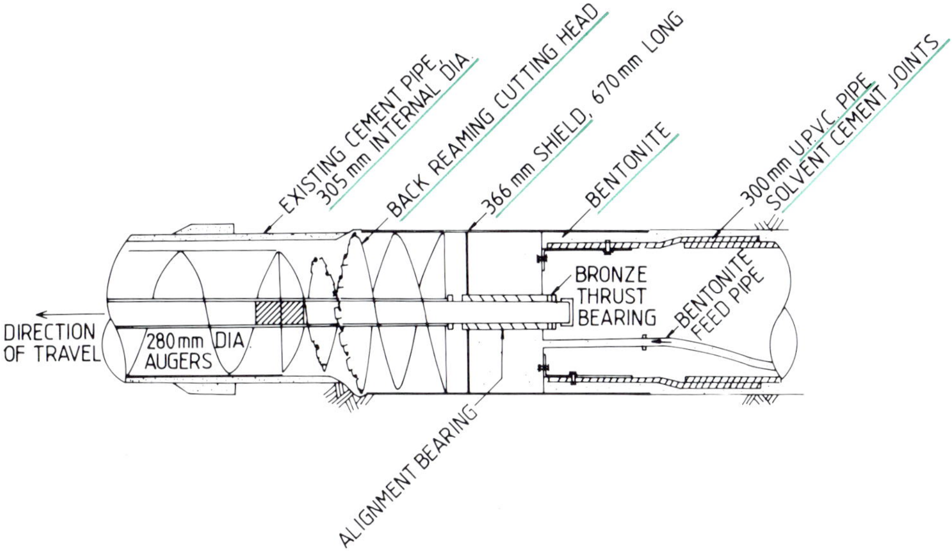

Figure 3.2.4. Back reaming general arrangement: sectional diagram.

Authors' introductions

Mr P T Etherton: Development of impact moling techniques in the UK

I would like to comment on a few points raised yesterday which I think are relevant to the use of impact moling techniques. Several things that were said sounded familiar — the call for increased Government expenditure, the illustration that we should be spending £X millions per annum on our decaying infrastructure and the fact that we are spending the equivalent of pocket money. I think we must all realize that there is a limited amount of money to spend, and probably will be for a long time, but we must not lose sight of the fact that many keen and inventive engineers from progressive authorities are working minor miracles on what little money is available. I suppose we should all hang on and hope that WRC and ERS will invent the cures for our ills which will be carried out at the rate of £1 per metre. In the meantime, engineers and planners should make use of the new techniques that are available and we should use our collective endeavours to make up with our brains what we lack in the purse. We have always, I believe, been an inventive nation that does things on a shoestring, and I regret that I can see little change in the future.

I have listened with great interest to the papers dealing with effects of moling on adjacent services, and the work carried out to date is most encouraging. It is, I believe, an area in which some guidelines are urgently required. I am left wondering how the moling operator will identify the type of ground through which he is moling and indeed the state of deterioration of the pipe that he is passing by. I would personally welcome a minimum standard to be set, i.e., do not mole closer than X to an adjacent service.

I would also like to comment on British Telecom's remark on contractors adopting new techniques. In my opinion contractors will, understandably, choose whatever method provides the greatest profit. I would suggest that you could pay them an extra £10 per metre to carry out a road crossing under local authority reinstatement by moling and pocket the difference in saving yourself, and I think you may be surprised at the response you will get.

For the benefit of any delegates who do not fully undertand the basic method of operation of moles and moling techniques, I will explain briefly the principles of operation.

Impact moles are available in two basic types, fixed- or moving-head design. Both have advantages, and disadvantages, and the end user should evaluate these accordingly. Figure 3.1.3 in the paper shows the simplest version available, complete with a screw reversing mechanism. Compressed air is applied via the tail hose which drives the piston to the front, whereupon it strikes the anvil and produces an impact bore. The machine burrows through the ground, compressing the surrounding material, and so forming the impacted bore. This particular model has a pull reversing mechanism, but the working principles are basically the same. The machines are available generally from 40mm to 250m in diameter. This particular machine is 40mm in diameter and has stabilizing flutes on the front end to prevent excessive rotation of the machine during operation.

Before commencement of moling activities, two basic guidelines must be observed. Firstly, the ground must be compressible to ensure that sufficient side wall friction is obtained, and secondly sufficient ground cover must be maintained over the machine during operations. As a rough guide this should normally be between 7 and 10 times its diameter. Do not attempt to form impact bores in extremely loose or wet material.

The procedure for forming impact bores can briefly be summarized as follows. First and most important of all, check the adjacent services with cable location equipment. Two, construct the entry and exit pits, normally 300mm wide by 750mm to 1.00m in length, at the required depth. The impact mole is then placed on a launching platform which can be adjusted for level and line. An Optisight or simple boning rod system ensures the correct position of the launcher in the entry pit. An oiling device supplies lubricant and anti-freezing solution to the mole and also enables the compressed air supply to be regulated by the operator close to the launch pit. When the operator is satisfied that the impact mole has been set on the right line and level the air supply is turned on, and the mole forms the bore. Ninety per cent of all services laid by impact moling techniques are below 40mm in diameter, and various items of ancillary equipment have been designed by manufacturers and are available to insert the services during or after the bore is formed.

A range of mandrels can be used to connect the pipe or cable to the compressed air hose, which can be then used in turn as a draw line to insert the pipe or cable. The service can also be connected to the front of the machine so that the mole can be placed in reverse and tow the service in where a blind end situation exists. Numerous attachments are available to be used with the basic mole.

A spinning mandrel has been specifically designed for use by British Gas to install 63mm polyethylene pipe whilst simultaneously forming the bore.

Small pipes can be ejected by attaching a ramming device to either 40mm or 55mm machines. This may be useful in ejecting old services in both the water and gas industries and is recently being introduced in field trials.

The use of the larger moles which are generally considered as mains-laying equipment requires a slightly different approach from the smaller service-laying machines. A greater emphasis must be placed on correct lifting procedures, and the use of mechanical handling equipment is recommended in order to avoid unnecessary hernias among operators. As a general rule a higher degree of expertise is required when forming larger impacted bores, up to 200mm in diameter.

I mentioned earlier the dangers of using impact moles in unsuitable ground conditions. The next process I would briefly like to mention is the fixed-head system, which is used and recommended when laying pipes or cables in waterlogged or loose materials, or indeed when steel carrier pipes are specified. I personally believe there is a strong argument for utilizing this procedure when crossing sensitive areas such as motorways, especially when ground conditions are unknown. This system is obviously more expensive than straight impact boring, but provides an extremely safe method of achieving a crossing.

A major advantage of fixed-head equipment is that it can be easily used for pipe pushing. The machine is attached to a steel carrier pipe by metal wedges, and the action of the hammer literally pushes the steel tube through the ground.

I would like to conclude by emphasizing a few do's and dont's for first time users or indeed those people who have limited experience of moles and moling techniques. I would emphasize that the first time user should ensure that the manufacturer or distributor he is purchasing equipment from can provide the necessary training and support his operators will require.

Do not underestimate the latent reluctance some operators will exhibit in refusing to abandon the pick and shovel in favour of this new-fangled equipment. Sometimes gentle persuasion is required, normally in monetary terms. An initial success by the operators in using the equipment will ensure they will readily adopt the new techniques. The cost-effectiveness of impact moling is, I think, beyond doubt, as illustrated by the British Gas experience. I am confident that trenchless techniques of all descriptions must become the engineers' and planners' first choice and therefore I would like to remind you as a parting gesture to "Think before you cross the road!"

Mr A G Munro: Auger boring for pipeline construction

In this presentation I will be briefly covering the development, explaining the system and extolling the virtues of an auger boring system developed by the Melbourne and Metropolitan Board of Works. Melbourne, the capital of Victoria in the south-eastern corner of Australia, covers an area of 2000 sq kilometres and has a population of over 2.5 million.

Services such as sewerage were not provided when land was subdivided until legislation in 1973 required this. By this time there was a backlog of 160,000 properties, many of them well developed. As most sewers are laid along the rear or side of properties, there was a need for a method of construction that minimized the surface disturbance. Legislation requires the Board or its contractors to restore properties to as near their original condition as practicable, and to compensate for any damage.

Hence development of a system to meet this need first started 27 years ago in 1958 using air-driven borers working on the core boring principle. By 1960 the development had switched to hydraulic-powered units and progressively the core boring was replaced by an auger system.

The current system consists of four basic elements, firstly a power pack which contains hydraulic pumps to provide the power to a boring unit. It is powered by a 6-cylinder diesel engine, weighs 2.8 tonnes and has a noise level of only 70 decibels. Secondly, a boring unit which consists of a boring machine and a base. The boring machine is principally a hydraulic motor to provide the rotational power for the augers. The base supports the machine and provides the forward and reverse motion by means of a thrust mechanism. There are two types of thrust mechanisms, the co-axial ram and the scissor ram. A thrust of 5 to 6 tonnes is applied to the augers. The beds are also designed so that a saddle bearing can be attached to either end. Thirdly, the augers and extension rods are used to transmit the thrust and the rotation to the head or reamer. The augers are also used for transmission of spoil, from the head or reamer to an adjacent shaft.

The last items of equipment are the heads and reamers. Basically the heads are used as the cutters for forward boring whilst the reamers are the cutters to enlarge the hole while being pulled back towards the machine. There are a variety of different heads and reamers for different grounds.

The procedure used to bore a line is to sink shafts where required to suit connections and surface obstructions whilst striving for a space of 10 to 13 metres between shafts. The machine shaft, which is only 2.4 metres long by 0.9 metres wide, is slighly larger than the intermediate shafts, which are 1.8 metres by 0.8 metres. The boring unit and base are set up in the machine shaft on levelling jacks and aligned and fixed in position by side screw jacks. Boring commences using screw-jointed augers, the first three or four being supported by the saddle bearing fixed to the boring base. The fifth auger is commonly a clearance auger and, following another three screw-jointed ones, pin-jointed spoil augers are used.

When an intermediate shaft is reached minor corrections are made using a steady bearing in the shaft. Once the number of augers is equal to the greatest distance between any shafts in the line extension rods are used, each third one having a wheel on it for support and to reduce friction. Spoil removal is from the last shaft passed through. When the bore is completed the line can be enlarged by back-reaming, or the rods and augers are removed at the machine shaft. By turning the machine round on the bed, boring can commence in the opposite direction. When the borehole is completed rubber ring joint pipes are jointed in the shaft and pushed along by the boring machine or by hand-operated jacks.

This system, which is based on the principle of achieving a high degree of accuracy from a small shaft, is most applicable in steeply sloping ground, highly developed and treed areas, under or close to structures, under roads, railtracks etc. Its use also minimizes the need for blasting. The auger boring system can full-face boreholes up to 300mm in diameter and over 80 metres in length in self-supporting ground up to hard reef. Holes can be reamed to 815mm in diameter.

The system has limitations including loss of relative accuracy on flatter grades, excessive ground hardness, running ground and significant variations in ground hardness which cause deflections, and underground obstructions. The flattest desirable grade is 1 in 80. Average production rates of 2 metres per working hour are achieved with rates up to 4 metres being achieved under favourable conditions.

Auger boring is not a low-cost construction method; however, in its correct application it is an economical method. It cannot compete with open cut in clay where this is practicable. It can, however, compete with open cut in reef, particularly where blasting is restricted. On an average reticulation area auger boring costs approximately 25 per cent more than open cut. This system has been progressively developed and used on what we call the backlog sewerage construction works. We have bored of the order of 250km with a peak production of 21km in 1978.

As this type of work is reducing, some of the equipment is being modified for sewer rehabilitation works. In the rehabilitation system small diameter augers are threaded through the pipe to a shaft where a reamer and shield assembly are attached. The new pipe is attached to the rear of the shield and pulled into the hole as the old pipe is reamed out. Same-size replacement has been proven and we are now moving into up-size replacements.

Trenching as a method of installing or replacing conduits is becoming less acceptable, particularly in metropolitan areas. The auger boring system was developed by the Board to satisfy a need which could not be met by commercially available equipment at the time. It will bore holes accurately from a small-size shaft and is hence particularly applicable in metropolitan areas.

Discussion on papers 3.1 and 3.2

G Woods (Naylor Bros. (Clayware) Ltd.): I see no mention in Mr Etherton's paper of the maximum size of mole. A recent report has indicated a maximum size of around 400mm, and I would welcome Mr Etherton's comments on that. Would he also comment on the question of accuracy, particularly when installing sewers. Munro and McMurdie refer to a design grade of 1 in 80, with a resulting worst case of 1 in 200.

The auger boring technique described in the second paper has been used predominantly in self-supporting clays. Has there been any experience of other ground conditions, particularly non-homogeneous soils? What classes of uPVC and polyethylene pipes have been used with the auger method?

Although the two papers deal with different techniques, I find a common thread running through them. This is the sensitivity of the methods to ground conditions — a sensitivity far greater than one gets with open-cut construction. Do the authors consider that more thorough site investigation is necessary when using these methods than with open cut? That would clearly affect the economics of the operation.

Mr P T Etherton: The maximum size of holes that can be formed by impact moles is about 200mm, owing largely to the fact that the ground one is trying to compress has a limited elasticity. But more important than that is the practical aspect that cover of 7 to 10 times the diameter of the machine must be maintained to prevent it from rising. To form a 200mm bore, one would use an impact mole of, say, 130mm, with a 200mm expander, requiring ideally 2 metres of cover. However, most gas and water mains of this size are laid at shallower depths than this, and this can be done — and is done every day with polyethylene gas mains of 180mm internal diameter — quite successfully where the surface is of good construction, e.g., a concrete road.

For larger bores, pipe pushing is adopted. Machines are available up to 250mm in diameter, and research is being carried out on a 400mm machine for pipe ramming. In this method the soil passes inside the tube, and there is not a compression problem. Pipe ramming is showing signs of progress in the UK after a slow start.

With regard to accuracy, the majority of impact moles are used for installing house service connections under 40mm diameter over lengths of perhaps 7 to 10 metres. Many thousands of these bores are being made with accuracies within ± 150mm, which is sufficient for this purpose. When using a larger machine to install a 200mm pipe, we would recommend digging an inspection pit every 20 metres and launching between pits.

Impact moles are not recommended for laying drain connections to accurate falls.

Site investigation is expensive, and is not practical on the majority of moling jobs, since soil conditions are liable to vary along the length of a bore. I would only recommend it on a large job, where the volume of work is enough to justify the expense. In general it is up to the contractor to evaluate the risk.

Mr A G Munro: The ground in which we prefer to use the auger boring system is the mudstone or siltstone referred to in the paper as the reef, which allows higher rates of progress than clay. The augers can handle clay quite readily, but the clay tends to clog and does not clear the auger as quickly as the reef. The auger will not handle running sand, but can manage fairly soft clay because the flights of the auger support the hole to a significant extent.

The railways in Melbourne insist that any boring under their tracks be fully supported at all times. They accept that the flights on the auger are supportive to the hole. Instead of withdrawing the auger and then threading the pipe through, we attach a back-reamer to it, hook the pipes behind it and then draw them back through the hole. In this way, the hole is never left without some form of support.

For the size of sewers I am referring to — mainly 150 to 300mm, although we do go beyond this — the favoured pipe material in Melbourne is earthenware or vitrified clay. We do, however, lay uPVC (class Extra Heavy) and polyethylene pipes at times. The advantage of using uPVC is that we can bore a smaller hole for the same size pipe.

I must agree that auger boring is more sensitive to variations in ground conditions than open cut, and more site investigation is sometimes necessary. Before we do any work, we have a certain amount of geological test-bore information, and we have a fair idea of whether the ground is suitable for boring or not. We then have to sink our working shafts, and these give us more information about the ground.

As regards other services being in the way, the Board has very strict rules that we must study the plans of all the authorities before we start work, to find out where their services are laid. We must call on an officer of each authority to tell us where their services are, and we mark these locations on the ground. If they are going to be in close proximity, we dig down and expose them.

I would like to amplify the statement in the paper that "the equipment can *readily* bore rock with a seismic velocity of 1070m/s." We do, in fact, bore rock with a seismic velocity up to 1500m/s. At the top end of the scale, we have bored 2000m/s rock, but only where no alternative method was practical.

P G Vale (Chairman, George Law Plant Ltd.): Three and a half years ago our company purchased several moles from Merstan, and since then we have acquired a great deal of experience. The main condition for success is a knowledge of the type of ground you have to go through. We contract on the basis of a fixed-price quotation for a job, and the majority of our work uses a system of pipe thrusting. So far, the maximum diameter of pipe we have thrust is an 800mm steel pipe over a distance of 20 metres. We believe that is the longest pipe thrust of that diameter in this country so far. We have surprised ourselves and our clients on the degree of accuracy we can obtain.

Acceptance of the method has been slow, and it is only in the last year that we have been working full-time. We have worked with British Rail and the Department of the Environment, and we find that, once people have had experience of the moling method, they accept it and ask for more.

V A Finney (Principal Assistant Engineer, City of Manchester): On the question of risk-sharing, raised by Donald Rees yesterday, Mr Etherton said we should make up with collective endeavour what we lack in the purse. I think that is an admirable objective, but I wonder whether the industry will rise to this challenge, or is it more likely that the contractor and the developing manufacturer will still want to secure that side of the market for themselves?

I am very interested in an auger system for on-line replacement, but would like to know how the problem of side connections is overcome with the system currently in use?

K Reed (Water Research Centre): Would Mr Etherton clarify his statement that British Gas are installing 200mm diameter pipelines in virgin ground?

How many intermediate shafts are required for auger boring?

Is any development work being carried out on steerable cutting heads for augers?

Mr Etherton: British Gas are installing 180mm pipes (not 200mm) in virgin ground on a daily basis, but obviously the sites are chosen with care. I do not know what percentage of new 180mm polythene pipes are installed by the moling method. No doubt British Gas could supply these figures, which I would be interested to see.

Mr Vale's comments support my own views on the inherent conservatism among many authorities and engineers, who are reluctant to try a new method.

I have personally been constantly frustrated by the lack of development money that is available. The will may be there, but the money is not. I think the contractors can be left to invest their money where they see a commercial return, but I would like to see a change in attitude as between authorites and contractors. Authorities still seem to suspect the contractors of trying to extract more profit than they have earned. The Department of Industry guidelines on development contracts, especially in respect of sewer renovation, should encourage a much more healthy relationship between the local authority and the contractor, so that, when things go wrong, the authority does not run for base and put the blame on the contractor. Perhaps more influence is required from Directors of Engineering.

Mr Munro: We have not yet developed any good systems for making connections on replacement or rehabilitation work, and still have to sink a shaft for each connection. In many places we can achieve two connections with one shaft, where properties on both sides of the sewer line have been connected at the same point on the sewer. We are very interested in the developments in Britain on minimum excavation methods to improve the making of connections.

It is desirable to have shafts every 10 to 13 metres or thereabouts along an auger line, although this is not always convenient. The longest bore we have ever done without a shaft was 36 metres, but the longer the distance between shafts, the less accuracy can be expected. There are ways of remedying inaccuracy of grade or line, e.g., by pushing the augers off-line while back-reaming, or by back-reaming to a size bigger than is required. But this is still not desirable.

We have not done any work on a steerable head. We believe the main secret of getting accurate holes is to have the first few augers relatively rigid, which is why I mentioned screw-jointed augers. The difference in diameter between the head and the lead augers should be as small as possible, to create a gun-barrel that sets the line. By this means we have been getting a satisfactory degree of accuracy without steerable heads, and now that we are doing more rehabilitation work and less new construction, there is little reason for investing funds in the development of a steerable head.

3.3 Drilling of short horizontal microtunnels in rock and rock and soil combined

K Olimb *Managing Director, Kristian Olimb A/S, Råde, Norway*

The drilling of horizontal holes of a size suitable for water and sewerage utilities in difficult or rapidly varying ground has, until recently, involved the use of very expensive equipment or costly civil engineering practices. However, by adapting the equipment and techniques used in vertical or inclined drilling in mines, whereby a pilot hole is back-reamed to the required size, it has been possible to achieve inclinations of less than 1° to the horizontal with good directional control. The paper describes the development of the technique, lists the sites on which it has been used, and gives a brief case history, with costs and savings, of some of the contracts undertaken.

INTRODUCTION

In Scandinavia, where the common underlying strata are rock or glacial moraine deposits, any underground construction is likely to be expensive and often involves drilling or blasting. In addition there is the problem of maintaining line and level in difficult or changing ground, especially in moraine deposit areas where large rolling stones can be encountered, and in coarse rockfill and swelling shale areas.

The solution has been to adopt a method known as raise boring, commonly used in coal mining to connect different strata levels by inclined or vertical shafts, so that nearly horizontal bores are achieved.

RAISE BORING

The principle behind raise boring is that a pilot hole is drilled downhill and the hole is subsequently enlarged by back reaming uphill so that the resulting tailings fall away from the drill head. The normal process is to drill the pilot hole with a tricone bit to give a diameter in the range 250mm-350mm (9⅞″–13¾″). The drill casing to follow the bit will have an outside diameter of 203mm (8″) up to 305mm (12″). The direction of the pilot hole is controlled by three or four stabilizers following the pilot bit. It is essential to set up the drill direction correctly and a trained operator can judge how the pilot hole will tend to deviate in a certain formation and can generally guarantee that the deviation will be within one metre of the target on a 100-metre length of hole. Raises of this type have been drilled up to a length of some 500 metres.

In a few cases the pilot hole is large enough for the purpose intended, but in most cases the hole is reamed up to the size required. Where this is done, it is essential that there is access to the target area to enable the bit to be changed and the cuttings to be dealt with. The pilot bit is removed at the target area and replaced with the raise head (Figure 3.3.1). This is dressed with cutters that will crush the rock when pulled back and rotated in a similar fashion to a conventional tunnel borer. Enlargements up to 5 metres diameter have been made using this equipment.

Originally it was only possible to carry out this technique on an inclination of 30° or more. During the reaming process the cuttings would then fall down behind the raise head by themselves.

DEVELOPMENTS FOR HORIZONTAL USE

The introduction of water-flushing techniques have made it possible to drill and back ream 'microtunnels' down to an inclination of less than 1°. The finished hole will be lined with a steel casing if the rock shows signs of weakness and this can easily be pulled in by the machine and welded on in sections, followed by annulus grouting. The microtunnels can be used for a variety of applications, sometimes in combination, such as water pipes, electric cables and sewers.

Drilling in changing ground

It was found that no suitable equipment was available to use this technique when the line of bore went through a combination of rock and overburden and so a machine was developed in-house and is illustrated in Figure 3.3.2 with details of the head unit shown in Figure 3.3.3.

The drilling is still carried out by a tricone bit, but between the bit and the drill string, an adaptor incorporating a thrust bearing is introduced. The casting is welded onto the bearing so as to allow the casing to be non-rotating, being only pulled into place by the bit.

When the drilling is complete, the bit is cut away from the casing and the string is withdrawn. The length of hole that can be achieved in one set-up is dependent on the profile and combination of the overburden and rock. An example of the type of problem that can be overcome is illustrated in Figure 3.3.4.

It is most important to achieve a good line with the pilot hole and it was necessary, in the complex case illustrated, to withdraw several times to control the deviation. Indeed there may be times when it is necessary to start all over again if a particularly difficult combination of circumstances is encountered.

Drilling, starting in overburden

Where the profile of the ground starts with overburden and then hits rock, as illustrated in Figure 3.3.5, then the technique just outlined has to be modified if the size of the finished bore exceeds 300mm diameter.

For the initial drive, a normal soil-drilling technique is used, whereby a steel tube is punched forward until it hits the rock, the soil being removed by means of an auger. The bore of this drive is normally one metre diameter, or at any rate large enough for a man to enter in order to attach the tube to the rock face by means of rock bolts drilled from the inside. When this is complete, a steering pipe, as illustrated in section A-A of Figure 3.3.5, is pushed inside the steel casing to allow a pilot hole to be accurately drilled in the rock section in a similar way to that previously described. The back reaming, to the same diameter as the initial steel casing, can then be carried out to complete the bore.

CONSTRUCTION RECORD

A number of successful contracts have been carried out and a list of some of these, with the diameter, length and inclination from the horizontal, is shown in Table 3.3.1.

A further list of most of the contracts carried out using the combined drilling technique, where a rotary bit has been used, simultaneously welding on a steel casing because of difficult rock, is shown in Table 3.3.2.

COSTS

Typical prices for combined drilling at 200mm diameter, including the protected steel casing, work out at £150/metre plus a setting-up cost of about £1000, the access pit and foundation being provided by the customer — the latter can often be designed to form part of the finished manhole or access arrangement. Further costs are involved if the site exceeds a distance of 100 miles from the depot. The conventional, practical, alternatives in moraine or rockfill are either to thrust a man-accessible steel pipe through the moraine or rock, digging out as the pipe advances, or resort to close-timbered or sheet-piled trenching. The latter, if buildings or roads are involved, can be highly disruptive and costly; indeed costs of five to ten times the combined drilling alternative are not uncommon.

Typical costs of thrusting will vary according to the nature of the ground, and costs, excluding initial setting up, will be of the order of £350–£450 per metre or two to three times the cost of drilling.

A further advantage, in the case of sewers, is the avoidance of pumping stations and rising mains, as a steady gradient can be achieved.

Detailed costs: NRK, Oslo

This contract, for the Norwegian Broadcasting Corporation (NRK), was to connect the water mains and sewers from their new building to the existing services. A single 300mm diameter hole, 21 metres long, was drilled to take both services which were later pulled through after construction. The pit for the drilling rig was provided by the main contractor and the small Rotamex rotary drilling unit was used. The bore was initially in rock and then continued in clay until the manhole wall was reached. Here a magnetic detector was used to locate the drilling head and the final breakthrough was achieved using a diamond core drill from inside the manhole. The rotary bit was removed in the manhole and the drillstring pulled back, leaving the casing in place to complete the job.

The cost of installing the corrosion-protected pipe in 1982 was only £6,200 and any risk of damage from blasting and clay movement was avoided.

Two alternatives were considered, one being to trench round the outside of the building and across the main access road and install a pump unit at an estimated cost of £23,000, some four times the final solution adopted.

The second alternative, to provide a gravity pipe by conventional excavation, would have involved deep clay excavation with sheet piles, as well as rock drilling from an existing cellar, and was rejected on the grounds of almost certain damage to the building from ground movement.

Detailed costs: Lily Tonga Service Station, Mastemyr

This contract is illustrated in Figure 3.3.4 and provided three parallel holes of 200mm diameter, 54 metres long, to serve a new service station on the main dual carriageway from Oslo. Traffic considerations were critical in the selection of the process and no disruption took place at all.

The access pit, which was blasted in rock and later used for the permanent manhole, cost some £4500 in 1981, transport and setting up being about £4600. The three parallel holes, each 54 metres long, passed under the dual carriageway and cost £16,500. The cost of conventional trenching and blasting would probably have been of the same order, but the disruption costs would have been many times greater and totally outweighed any other considerations.

SUMMARY AND CONCLUSION

The adaptation of raise-drilling techniques for provision of statutory services in difficult or rapidly varying ground has been successfully achieved and used on a number of contracts in Norway[1]. The technique shows considerable cost saving and avoidance of social disruption and is particularly useful when crossing railways or motorways or going under buildings.

REFERENCE

[1]Olimb, Kristian, Boring av korte, horisontale fullprofiltunneler i vekslende fjell og løsmasser, *Fjellsprengningsteknikk Bergmekanikk/Geoteknikk*, Norway, 1982.

TABLE 3.3.1 Modified rasie-drilling contracts

Place	Customer	Purpose	Diameter mm	Length m	Inclination from horizontal
Djupdalen	Oslo Vann og Kloakkvesen	Utility tunnel	1200	40	16°
Kastellbekken	Oslo Vann og Kloakkvesen	Utility tunnel	750	150	40°
Elvåga	Oslo Vann og Kloakkvesen	Water mains	1200	80	3°
Fagerliveien	Oslo Vann og Kloakkvesen	Water mains	250	60	5°
Setesdal	Ingeniørbygg	Tunnel	1200	85	90°
Setesdal	Ingeniørbygg	Cable tunnel	500	60	90°
Mo i Rana	Sydvaranger	Ventilation shaft	1200	262	90°
Bergen	Bergen Kommune	Sewer tunnel	250	160	2°
Skien	Skien Kommune	Sewer tunnel	250	180	5°
Karihaugen	RA2	Sewer tunnel	250	220	6°
Karihaugen	RA2	Sewer tunnel	250	70	4°
Voksen Skog	Oslo Vann og Kloakk	Water and sewer tunnel	600	120	16°
Voksen Skog	Oslo Vann og Kloakk	Water and sewer tunnel	600	140	18°
Fagernes	Nord Aurdal Kommune	Water and sewer tunnel	1200	120	2°
Stockholm	Siab	Water and sewer tunnel	660	45	8°
Stavanger	Stavanger Kommune	Water and sewer tunnel	600	62	10°
Bodø	Astrup & Aubert	Cable tunnel	250	137	2°
Tyssvær	Tyssvær Kommune	Water mains	250	170	2°
Klemetsrud	Oslo Vann og Kloakk	Water mains	600	95	3°
Klemetsrud	Oslo Vann og Kloakk	Sewer tunnel	250	95	3°
Ulla-Førre	N.V.E.	Hydroelectric tunnel	600	80	60°
Ulla-Førre	N.V.E.	Tunnel	2400	130	90°

TABLE 3.3.2 'Combined' drilling contracts

Place	Customer	Diameter mm	Length m	Type of Ground and Remarks
Time	Time Kommune	200	15	Moraine
Økern	Henriksen	200	30	Corrosive swelling shale
Strømmen	Knappkøyen	200	16	Rockfill
Østvangen	Oslo Kommune	200	8	Rockfill
Vestby	Arne Jensen	200	18	Rockfill
Rygge	Arne Lyshaugen	200	18	Underwater rockfill
Lunner	Waterworks	200	23	Rockfill
Norway	Bygg og Anlegg	200	17	Rockfill
Norway	National Railway	200	10×12	Rockfill
Lesja	Lesja Kommune	200	24	Rockfill
Oslo	Moderne Bygg	200	42	Very broken rock
Nordstrand	Oslo Kommune	200	15	Rockfill
Karmøy	Karmøy Kommune	200	15	Moraine
Oslo	Selmer	200	15	Rockfill
Mastemyr	Lily Tonga	200	3 × 54	Solid rock and gravel
Bårum	Bårum Kommune	200	30	Rockfill
Norway	Fetsund	250	27	Moraine
Norway	Fjordservice	250	30	Rockfill
Kjeller	Selmer	250	3 × 18	Rockfill
Oslo	Norwegian Broadcasting Corpn.	300	21	Rock and clay

Figure 3.3.1. Detail of raise head prior to back reaming.

Figure 3.3.2. Modified drilling machine with setting up frame.

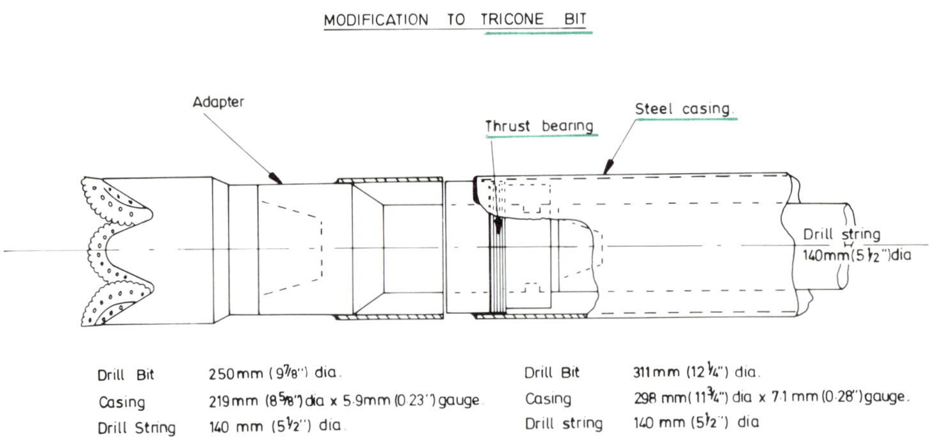

Figure 3.3.3. Detail of 'combined' drill-head modification.

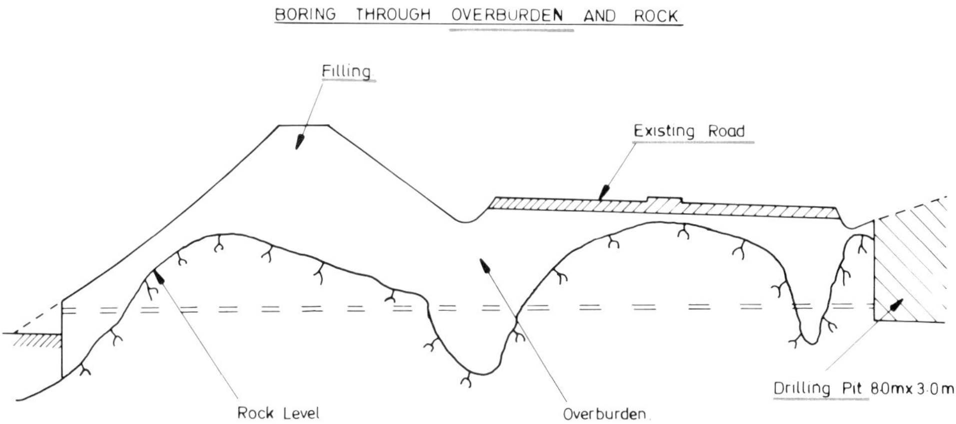

Figure 3.3.4. Typical profile (starting in rock).

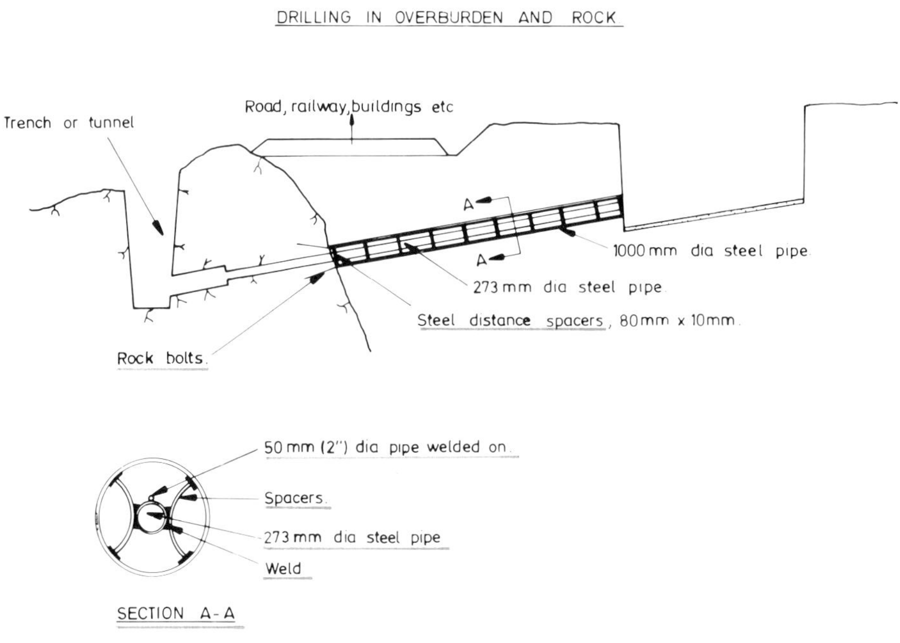

Figure 3.3.5. Technique when starting in soft material.

3.4 Small-diameter tunnelling by pipe-jacking methods

A F Moss *Past Chairman, Pipe Jacking Association*

R N Craig *Principal Engineer, Sir William Halcrow & Partners*

The paper briefly discusses conventional pipe jacking in an urban environment and enlarges on the recent development trends in the United Kingdom. The conventional pipe-jacking system has developed to suit the considerable restraints in urban sites and the need for small-size shafts. The jacking equipment has been designed to be versatile and to be accommodated in the small shafts. Recent developments have included the use of partial- or full-face machines using telescopic jacking systems.

British technology is being used in the development of pipes in new materials, including glass-reinforced plastic, glass-reinforced cement and vitrified clays. The paper discusses shafts, jacking equipment and methods of excavation and briefly covers jacking loads, alignment and secondary linings. The system is compared to other forms of tunnel lining and areas are identified for future research in the laboratory and in the field.

INTRODUCTION

Pipe jacking was developed in the United Kingdom in the late 1950s. At about the same time similar systems were developed independently in other countries in Europe and in the United States of America. The principles of the system had been used in isolated instances during the previous decades.

The pipe-jacking technique is a method of lining a tunnel by shoving a line of pipes through the ground, using hydraulic jacks, from a prepared jacking pit. Excavation is carried out at the face within a shield as the line of pipes is pushed forward from the jacking pit. After pushing a full pipe length into the ground a new pipe is lowered into the shaft and the process repeated.

Pipe jacking is the basis of three main systems currently being used in the United Kingdom:

- auger boring;
- full-face remote-control jacking;
- conventional pipe jacking.

The paper briefly discusses the systems and enlarges on the British developments in conventional pipe jacking for urban situations and outlines areas for future research.

PIPE-JACKING SYSTEMS

Auger boring

The auger-boring system is still commonly used in the United Kingdom for the installation of service ducts for pipes of diameters of 0.3m to 0.75m and occasionally for smaller diameters or larger diameters up to 1.5m. The linings are normally of steel tube with thicknesses varying from 6mm to 20mm, depending upon the diameter, ground loadings and the need or otherwise for corrosion protection. The system is usually used for relatively short lengths under roads, railways and other embankments. The size of the pit from which the pipe is installed depends upon the jacking system and the length of the pipes, and varies from 2.5m to 13m. The system usually achieves high rates of progress. Jacking lengths are limited to between 60m and 100m in ideal ground conditions with considerable reductions in poor ground conditions. The auger-boring system is discussed in more detail in the paper by A G Munro[1].

Full-face remote-control machines

In Japan full-face remote-control machines are in use for a range of excavated diameters from 450mm upwards. Both bentonite and earth pressure balance systems, as well as combined systems, have been developed commercially for excavating in difficult ground conditions below the watertable. The excavated material is, in all cases, removed as a slurry, rock crushers being used ahead of the machines to regulate the size of spoil.

The use of these machines in the United Kingdom has to date been limited, on account of the more variable ground conditions, and the high capital cost involved. In addition their use in urban areas is constrained by the need for long jacking pits, 7m–9m in length. The system is discussed in more detail in the paper by A Rooke[2].

Conventional pipe jacking

The original use of pipe jacking was mainly for lengths of drive where a sleeve or duct was required for another service or pipe, often for a crossing under a road, railway or other embankment. The system was only occasionally used for long lengths of tunnel. In the last decade or so the emphasis has changed and now the majority of the length of pipe jacks is for finished tunnels for the sewerage market. The techniques have therefore had to be developed to achieve higher degrees of accuracy in the construction[3,4].

The present market for pipe jacking with concrete pipes varies from 10 to 20 per cent of the tunnelling market, equivalent to between 6km and 15km a year. The diameters are relatively small compared to those for other tunnelling methods, with an average excavated diameter of about 1.7m. The recent developments of the system behind full-face tunnel-boring machines or semi-mechanized shields has increased the annual length and the average excavated diameter. The increase in the market has also partly been achieved by contractors offering pipe jacking as an alternative to precast concrete segments.

CONVENTIONAL PIPE JACKING AND DISCUSSION ON LATEST DEVELOPMENTS

Pipes

Until recently the pipes have been manufactured in steel or in concrete. Steel pipes have normally been used for a sleeve for other services; however, there is nothing to preclude their use as a pipeline, subject to the efficiency of the corrosion protection employed.

The concrete pipes have conventionally been manufactured using centrifugal spinning methods with joints of the rebated form (Figure 3.4.1). Whilst they have generally been used for relatively short drives, if interjack stations are incorporated in a line of pipes, long lengths of drive can be achieved[5].

In Europe and in some other overseas countries the vertically-cast butt-ended form of pipe is now commonly used. The details of the joint may include an attached or detached steel collar or dowels[5].

Butt-ended pipes with a steel collar are now available in the United Kingdom manufactured in spun concrete or vertically cast.

Pipes manufactured in new materials for pipe jacking are being developed in the United Kingdom and overseas. These materials include glass-reinforced plastic (GRP), glass-reinforced cement (GRC), resin concrete, vitrified clay and asbestos cement. British technology is being used in the development of pipes in all these materials, although they have not been used, except for trials for pipe jacks, in this country. Glass-reinforced cement pipes have been manufactured under a United Kingdom licence, for jacking pipes in Japan.

Glass-reinforced plastic pipes have been used for a pipe jack in the Middle East and vitrified clay pipes are being used in a research test project in Germany with encouraging results[6].

Shafts

Shafts for pipe jacking can either be trench-sheeted, sheet-piled or of segmental construction and may be of whatever shape or size is necessary to accommodate both the permanent works, such as manholes etc., and the pipe-jacking set-up used. The conventional form of pipe jacking has been developed to suit the considerable restraints of urban sites. Shafts of 2.4m to 3.0m internal diameter or 3m × 2m section are commonly used (Figures 3.4.2 and 3.4.3). These are the typical sizes for shafts for timber headings and can normally be fitted into congested areas with the minimum disruption to traffic. Whenever larger room is required at the level of the tunnel to accommodate the jacking equipment, a base heading may be constructed at the bottom of the shaft.

Depending upon the diameter-to-length ratio of the pipe it may be necessary to lower the pipe with its axis vertical and to turn it into the horizontal position at the level of the base heading. For this reason it is more usual to have the base heading at the front of the shaft.

Jacking equipment

For small shafts two jacking systems have been developed. The most versatile of these employs short-stroke jacks with spacers between the head of jack and the pipe thrust ring to provide the full shove for the pipe (Figure 3.4.4). This system requires a number of shoves of the jacks to push a full pipe length but it is a very economical method using small capital cost equipment adaptable to all sizes, which can be delivered to site and set up within a very short period. The alternative employs a combination of long-stroke and concertina jacks acting through a yoke. The long-stroke jacks are positioned outside the pipe barrel and act in the reverse direction onto the thrust wall. The system allows for a complete pipe length to be installed without spacers. This is particularly useful for machine drives, but is not adaptable for different sizes of pipe and has a high capital cost.

For mechanized long pipe jacks, where the rate of progress is high, long telescopic jacks acting directly onto the thrust ring are normally used.

Methods of excavation

The method of excavation for conventional pipe jacks is by hand. A number of pipe jacks have been excavated by partial or full-face machine. The first machine drive in the United Kingdom was in the early 1970s. More recent applications have included a variety of full-face, partial-face or remote-control machines. The use of such machines requires long drives or a series of long drives to recover the high set-up costs. The machines have been used in all types of ground conditions from soft ground, silts, soft clay, clays, sand and boulder clay up to hard rock of sandstones, siltstones and mudstones. The pipe-jacking system has now been used behind all types of shields and machines including those normally used for the construction of segmental linings. The pipe jack has therefore now become an alternative lining system for a tunnel.

Pipe jacks have been used in conjunction with most technical forms of ground stabilization including compressed air, chemical consolidation and dewatering[7]. Overseas the use of compressed air at the face, with the operator of the machine in free air, and of the slurry machine are commonplace. The slurry machine has to date only been used on a small number of drives in the United Kingdom.

Excavated material is normally removed by skip except in the case of slurry machines where the material is removed in the slurry. Skips are carried from the face on rubber-tyred bogies. Alternative methods, used overseas, include pneumatic transmission and long conveyors.

Jacking loads

The load required to thrust a line of pipes is the sum of the shield loads and the frictional pressures along the line. This load may be provided by jacks at the back of the shield, at any interjack stations and at the thrust pit. The loads at the thrust pit are normally resisted by a thrust block at the rear of the shaft. Alternatively they may be transferred, through the floor to the shaft lining, to ground anchors or any other suitable adjacent permanent structure.

The use of lubricants injected under pressure into the void around a pipe reduces considerably the friction between the ground and the pipe. For short drives bentonite is often brushed over the pipes prior to jacking to reduce the friction between the pipe and the ground. In this instance the lubricant seals the pores in the concrete pipes, thus reducing the drainage effect on the porewater in the ground and reducing the friction on the pipe.

In the United Kingdom the use of injected lubricants is generally restricted to long drives over 50m to 100m in length. The use of interjack stations is normally restricted to relatively long drives.

Alignment

Particular attention should be paid to the alignment during the construction of all one-pass lined tunnels including pipe jacks. Jacking loads, and subsequently construction costs, are significantly reduced as a result of keeping an accurate line and level for the drive.

In practice the overall alignment of a pipeline is the combination of the manufacturing tolerances of the pipes, the surveying tolerances and the excavation (shield) tolerance. Current UK specifications normally require plus or minus 75mm in line and 50mm in level, although much closer tolerances are normally achieved.

Secondary linings and reinverting

The effective diameter of the finished pipe-jacked tunnel can be reduced either by reinverting or by the installation of a carrier pipe or duct. For a sewer, a precast channel of suitable section for the dryweather flow can be installed leaving considerable cross sectional areas for storm conditions. In some instances it may be possible to accommodate both foul and stormwater sewers within the same pipeline (Figure 3.4.5).

Recently pipe-jacked steel sleeves have been installed in preference to timber headings. Timber headings are commonly used in urban areas to provide temporary tunnels for laying pipes or other services. After laying the service the timber heading is backfilled with concrete and grouted. Difficulties occur in the quality of concrete placing and voids may be left which, if ungrouted, may percolate to the surface. In time the timber may rot, causing additional settlement.

A thin steel tube pipe jack may be pushed under an embankment and a sevice laid inside with the void between the inside of the pipe and the sevice backfilled with a weak cement-based grout, thus avoiding possible future problems with the timber heading.

COMPARISON OF A PIPE JACK WITH OTHER FORMS OF TUNNEL LINING

When considering the alternatives of a pipe jack with other forms of tunnel lining, the following advantages should be taken into account:

- clean to install;
- requires no secondary lining;
- characteristics of a flexible jointed pipeline;
- fewer joints and a higher degree of watertightness;
- a factory-manufactured pipe.

In addition the size of the tunnel may preclude the use of proprietary segments. Bolted segments are available down to 1500mm internal diameter, and unreinforced smoothbore segments down to 1000m or 1200mm internal diameter, whereas pipe-jack linings are available in conventional construction down to 900mm internal diameter and with remotely-controlled machine excavation down to 450mm internal diameter and below.

In urban situations, at shallow depths, where future development or redevelopment may involve excavations or piling operations to be carried out in close proximity to the pipeline, the rigid nature of the pipe cylinder would be an advantage compared with a segmental lined tunnel which could collapse if the lateral pressure were removed.

The main limitation, often given, for a pipe jack is the obvious risk of seizing up through high friction loads. If a conservative estimate of the likely friction loads is made, and sufficient allowance given for lubrication and/or intermediate jacking stations, and the tunnel is constructed accurately to line and level, then the risk is very small. However, in the event of unforeseen conditions giving rise to substantially higher friction loads then, subject to compatability of size, the pipe jack may be converted *in situ* to a segmental tunnel in order to complete the drive.

FUTURE RESEARCH

The Construction Industry Research and Information Association (CIRIA) has carried out a preliminary study into the pipe-jacking technique, both in the United Kingdom and overseas. A technical note on the state of the art of pipe jacking in the United Kingdom was prepared in 1982, and a second phase of study is already under way[5]. Following those initial studies, research objectives have been identified by the industry in the field of pipes, joints and lubricants, aimed at increasing the knowledge and understanding of the design aspects of the technique, particularly those related to the stresses in the pipe during construction.

A research programme is being developed by the Construction Industry Research and Information Association in conjunction with the Pipe Jacking Association and other commercial sponsors for research projects at the University of Oxford and Brighton Polytechnic, to study these research objectives both in the laboratory and in the field.

REFERENCES

[1]Munro, A G, Auger boring for pipeline construction, NO-DIG 85, Paper No. 3.2
[2]Akesaka, T and Rooke, A, Development of the Iseki Telemole and Telemouse and their application to trenchless pipelaying, NO-DIG 85, Paper No. 4.4
[3]Pipe Jacking Association, *Jacking Concrete Pipes,* Technical Bulletin No. 5
[4]Pipe Jacking Association, *A guide to Pipe Jacking Design*
[5]Craig, R N, *Pipe Jacking: A State-of-the-art review,* Technical Note 112 CIRIA 1983
[6]Kuntze, E h E, Microtunnelling in Germany, NO-DIG 85, Paper No. 4.3
[7]Pipe Jacking Association, *Case histories in Pipe Jacking* (in conjunction with Tunnels & Tunnelling).

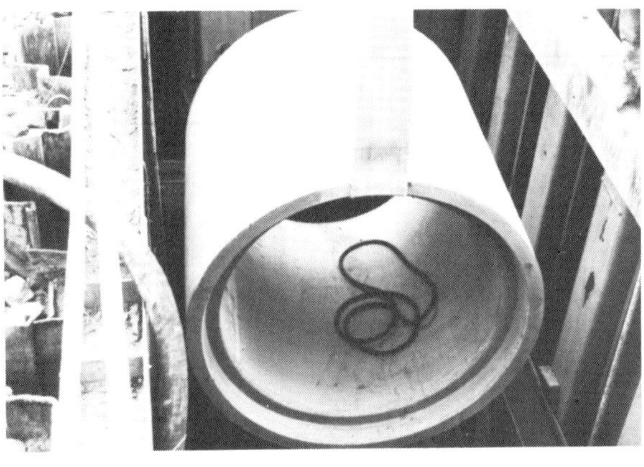

Figure 3.4.1. Rebated pipe in small shaft.

Figure 3.4.2. Segmental shaft with base headings.

Figure 3.4.3. Small rectangular shaft.

Figure 3.4.4. Shaft bottom

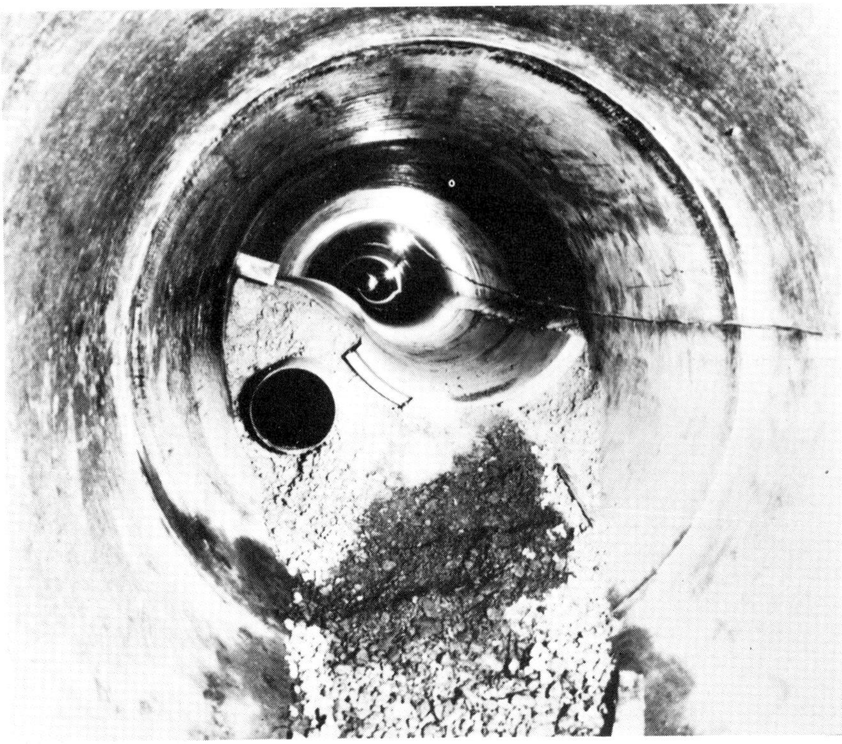

Figure 3.4.5. Storm and foul sewers in pipe jack.

Authors' introductions

Mr K Olimb: Drilling of short horizontal microtunnels in rock and rock and soil combined

My paper deals with the drilling of short horizontal microtunnels in rock and rock and soil combined. The topographical situation in Norway does on many occasions make it very profitable to drill instead of trenching. Our firm started drilling shafts in 1975 and very soon we got enquiries for horizontal tunnels in rock and rock and soil combined.

Figure 3.3.3 in the paper shows the tricone bit modified for rock and soil combined. The adaptor screws onto the bit. We have a thrust bearing to which we weld the steel casing. The question is why do we not rotate the casing together with the bit. On the way through it will always gyrate and then we need very much torque. It is also to avoid joint bursting. When drilling through rock and soil combined we go through the whole way with the steel casing, which is centred on the drill pipe before jointing. The centring ream has a smaller diameter than the casing so that when we have finished the hole we pull the steel rod and also the centring ream back together with the steel rod.

Figure 3.3.5 shows a place where drilling is superior to trenching. Trenching here would cost ten times the drilling cost. Where the profile of the ground starts with overburden and then hits rock and a larger diameter than 300mm is required we use an auger borer whereby a steel tube is pushed forward until it hits the rock. With bolts we fasten the tube to the rock and then drill the pilot hole, connect the reaming head and ream back. Further details are given in the paper.

When drilling holes between horizontal and 45 degrees it is very important to flush the cuttings away from the cutters, otherwise the cutters will soon become worn and we will have low penetration. When we are drilling horizontal tunnels we have to use a high-pressure cleaner close behind the reamer head to flush the cuttings away.

Very often in Norway, especially in the western area, drinking water is drawn from lakes up in the mountains. At one such place we drilled a hole through the hill and saved the municipality the cost of a pump station. When we had finished the hole and pulled back the drill rod we also took with us their polyethylene pipe.

We have developed a system where with a middle-sized drilling unit we are able to drill larger holes by going through two or three steps. When drilling an 8ft tunnel we first drill a 5ft and then the 8ft. It is also necessary to stabilize the steel pipe and we use non-rotating reams.

[Mr Olimb showed and described a number of slides illustrating work undertaken by his company.]

Mr R N Craig and Mr A F Moss: Small-diameter tunnelling by pipe-jacking methods

Mr Craig: I will introduce the paper with a few words on small-diameter pipe jacking and then Arthur Moss will describe a particular project in Bristol

There are three main types or systems of small-pipe jacking currently in use in the UK: auger boring, full-face remote jacking and conventional pipe jacking. Auger boring has already been discussed by Mr Munro and this afternoon Mr Rooke will discuss remote jacking.

Auger boring is generally 0.3 to 0.75 metres in diameter, with steel tubes, in relatively short lengths and with shafts varying from 2.5 metres to 13 metres in length. Remote jacking is generally for 450mm diameter upwards. It is used generally in difficult ground conditions and the shaft sizes are relatively large, 7 to 9 metres in length.

Conventional pipe jacking in the UK has a market of 10 to 20 kilometres a year for concrete pipes and up to 30 kilometres a year overall for all forms of pipes. The average internal diameter is about 1.4 metres and it is therefore at the top end of what we are discussing at this conference.

The pipes are manufactured in concrete, spun or vertically cast, and in steel, although other types of pipes are used overseas and many of these are with UK technology, in particular, glass-reinforced plastic, glass-reinforced cement, resin concrete, vitrified clay and asbestos cement. Joints are normally of the rebated type, although more recently other forms of joints have been used in the UK including the butt-end pipe with a steel collar. A contract will commence shortly using a butt-end pipe with dowels.

The conventional small-diameter pipe-jacking system is from 0.9 to 1.35 metres in diameter, although pipe jacks of 750mm diameter have been successfully constructed. The system has been developed to suit the considerable restraints of urban sites.

The minimum diameter of shafts is about 2.4 to 3.0 metres, or 2m × 3m rectangular, but for deep shafts base headings may be required to accommodate the jacking equipment. For shallower tunnels, larger shafts may be more economical and therefore lengths of up to 6 metres are sometimes used. For small diameter shafts, short jack systems are necessary, but for larger shafts full-size or telescopic jacks may be used.

Excavation is normally by hand in these small diameters although more recently one or two back-acter machines have been used in faces down to 1.35 metres internal diameter. The system can be used in all types of ground conditions, as with other tunnelling methods.

Jacking loads have to overcome the frictional resistance between the pipe and the surrounding ground and also the face resistance. To reduce the friction bentonite or some other slurry can be injected between the excavation and the pipe. Alternatively, intermediate jacking stations can be incorporated in the pipeline.

For the smaller diameters, reinverting of the pipe can be carried out to get the required diameter.

The system needs further developing and research is programmed to be carried out under the guidance of CIRIA and the Pipe Jacking Association.

In summary, pipe jacking is clean to install, gives good safety during construction, requires no secondary lining, has very few joints to seal, uses factory-made pipes and can be reinverted to small diameters.

Mr Moss: The business of NO-DIG; the business of non-disruptive construction of utilities in urban situations; the business of risk. Risk management is the true way to assess whether a project has been carried out correctly. Has the risk management been sorted? Do we know when we go in, that our source investigation in the city centre is realistic? We cannot consider topography — there are blocks of flats, other high-rise buildings, loads of excavations carried out in the past, all there to mislead us. So we take a few boreholes. We never take enough. The thing with boreholes is that if you don't take them you still pay for them in the project. So risk is important. We might meet all sorts of other things below ground, buried debris of all descriptions. I have had in one sewer length, a bomb, a thermal spring and permafrost, and that is in this country. The major difference between hand-driven pipe jacks and the other systems which are being discussed today, is that we put a man up front, and we can cut down the risk by having him there because when we meet unforeseen conditions, we can see straight away what the best solution might be. That is why, on the example I am going to describe, I chose pipe jacking hand work in very difficult conditions.

The job concerned was 34 metres long, in a congested situation in a gasworks. The ground conditions at one end and the borehole showed us that we had cyanide levels some 1000 times greater than those acceptable normally, high phenol content, and all sorts of other rubbish. Typical gasworks floor, and alongside these some 19 services, ranging from a 1200mm high-pressure gas main, down to a GPO cable that was supposed to be connecting the President with Mrs Thatcher. So we had no choice but to be sure of the construction method, hence pipe jacking, small diameter required, short length pipes to fit into a very tight pit situation, hence the steel tube jack. In the leader shield there was a curtain air system for the miner, and just to keep his atmosphere down to a realistic level, he was wearing a mask as well so we could keep him safe.

The thrust pit was very small, 2.7 metres long by metres wide. We can go even smaller in special circumstances. The pipes were a metre long (we can use shorter pipes if necessary). Excavation was carried out manually, with a face board, across, and there was room in 1000mm, which was the size of the tube, to work a lot of the usual safe systems for manual excavation. You can put in boarding, you can put in headboards, you can put in face rams, all these things are possible down at this size.

Because we used the steel tube jack we did not need a shield as such — the shield stayed in, so the reception pit was very small, only 1.2 metres long.

Muck came out with standard hand carts, pulled out with a winch, and because this was toxic material we put it into a hopper first and then took it away in a lorry.

When we finished we ended up with a pipeline 34 metres long to within 40mm of line and level at every point, all ready to take a slimline concrete pipe, to grade up afterwards to give us a finished 750mm tube in the most difficult conditions I have worked in for a while.

Discussion on papers 3.3 and 3.4

D G Clow (External Plant Division, British Telecom): These two papers detail techniques which are becoming increasingly attractive as a means of providing the larger NO-DIG tunnels. However, in British Telecom's experience, the high setting-up costs on short lengths of tunnelling tend to restrict application *vis-a-vis* timbered headings. BT is probably the last major user of timber headings in the world. They are cost-effective, and unanticipated obstructions can be readily circumvented.

However, any spare space in the completed heading has to be backfilled, whereas with mini-tunnels the space does not have to be backfilled. The question then arises: why not use the spare volume for other utilities' services? This has been discussed for many years, but there are quite severe problems of security and safety in view of the incompatible nature of the services in fault conditions, or potential unwitting interference with one utility's plant by another's operatives.

Mr A F Moss: In a manually-excavated tunnel, the main cost is in digging the muck. The lining is a secondary cost. All the client gets for his money is the lining. But when you put in a pipe, you end up with a nice concrete pipe or steel tube which will stand there for a hundred years.

Mr R M Craig: The main concern with a timber heading is the long-term possibility that the timber might rot, causing a settlement problem in the future.

D B Dawson (Design Assistant, British Rail, Western Region): The paper by Moss and Craig refers to the need to reduce friction on the pipes to minimize jacking loads, and the use of bentonite is mentioned. But the paper does not refer to the importance of reducing friction in the ground to safeguard property, such as a railway or highway, above the pipe jack. Jacking forces transmitted to the soil by friction can cause severe settlement.

In the UK drag sheets have been used on top of rectangular units to reduce friction in the ground. During the discussion on paper 1.3, Dr Tohyama referred briefly to the Japanese development of a "membrane-wearing pipe-jacking" system which appeared to provide a friction-reducing envelope around a circular pipe. I would like to know more about this system, in particular the material used and the method of feeding the envelope to the face of the pipe jack and onto the outside of the pipe.

The paper mentions new materials such as GRP, GRC and others being used for pipe jacks. Could the authors explain why they are needed (e.g., for chemical resistance?) and how they behave under jacking loads?

Mr Clow said that public utilities could not share tunnels because of the risk of accidents in fault conditions such as leaking water mains, arcing electric cables and leaking gas. This cannot be an acceptable reason for not sharing tunnels. If problems exist with the safety and security of services, they need to be rectified whether the utility is in a tunnel or not. Last year there was a case where a water main leaked and caused ground movement which fractured a gas main. The gas system started to fill with water and a large part of a town had its gas cut off for several days. If the utilities had been in a shared tunnel the accident could not have happened!

A Coulthart (Public Health Engineer, Property Services Agency): Mr Craig stated that butt-jointed pipes, both collared and dowelled, had recently been introduced in the UK. Was this related in any way to problems being experienced with rebated joints? I would like to hear the authors' comments on the performance of GRC pipes.

C E G Bland (Director of Engineering Services, Clay Pipe Development Association Ltd.): Mr Moss described a pipe jack in a disused gasworks. What about the corrosion resistance of the steel pipe liner in the very aggressive ground conditions and the penetration of phenols? What grout was used in the annular space between the lining and the pipes? Could he please comment on the corrosion resistance of slimline concrete pipe to the various chemicals to which it is exposed? Could he also tell us how many pipe jacks had become stuck, never to move again?

Mr Moss: In reply to Mr Dawson, when pipe jacking is carried out in cohesive soils the line loads are likely to be relatively low and we are not much concerned about drag, though we may use bentonite for other reasons, e.g., to increase the length of drive. But in sands and gravels the line loads are high and we would normally employ bentonite to reduce these loads, including drag. My experience is that, with overbreak kept to a very small level — say, 10mm on diameter in sizes of 1200mm o.d. and upwards — there are enormous benefits in using bentonite.

The use of drag sheets is very new in circular sections, and a lot of research will need to be carried out in the next few years. Whether it will be viable commercially for long pipe jacks is another matter.

We did consider corrosion when pipe jacking in a disused gasworks. First we thought of corrosion of our men! That is why we used a curtain air system at the front of the face, and used steel tubes to provide a seal. For the pipe itself, we used a glass-reinforced spun concrete pipe in sulphate-resisting cement. This gave a high-density pipe, which, being thin, enabled us to reduce the excavation diameter. We grouted the annular space with something we firmly believe to be an ideal tunnel grout medium — pulverized fuel ash (PFA). We added Portland cement to it because that gives a sulphate-resisting mix. It is easy to pump, and we can almost guarantee that it will fill every void along the pipeline.

The first butt-jointed pipe in the UK — dowelled butt-joint pipe with no sleeve collar — is being used by my company, for reasons of cost. It will give the same type of joint as segmental construction, but, with only 10 per cent of actual joint length being involved, there should be a realistic low leakage factor compared with a segmental tunnel. So we have a one-pass lining that is very realistically priced.

Yes, pipe jacks do occasionally get stuck — twice in 25 years of pipe jacking experience! On both occasions the pipeline was completed from the other direction. It would be very rare for such a difficulty to result in a completely abortive pipe jack.

Mr Craig: Mr Dawson mentioned settlement as a result of pipe jacking. Some people think that pipe jacking causes more settlement than conventional tunnelling. Others say that, because the overbreak is less,. and as long as that overbreak is grouted afterwards, there is less settlement. My own view is that a well-constructed pipe jack should cause less settlement than a well-constructed tunnel. Whatever form of construction is used, there is likely to be a small amount of settlement as soon as excavation starts.

With regard to materials for pipe jacking, steel pipes are generally used as ducts to carry pipes of other materials such as GRC or GRP. GRC (glass-reinforced cement) pipes have been used for pipe jacking in Japan and Singapore, but not, so far as I am aware, in the UK, except as a secondary lining. I think they should be very suitable for jacking. GRP (glass-reinforced plastics) externally lined pipes have recently been used in Newcastle. They have also been used in Baghdad, mainly to resist corrosion, where the extra cost is justified. GRP can be applied to the outside of a concrete pipe to protect it against corrosive soil, or internally to protect it from the effluent.

To add to Mr Moss's answer on butt joints, the cost of a steel collar is probably equal to the cost of between a half and one metre of pipe. If dowels are used instead of sleeves, there will be an appreciable reduction in cost. Dowelled joints have been used quite widely on the Continent, but probably in ground where a water-stop is not needed. It is possible that dowelled joints would allow more joint movement, but this could be overcome by joining the first few pipes together during installation, with straps across the joints, to ensure they start the jack in the right direction.

If a pipe jack gets stuck, one can complete the drive by continuing with the shield only and building segmental linings behind it.

Mr K Olimb: Reference has been made to setting-up costs. These are really the same, whether for a long or a short tunnel. If the setting-up costs include the access pit, the cost depends on the thickness and nature of the overburden and whether steel sheeting is required. The cost of driving a 300mm casing in rock and soil combined would be about £140 per metre, including the casing. For a tunnel 1 metre in diameter, in rock and soil combined, the cost would be approximately £3500 per metre. The reason for this high cost is that, as described in the paper, we have first to push a casing through with an auger bore, and then centre a pilot awl.

How are we able to arrive at the correct point at the far end? There will always be a tendency for the bore to deviate, but we have a simple system for checking the position of the awl at any point. We fill the drill pipe with water and, by means of a manometer, measure the water pressure for comparison with the starting point. This gives us the relative levels, and we are able to control the finishing level within 30 to 40cm.

J Boon (Technical Officer, ARC Slimline): Slimline pipe has been extensively jacked in Japan. The lack of steel means that there is no stress concentration in the pipe wall, and therefore greater distances can be jacked.

T Brinch (Kristian Olimb A/S): I would like to mention a system that has just been developed in Norway to utilize old telecommunication cables that have been buried. A gas mixture is introduced into the old cable duct and is ignited to burn away the insulation. The old copper conductors are then pulled out and the duct is available for a new fibre optic cable to be installed. The cost would be about 30 per cent of the cost of buying a new duct and cable. Mr Olimb can give further details if anybody likes to write to him.

Professor H P S van Lohuizen (Delft University; Session Chairman): It may be of interest if I describe a few of the problems experienced in Holland, which are much the same as in the UK.

Recently there has been so much disruption of traffic in Amsterdam as a result of the restoring and relaying of pipes and cables that the local authorities had to stop all work on these utilities. This is purely a problem of bad co-ordination between the utilities, responsibility for which may vary from one city to another. In Amsterdam, the waterworks, sewage works and electricity supply are municipal services, but telecommunications are owned by the Government. The gas works are partly owned by the municipality, and partly by a private company which distributes gas to cities all over the country. In some areas electricity is controlled provincially.

There has been a lot of discussion about combined utility tunnels, and it is agreed that this would be a good idea, so as to avoid the kind of problems I have mentioned in Amsterdam. But such schemes would involve heavy pre-investment and higher charges to the public, which would be politically disastrous.

Nevertheless, the Dutch Tunnelling Group in the Royal Society of Engineers has been trying for ten years to get a clearer idea of the technical implications of integrated services, especially in city areas. It has produced two reports, and in reaction to these the city of Lelystad which is a completely new city in one of the recently-reclaimed polders, showed interest in planning one of its new extensions from the start with integrated services. The utility companies were also quite willing to co-operate as long as they could see that the scheme was economically viable. They asked us at the University to co-operate with them and we got the Minister of Economic Affairs interested. He voted money for a first-stage research project, on which we started work. The tunnels would give enough room for maintenance, and it would no longer be necessary to use the street as a means of putting in the utilities. From this would come a wholly new concept for a city plan laid out with these integrated utilities.

For the second stage of our research money had to be voted again, but the Minister decided that in future he would only agree to funding if 50 per cent of the cost was paid for by the industries themselves. Through this very project we wanted to prove both to the industries and to the utility companies that the whole concept was feasible, not only during the construction phase, but for the whole of its life, including maintenance, etc. The funding was not agreed because the industry did not want to contribute their 50 per cent in the second phase of our research project, so the city of Lelystad goes on in the traditional way, and a big chance was missed.

However, some Dutch contractors have become interested in NO-DIG methods recently, and are working with British companies on the new techniques. I shall be able to report back on this Conference, and will look at the application of the methods to Dutch conditions, where we have a high water-table with very soft ground conditions. Rock is found only at 1000-metre depths.

N L Harrison (Laboratory Manager, Humes Ltd.): Humes design and manufacture concrete pipes in Australia. We have been designing jacking pipes for many years, and the pipes have been duly installed, sometimes in very long lines. I have never heard of any problem resulting from stress raisers caused by the difference in stiffness between steel and concrete. Is Mr Moss aware of any problems due to stress raisers in steel-reinforced concrete pipe?

Mr Moss: Once the safe load that can be applied to the end of a pipe is known, that can be used in calculations when planning a jacking operation, e.g., the load to be applied to the pipe, the load to be taken by the back wall, how many intermediate jacking stations will be used, etc. Reinforced concrete pipes do have certain problems where the steel is close to the end, but we know from experience what the safe load is. If our calculations show that a particular pipe cannot offer an economic solution, we try another pipe until we do reach a satisfactory result.

Mr Craig: I should point out that when jacking one pipe against another it is possible to get tensile stresses in a zone a metre or so from the end of the pipe. It is therefore necessary to ensure that the steel is adequate to take that stress.

K G Stanley (Head of Underground Construction Group, British Telecom): As far as BT are concerned, the purpose of underground construction is to provide a facility for laying and replacing cables. We want to buy a space rather than a lining. We use various forms of NO-DIG construction where this is seen to be cost-effective. This use, however, will surely only increase if the comparison with conventional trenched construction on cost grounds becomes more attractive, and system reliability is such that line of route, negotiation of obstacles and potential damage to other services are not problematical.

BT are generally prepared to try any new or improved method of NO-DIG construction, but potential providers of such techniques must be able to market their systems as commercially viable alternatives. Our experience to date suggests that the use of NO-DIG techniques is not always simple and straightforward, and we feel that much work remains to be done, particularly as regards underground mapping.

P J Everett (Executive Engineer, British Telecom): Some time ago we attempted to thrust a 600mm pipe under railway tracks over a distance of approximately 50m, at a depth of about 2.5m. The pipe stopped at 35m and could not be thrust any further. Attempts to locate the pipe head by normal location equipment were unsuccessful because of interference from the railway electrification. We then used a discriminatory borer, developed by our research branch, to trace the line taken by the pipe from the drive shaft. Boring (0.875 inch diameter) was carried out above the drive line until the pipe was struck. This was confirmed by feeling the vibrations given out by the vibratory probe through the pipe at the drive shaft position. We found that the pipe had cleared the railway, but was 1.5m off line in both the vertical and horizontal planes. We were able to excavate to the pipe and complete the thrust. The reason for the stoppage was merely ground friction combined with the length of the thrust. Our conclusion from this experience is that careful consideration needs to be given to soil characteristics, length of drive and diameter of pipe to be driven.

Professor van Lohuizen: The point that Mr Stanley raised is very important. The problem is always how to contain the costs that are generated. In general, in Holland the authority will not repay any losses incurred by people as a result of streets being broken up, traffic delays, etc. That attitude may have been valid in the past, but I think it is not right any more to look only at the costs of the trenching or pipe jacking. We must take into account all the other costs — economic or social costs. If we do not, then the new methods will probably always be more expensive.

Mr Moss: Mr Stanley said that British Telecom were concerned with buying space as opposed to a lining. That is true, but the space still needs to be lined, and a timber lining occupies a large volume relative to the excavated cross-section. If, instead, he used a steel tube jack, with a 12mm thick steel tube, 1 to 1.2m diameter, he would end up with a lot more space for his money.

But I would stress that all these processes — pipe jacking, auger boring, impact moling — are specialist processes. When done expertly, they look simple. But the client must choose a well-experienced engineer to design his work. The engineer, in turn, must invite experienced contractors to tender for the work, and the contractor must employ properly skilled labour.

R Tharby (Chairman, National Joint Utilities Group): The idea of shared tunnels for all utilities, as mentioned by the Chairman, is an attractive one, but it has been tried in at least one of the new towns in the UK with very little success. One of the reasons for this is, I believe, that the difficulties are not clearly understood. It is easy enough for two or three utilities to share a pipe put under a railway track. But in town streets, each utility has its own distribution network, with services to all of the houses, offices and other buildings. From time to time, in order to remain cost-effective, each utility may need to change the form of its network, e.g., from a T-system to a ring system, or to introduce improved cables or equipment. The shared facility must therefore allow for this flexibility, and provide access for people to repair joints, carry out maintenance or replace plant. This makes the shared tunnel a rather more difficult matter than sharing a tube under a railway.

Professor van Lohuizen: I am sorry nobody from Belgium has spoken about the experience in North Brussels. The old part of the city has been replaced by a "new" city of high-rise office blocks and apartments. The Brussels Municipality was able to find the money, and it decided to finance the construction of utility tunnels in this new area. The tunnels run along the sides of the streets, and combinations of services go from the tunnels to the buildings. There is room to walk through the tunnels, and enough space to carry out any necessary work on cables and pipes. By municpal law, the utilities are obliged to use these tunnels and to pay for doing so. I do not think that forcing people to use the tunnels is necessarily the best way, and prefer the idea of offering people a system that is so attractive that they need no coercion.

SESSION 4. Chairman: Dr. L. S. Blake, Director, Construction Industry Research and Information Association, UK

4.1 Replacement of small-diameter pipes by pipe bursting

A G Poole — *Divisional Manager, Avon Lippiatt Hobbs Ltd*

R B Rosbrook — *Development Engineer, Southern Water Authority*

J H Reynolds — *Senior Engineer, Southern Water Authority*

The replacement of underground plant without excavating is arguably the most significant development the utility industries have experienced this decade. ALH, from its experience of marketing the system in the UK and overseas, will describe the fundamental differences between the replacement of gas/water cast-iron pipes and the renewal of clayware sewerage systems. The innovative designs relative to both are extremely important and the paper will clearly define each aspect of the improvements made and the necessity for same.

The paper will also discuss the more detailed work that has been undertaken in the Southern Water area for the replacement of clay, concrete and iron pipes. From the early development work in Winchester on size-for-size replacement of 225mm clay sewer pipes, a system has evolved which can deal with the replacement of iron pipes and other materials in difficult ground conditions, including water-bearing strata, without recourse to expensive stabilization techniques. A further stage of the development has resulted in the oversizing of pipes to remove hydraulic deficiencies.

INTRODUCTION

Pipe bursting is the method of inserting a new pipe into an existing pipe of an internal diameter equal to or smaller than its replacement by means of fragmenting the existing pipework and forcing the material into the surrounding soil to form access for the following replacement pipe.

The concept was originally developed in the UK some four years ago by British Gas in conjunction with D J Ryan & Son to replace cast-iron gas distribution pipes. A little later a similar concept was developed by the Water Research Centre together with Merstan Impact Moling Ltd. to replace glazeware and other sewerage pipes.

The concept of replacing gas, water and sewerage pipes is protected by a British Gas patent application. However, in the UK, its use is allowed on a licensed basis.

Overseas the application has been patented and the system is marketed by Avon Lippiatt Hobbs and also PIM Division, National Utilities Industries, New Jersey, in the USA and Canada.

At present, the concept is operational in North America, Europe and Australia.

The bursting of the pipe material is accomplished by the use of a pneumatic burster which is derived from the "mole" machine originally designed to insert pipes into "virgin" ground for road crossings, etc. Whilst the replacement principle is the same for cast iron and sewerage pipes the installation criteria necessitated by the working environment have caused the concept to be separated into two very different equipment packages. They are:

The cast-iron system The technique of replacing gas, water and sewerage pressure pipes.

The sewerage system The replacement of all non-pressure sewerage pipes.

Figure 4.1.1 shows an 'exploded' cartoon of the burster unit and replacement pipe being inserted into an existing clay sewer.

THE CAST-IRON SYSTEM

The method of replacement developed by BGC/Ryan entails the excavation of insertion and exit pits at each end of a section. All service pipes are located and disconnected prior to commencement of the operation.

A constant-tension winch is positioned at the exit pit, the wire being passed through the existing pipe and connected to the bursting unit.

When required, a polyethylene liner pipe equal in length to the section to be replaced is connected to the unit which is positioned in the insertion trench.

The bursting unit is powered by compressed air which is supplied by a hose situated within the liner. When the air is applied the burster starts compacting and with assistance from the winch, passes through the section at speeds of up to 2½ metres per minute. On completion the liner is cut at service positions and the new pipe is winched through the liner. Service pipes are reconnected and excavations backfilled. The total replacement of sections in excess of 100 metres can now be accomplished within a working day.

System development

Constant-tension winch

The bursting system requires a directional force and whilst standard line winches were acceptable in the short term, it soon became clear that the most satisfactory operation was provided by hydraulically-operated constant-tension winches. Irrespective of the unit's speed, the tension force remained constant at approximately 2½ tonnes.

The winch can also be the motive force for the insertion of the carrier pipe into the liner.

PE liner

The introduction of a liner pipe was necessitated by the need to guarantee against the possibility of abrasion of the polyethylene. The new pipe, by virtue of its service requirements, carries liquids and gases at pressures up to 12 bar. Its long-term life would be seriously affected if its external surface were damaged. The abrasion characteristics of fragmented cast iron, particularly when subjected to traffic vibration, was found, through considerable investigation, to be potentially detrimental to the PE carrier pipe. Therefore, the use of a thin-walled pipe of a diameter slightly larger than the carrier pipe provided the necessary protection.

Hydraulic Breaker Arms

Early experience proved that the cracking of joints, clamps and fittings occasionally presented a greater problem than the force required to crack the barrel of the pipe. The situation was accentuated when attempting the breaking of water pipes of a greater wall thickness than gas pipes. Considerable time and expense were spent by the inventors in the search for an acceptable answer. However, the final result more than justified their commitment. Hydraulic breaker arms were fitted to the nose of the burster unit. These arms work independently of the drive unit. When a difficult obstruction such as a joint or collar is encountered, hydraulic pressure is applied to the arms to force them diametrically into the obstruction and crack the problem area. Figure 4.1.2 shows the burster unit being inserted into a cast-iron pipe. The breaker arms can be seen in the open position.

Oversizing

The early trials of the system were based purely on size-for-size replacement. Therefore, the external diameter of the existing pipe required a movement of little more than the wall thickness to accept the replacement pipe.

However, it soon became clear that the motive force provided the power to expand existing pipes of diameters up to 150mm, to the extent of enabling large diameters to be installed. Now the ability to increase cross-section areas by over 180 per cent is a widely accepted practice.

The cast-iron system has evolved during the last two years into an extremely efficient replacement alternative and is accepted as a means of reducing relay costs by an average of 20 per cent.

It should be stressed, however, that its success has been achieved only as a result of considerable financial backing and development by the investors. Another extremely important factor is the provision by British Gas of not only a considerable amount of trial sites, but a regular supply of contracts, amounting to over 500 kilometres to date, on a competitive basis, to provide a return for past and future development investment.

SEWERAGE SYSTEM

There are differences between replacing a gas or water main and a sewer and they can be summed up as follows:-

(a) Sewers are generally the deepest and largest of all public utility plant.
(b) Sewers form part of a gravity network and cannot thus be readily bypassed by closing valves and re-routing supply.
(c) Sewers have house connections which are infinitely variable in their size and angle of entry relative to the existing pipe.
(d) Sewers offer an unpleasant working environment.
(e) Sewers are usually at a greater distance from other pipelines.
(f) Since sewers are generally deeper, there is less risk of surface heave.
(g) Clay pipes, etc. are easier to break out (except when surrounded by concrete).

The hundreds of kilometres of sewers requiring renovation or replacement are well documented. In recent years, many statements have been made both in the form of papers at conferences and by the media requesting the funds for development to be made available.

Until the recent initiative by the British Government to provide trial sites, there was inadequate opportunity for innovating new systems. The idea of joint initiative between client and contractor to solve the UK problem was not entirely a new concept but with government approval Water Authorities were able to create a new environment with actual practical application being made in the field at an early date. This avoided the extensive "off-site" proving programme which does not provide an early enough return of capital for the innovator in a depressed market which he frustratingly knows will have to upturn in the not too distant future.

It was against this background that Avon Lippiatt Hobbs and Southern Water Authority decided to explore the commercial realities of size-for-size replacement of sewerage pipes. At this point the Water Research Centre at Swindon joined in the project. They had been given the task of monitoring this new initiative and their normal role of giving advice to the industry, particularly on materials, was of considerable help.

Preliminary discussions soon identified the important areas for development requirements within a contract:

Accurate lateral location;
Effective connection of laterals;
Material suitability;
Pipe bursting.

Accurate lateral location

Minimum excavation techniques for lateral reconnection would form an essential part of any such project. If the work was to be properly planned the exact position, depth, and mode of entry of the lateral at its junction with the sewer would need to be known. It would also be necessary to plot the exact position of all surrounding public utility plant so that lateral excavation could be sited to avoid these services or carefully positioned if avoidance was not possible.

In the early project work the position of all services was identified using electrolocation techniques provided by Subtronics Ltd. Similarly, the position and depth of lateral connections and of the sewer itself were also plotted.

Seer TV Surveys Limited were then employed and given a four-point survey brief:

1. Confirm the number of laterals.
2. Provide an accurate linear dimension from a manhole to the point of lateral connection or junction. It should be noted that the UK specification for CCTV work in sewers allows for an error of ± 2 per cent in linear dimensions. System accuracy reduced that to 0.2 per cent.
3. To view each lateral from within the sewer using a rotate-head CCTV camera on its first commercial contract. The camera has the ability to look at right angles to the line of the sewer and so identify those laterals which are no longer in use or are cut off. 28 per cent of the connections to the Edgar Road sewer were in fact dead.
4. To survey every lateral possible using a small camera which travelled down the lateral. This confirmed the method of lateral approach to the sewer and checked the condition of the lateral itself. Obviously it would not be correct to reconnect a lateral which was in poor structural condition without taking other remedial action.

Thus, a considerable amount of information about the underground services was available. The effort and cost of carrying out these surveys was more than repaid by not having to excavate dead laterals and the benefits gained by being able to plan excavations with a fair degree of confidence.

Effective connections of laterals

When a sewer is sliplined the liner blocks off the lateral connection. Until the laterals are connected to the new pipe, the annular space between the liner and the old sewer is used to convey sewage. Wire mesh filters were installed in each inspection chamber to prevent solids entering the annular space. The lateral connection to the clayware pipe was then broken out and reconnected to the polyethylene pipe. This was achieved by using a unique lateral connection system developed by ALH specifically for sliplining and pipe bursting. It consisted of a polyethylene stub pipe which was machined to suit the outside diameter of the liner. The stub was fixed to the liner by electrofusion and a coupon was cut out and removed from the pipe wall. Connection of the stub pipe to the lateral was achieved by the use of conventional drain pipes including the use of standard bends, to accommodate any angle, terminating in a rubber double collar joint supplied by Naylor Bros. to ensure flexibility between lateral and sewer.

The lateral/sewer joint was part concreted on completion but care was taken to maintain flexibility between the two. The connection requirement for pipe bursting is precisely the same but the lateral must be disconnected prior to insertion to avoid damage.

Material Suitability

Due to the shallower depth of gas and water mains, it is normal practice for the driving unit, or mole, to expand the pipe and at the same time tow in a continuously-welded length of replacement main. However, the use of long pipe lengths is often rendered unsuitable in sewers because of the size of insertion trench required (length approximately 5 × depth). By utilizing short-length pipes from manholes or small excavations it was felt that the system would receive wider acceptance, particularly in deep bad ground situations or heavily-trafficked areas.

It was also considered that the design of the pipe should take into account the following:

1. Structural adequacy.
2. Flush joints both inside and out.
3. Quick and simple jointing.
4. Withstand end jacking loads.
5. Joints not adversely affected by jacking.

6. Resistant to surface abrasion.
7. Normal testing procedures to apply.

Modified HDPE pipes with 'Snaplock' joints as provided by Stewart & Lloyds Ltd and polypropylene pipes with screwed joints and a rubber "O" ring seal provided by Demco Ltd., both of 6-bar rating, satisfied these criteria. Figure 4.1.3 is a photograph of a number of the 'Snaplock' pipes being threaded over the compressed air line.

PIPE BURSTING: CASE HISTORIES

General

Through on-site development the system that evolved allowed a completely free slip-joint between the driving unit and pipe so that the only load on the pipe was effectively pushing from the rear. The push was primarily achieved with a hydraulic pushing machine set up in the entry pit. Co-ordination of the push with the progress of the driving unit proved to be essential to prevent its separation from the front of the pipe. This was achieved by physically tying the driving unit to the pushing machine by a chain which, although simple and effective, required the chain to be disconnected and retensioned each time a new pipe was added. Pre-threading strings of pipe with chain and air supply hoses has reduced the joint cycling time for these pipes to approximately two minutes.

Case History No. 1: Edgar Road, Winchester

A quiet residential road in this historic City was the site chosen to put theory into practice. A survey has shown the 229mm diameter clay foul sewer to be in a very poor structural condition over the whole 800m length.

Hydraulic calculation showed that the first 230m of pipeline could be renovated by sliplining. The method was adopted so that initial efforts could concentrate on lateral connection methods without the problems of proving the pipe-bursting technique.

Work started on site in November 1983. The sliplining and lateral connections went smoothly, with the length being completed within three weeks. Efforts were then to be concentrated on the main task of size-for-size replacement of the remaining 570m of sewer for which the Merstan bursting unit was to be used.

The first length to be moled measured 74m and was replaced using Snaplock pipes. Progress averaged around 20m per hour although the combined moling/jacking operation was at a rate of only 20 seconds/metre. The majority of working time was spent manoeuvring pipes into position and performing the jointing operation.

The second length to be moled measured 116m. Again 1m long Snaplock pipes were used but as an experiment approximately 20m were installed in pairs as 2m long lengths. This did not increase the driving rate, however, as the stroke of the jacks in the hydraulic pushing machine was a maximum of 1.4m and the frame had to be moved forward to complete each 2m length.

The third length was for a distance of 101m and Snaplock pipes were used. Throughout the second and third lengths, the evolution of better equipment came about by a number of events, not least by the increasing operating experience and more specifically, improvements to the mole to rapidly exhaust the air supply. The exhausting improvement reduced the 'running on' when the machine was required to stop. Further development which included a remote air valve provided for further improvement to efficiency and reliability of the system.

The fourth section of the sewer to be replaced was 104m in length and on this occasion Demco polypropylene pipes were installed from the base of a 4m deep manhole. The main purpose of working through an existing shaft was to develop the system for tight urban locations where both traffic and deep sewers might demand the restriction of additional excavations. Figure 4.1.4 shows work in progress. At the bottom of the 1200mm diameter manhole shaft, which had had its cover slab removed for easier access, a timbered heading 1.5m long was constructed. This gave sufficient, if restricted, working space for the jacking frame and two operatives. Progress for the mole was similar to previous occasions but the rate of installation of the pipe was significantly slower at around 10m per hour. Progress was hampered by having to lower each pipe individually down the manhole shaft on a rope. Also aligning the pipes to engage the joint screw threads took longer than anticipated.

Completion of the fifth and final length was delayed for seven weeks to allow further development to the mole. As the driving unit is inaccessible whilst operating, any failure had required an excavation to correct the fault. This was both time-consuming and expensive, and so the mole was modified to allow the driving unit to be withdrawn to the launch pit, repaired and reinstated into the pipeline. The initial expander head at the front of the mole enlarged the existing sewer to 260mm diameter, ie 10mm more than the 250mm o.d. pipe being installed. It had been proved on the earlier sections that the surrounding ground recovers quickly onto the new pipe, and within a period of two days the pipe became impossible to move. The expander head was therefore increased in size to give a 25mm overbreak. With these modifications the final length of 110m was completed by mid-April 1984.

Laterals on the moled length were dealt with in a similar manner to those on the sliplined length, except that they were disconnected from the sewer before moling. This ensured that no damage was caused to the lateral outside of excavation beyond the reconnection pit. Over the final lengths, the depth of the sewer increased from 3 to 5m, and so an alternative method of uncovering laterals to hand excavation was sought. After excavating a 1m deep trial hole to ensure there were no service connections in the way, a lorry-mounted auger was used to form a 900mm dia shaft on to the lateral. To ensure safety of the site operatives who would carry out the reconnection work, these shafts were immediately lined with steel shields of an appropriate diameter. The speed of this operation was such that it is now intended to make much more use of the auger in future contracts. Figure 4.1.5 shows the liners *in situ* and also the connection under construction.

To investigate the effect of moling on adjacent services the Water Research Centre installed a monitoring pipe approximately 20m from an insertion pit and approximately 1.5m away from the sewer. The monitoring pipe recorded an increase in strain as the mole passed by but this was so small as to be negligible. Further details of this experiment can be found in Reference 1.

In conclusion the contract was essentially successful with 800m of sewer being replaced for much the same cost as the open-cut alternative. Although the road was narrow there was no need to close it to traffic, demonstrating that the system has significant social advantages. Sufficient confidence was gained in this system by Southern Water for ALH to be offered a further sewer replacement contract in Tonbridge, Kent.

Case History 2: Tonbridge

A badly corroded 230mm cast-iron sewer running to a pumping station had to be replaced in very difficult conditions. The pipeline ran at a distance of just 3m between houses at a depth of approximately 4.5m in water-bearing gravel. The pipe was in fact within the zone of influence of foundation pressures from one of the properties and this made conventional dewatering impractical. The only manner in which replacement by conventional means could have taken place would have been to pre-grout the gravels prior to excavating. The total length of the sewer was approximately 100 metres with a 45° bend part way along its length. It was realized that a successful pipe-bursting project would show considerable savings.

The contract for replacement of this sewer was placed with ALH who subcontracted D J Ryan to undertake the moling work because of their considerable experience of cracking and replacing cast-iron mains for the water and gas industries with their PIM system.

A shaft was excavated at the location of the 45° bend in the sewer and moling carried out in two lengths from this position. On completion of the moling a new manhole was constructed in the shaft using precast concrete. Overpumping of the full 100m of the sewer was required and suitable safety precautions needed to be taken to permit safe working at depths in the water-bearing gravels. The cast-iron sewer was replaced by a 250mm o.d. HDPE pipe of 10-bar pressure rating. A 10-bar pipe was selected in preference to the 6-bar pipe used at Edgar Road because then its increased wall thickness provided an additional factor of safety against scoring by the broken cast iron. In this instance the HDPE pipe was supplied in standard 12m continuous sections and was butt-fusion welded into a continuous length. Debeading was carried out both internally and externally.

The mole used was of 155mm diameter, modified by Ryans and fitted with a 300mm diameter outside steel expander. The nose of the expander was pointed, with two sharp-edged hydraulically-operated retractable breaker arms. The first 65m of sewer was replaced first on the 15 May 1984. The mole started off at approximately 40m per hour but this progressively slowed down to around 15m per hour before completing the length. An explanation for this would be the increase in pipe frictional forces together with the possibility that the ground was re-exerting itself onto the pipe during installation and adding to the frictional forces.

This rate of ground re-exertion was faster than that seen in Edgar Road and was due to movement in the water-bearing gravels. Overall progress averaged 24m per hour which is considered to be very promising for a first attempt at such a large cast-ron sewer. As the mole passed each joint (spigot and socket) in the original cast-iron pipe it slowed down momentarily until it had broken out the extra thickness metal. At no time was it necessary to use the hydraulic breaker arms on the mole. A jacking frame was used to assist the mole when it was at its lowest speed, but it was unable to generate sufficient grip on the HDPE pipe to have a significant effect. The mole had to pass within 5m of the foundations of a house adjoining the pumping station. The possibility of vibrational damage to the foundations was a cause of concern but although some vibration was felt in the house, it was considered to be insignificant.

The second section of the pipe was completed on the 23 May 1984. Progress of the mole over this short length was rapid, with the full 35m of pipe being installed in just 35 minutes. As with the first length of pipe the installation rate was starting to slow down as more pipe was installed.

Apart from a small problem with a proprietary hose fitting, the equipment used performed well and showed no signs of inherent unreliability and the contract was successfully concluded.

Case History No. 3: St. Cross Road, Winchester

St Cross Road is the main road from the south into the centre of the City of Winchester. Although it is a busy thoroughfare with traffic peaks reaching 1600 vehicles per hour, it is also by modern standards a narrow road, only 5 metres wide over certain lengths. The road is fronted by a number of historic buildings and commercial premises. When a CCTV survey of the main foul sewer demonstrated that the sewer had deteriorated structurally to a point where urgent remedial works were required, it therefore presented the City Engineer with a major problem. Discussions with the contractors who specialize in drainage works indicated that the road would have to be closed for up to 30 weeks to allow replacement of the sewer by conventional means. A large number of tourists visit the City in the summer months and so it was felt that any major drainage works, which would inevitably affect City life, should be carried out in the winter, even though this might prolong the duration of the work.

The 9-inch clayware foul sewer was over 100 years old, and laid at a depth of up to 3.5m under the centre of the carriageway, for a total length of 1100m. In common with the Edgar Road sewer (Case 1), the pipe also carried a portion of surface water drainage. Calculations have shown that the sewer was of adequate size over its length, and could even accept a slight reduction in size for 200m on the upper section. The benefits of reducing the contract period using size-for-size replacement methods of sewer renewal led Southern Water and Winchester City to discuss the project in more detail with Avon Lippiatt Hobbs Limited. ALH were confident that the works could be completed in a much shorter period than the open-cut alternative and agreed to a 12-week contract period. The contract thus involved 200m of sliplining the sewer with 180mm o.d. polyethylene 10-bar pipe, 835m of size-for-size sewer replacement with 250mm o.d. 6-bar polyethylene pipe, the reconnection of 50 laterals, 65m of open cut with 225mm diameter clayware pipes in an area where re-routing of the pipeline was necessary, and the construction of six manholes.

The specialist contractors were again employed as specialist subcontractors for the pipe-bursting operation. Because of the relatively shallow depth of the sewer on this contract, it was elected to use continuous butt-fusion welded pipes instead of the short-length pipes used at Edgar Road. This was also made easier because space was available as a result of eliminating all through traffic. A new design of pushing frame was provided which gripped the outside of the pipe rather than the pipe end-pushing machine used on the previous Winchester contract.

Before the contract commenced, an intense publicity campaign was launched by the City of Winchester, with 3200 leaflets being distributed and full-page advertisements being taken in local newspapers. It was realized that due to the narrow road width and the position of the sewer, even with this method of work, heavy goods vehicles would not easily be able to pass the operations, and so a road closure order was enforced. However access was maintained at all times for residents and commercial premises.

Work started on site on the 8 October 1984. Laterals were excavated and reconnected in much the same way as on the earlier contract, but this time more use was made of the lorry-mounted auger. Altogether six sections of pipe were replaced using the mole in lengths varying from 50m to 130m. Pipeline insertion rates were fairly consistent throughout the contract, averaging approximately 1 metre per minute, and no unexpected problems were encountered.

Such was the pace of the operation that the work was completed in nine weeks and the road opened to all traffic three weeks ahead of schedule, and up to 21 weeks sooner than the open-cut alternative. Additionally it was possible to open the road to through traffic in sections and thus alleviate the City's traffic management problem.

The final contract sum of £149,000 compared very favourably with the alternatives, and to this should be added the considerable saving in social costs and commercial losses and lack of diversion and replacement cost of other service pipes and cables.

A SUMMARY OF THE ADVANTAGES OF PIPE BURSTING METHODS

Pipe bursting has been described — its evolution and development. The paper so far has discussed in detail work by British Gas/Ryan in the replacement of iron mains which naturally had application to the UK water distribution system. Many kilometres of cast-iron main exist with varying degrees of corrosion in both the gas and water industries. The ability to uprate the capacity of existing pipelines by oversizing the replacement main has also been noted. Many of the iron mains to be replaced are in the size range, 150mm and below. Some 80 per cent of the UK water mains are of this size. The largest known cast-iron pipe to be replaced to date was the 225mm internal dia. sewer which has been described in the Tonbridge project. (Case history No. 2).

There are greater benefits to the replacement of sewerage systems as these are generally deeper than water or gas mains. The cost of the process described increases disproportionately with depth so a clear advantage in applying the technique to sewers emerges.

There are in addition many other cost advantages for on-line replacement of sewers by the method which has been illustrated in the preceding text: Overcoming bad ground (Tonbridge project). Avoidance of damage to other pipelines and cables, particularly gas mains (Winchester projects). Speed of operation is extremely important in any comparison of methods.

The reconstruction of carriageways, in total and in part, can occur after trench works have been undertaken and this is an extremely expensive operation. Depending on the trench position, greater reconstruction may be necessary particularly where concrete slabs are interfered with.

The disruption to city or town life is clearly apparent and is the subject of the paper to this conference by Ken Reed and Edward Glennie of the Water Research Centre, entitled 'Social costs: trenchless v. trenching'.

Today there is a greater awareness of the safety of the work force and clearly this method is considerably better in this respect for deep sewers. Obviously there are still some risks but these have been minimized by the limited excavation.

A further often unrealized benefit is for public utilities to take advantage of each other's redundant mains. Where there is a hole in the ground then it should be used to full advantage. An illustration of this was the recent purchase of an old redundant gas main by Southern Water Authority, which has been converted, after renovation, into a sewage pumping main. There is no reason why the space occupied by a redundant main should not be recovered for a cable duct for electrical or telecommunications. The trouble seems that in developing this new technology a public service undertaking may guard jealously its redundant pipes for future use!

THE FUTURE SCOPE FOR THESE SYSTEMS

The paper by Reed and Glennie, as mentioned above, covers the potential for this type of operation. Clearly there is much scope for future investment in such methods not only within the UK market but in worldwide situations.

Much of the UK system is more than 100 years old and although a great many lengths are still serviceable at this age there will be a growing demand for such replacements. It is difficult to predict a level of investment for pipe-bursting techniques but it is a fact that a high proportion of existing pipelines are in the size range for which this equipment can be used.

The situation in hotter climates is not the same in terms of pipeline life, particularly where cementitious products, concrete and asbestos cement have been used for sewers. These products can be severely attacked in an environment where hydrogen sulphide is readily generated, leading to acidic formations on the soffit of the pipes.

A REVIEW OF THE SYSTEM

In the final section of the paper it is necessary to draw conclusions from the works so far and summarize the achievements to date. The state of the art for gas and water distribution is well advanced with some 500km experience in the UK. To date the work undertaken on sewers amounts to some 10km and although confidence has been gained of the commercial viability of the system, partly because of work at Winchester and the previous gas/water main experience, there are undoubtedly some areas which need to be further explored. Further projects are current in the UK and further information will be available at the Conference.

Briefly it is possible to summarize the situation to date as in Table 4.1.1.

Other replacement materials

Clearly the development of the short-section pipes (1 or 2 metres) will evolve further. There are circumstances in the UK and elsewhere where the desirability for insertion through small holes will be necessary. This may well be in heavily-trafficked areas or for deep works, say beyond 3 or 4 metres deep. In situations where the ground is in sandy, gravelly soils with very high water table, then a small entry pit will reduce the dewatering problem.

Interest is now being shown in alternative materials to polyethylene/polypropylene, those being clay, concrete and GRP. Developments of small jacking pipes in Germany[2] have led the way forward, although these are somewhat heavier in construction than necessary for moling, the end thrust being considerably less. There is also an obvious penalty for the thicker pipe wall for this takes up some of the oversize capacity. Clearly it is necessary to contain the joint within the wall thickness and this has a contrary effect.

Development in moling devices

To date compressed air-driven and hydraulic models have been developed. Moles have evolved initially with a length as long as 3 metres but shorter designs have been developed for insertion through small holes or existing chambers. Oversizing is a further application of the system and is proven in small-diameter gas and water mains. Some early work in sewers for oversizing from 225 to 300mm has already taken place and more projects of this nature will be undertaken in 1985. It is not clear yet as to the ultimate size for which this process is capable of development. It may well transpire that pipes up to 450mm can be replaced.

TABLE 4.1.1 Works successfully carried out (Known contracts in UK)

Existing material	Size range	Replacement material	Remarks
cast iron	up to 250mm	polyethylene with PE liner	up to 10-bar fusion-welded strings
ductile iron			untried
pitch fibre	150mm	polyethylene	completed project but some difficulty with rubber ring joints
clay	225mm 300m oversize	polyethylene polyethylene	welded strings and short-section polyethylene and polypropylene have been used
concrete	225mm	oversize to 300mm	welded strings
asbestos cement	150mm	polyethylene	joint sockets may cause problems
fibre-reinforced pipes	250mm	polyethylene	difficulty in splitting material; special cutting devices required
PVC			acceptable in terms of brittle fracture; specialist tooling requirement
polyethylene polypropylene			not thought possible with present technology

Vibration

Obviously the driving mechanism is important as the mole could excessively affect adjacent structures. The shape and progressive taper of the outer case of the moling device can be an important factor.

Experience has shown that in gas main cracking in certain circumstances other pipelines have not been damaged where they are as close as 300mm.

Work by WRC has been described in reference 1. This paper refers to the experiment which was carried out in Case history No. 1 by monitoring 'dummy' pipes instrumented with strain gauges. Sophisticated monitoring of the effects of the mole was carried out as this passed along the existing pipeline adjacent to and below the monitoring stations.

Further site monitoring is to be undertaken by WRC and in addition during the winter 1984/85, a test site is being made available at Swindon, UK, where a number of moling devices will be monitored as they perform in specially-laid pipelines. Not only will the vibration effects be studied but also the power and forces required by the differing devices.

Lateral Connections

The state of the art has been described earlier in the paper. A number of UK firms are evolving systems for reconnection of renovated pipelines from within the pipe. The added problem with pipe cracking is that more of the lateral or service connection is affected because of the expanding action of the mole. Present practice is to sever the connection point prior to cracking. The challenge clearly exists for innovation in this area and no doubt many minds are busy on this subject.

CONCLUSIONS

The ability to replace cast-iron mains on a size-for-size and upsize basis by pipe-bursting techniques has now evolved into a full operating system.

The case for replacement of sewers by this method is also proven but the system will undoubtedly evolve to cover the many varied circumstances that are found in deteriorating sewerage pipelines.

The concept now shows obvious advantages over conventional means of pipeline replacement and if there is not a clear saving in some circumstances to the client, particularly in urban areas, then the other advantages described, not forgetting the savings in social cost, will undoubtedly tip the scales in favour of this method.

REFERENCES

[1]Dorling, C, *Pipe strains and vibrations caused by percussive moling*. Site Experiment No 2 Water Research Centre External Report No 154E.

[2]Stein, D, Bielecki, R and Maass, U, Small diameter jacked pipes: development and experiences from Hamburg. *Europipe '83 Conference*, Basle, Switzerland.

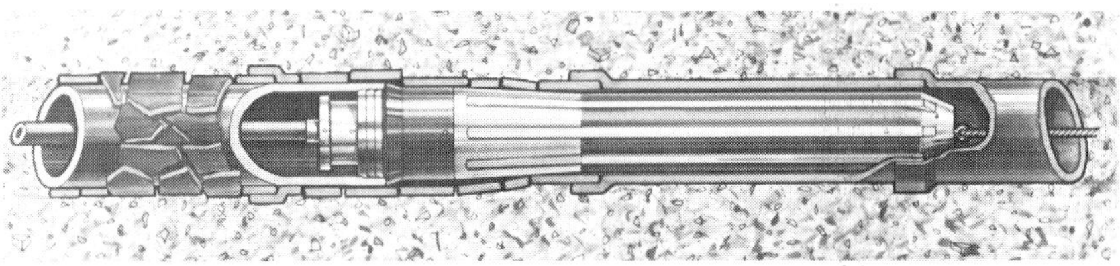

Figure 4.1.1. Burster unit and replacement pipe being inserted into a clay sewer.

Figure 4.1.2. Burster unit being inserted into a cast-iron pipe.

Figure 4.1.3. Snaplock pipes being threaded over compressed air line.

Figure 4.1.4. Edgar Road, Winchester: installation of equipment in manhole.

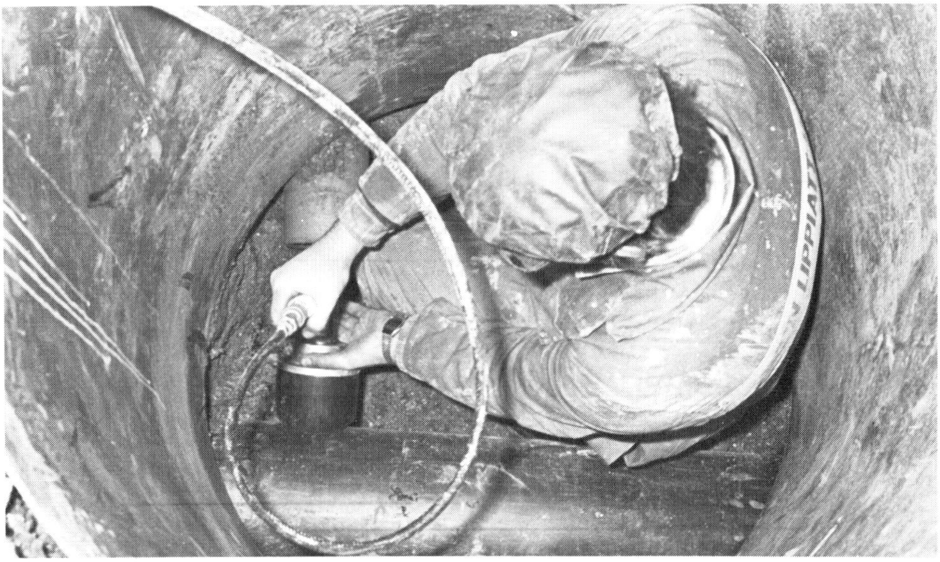

Figure 4.1.5. Lateral connection being made in 900mm diameter access shaft.

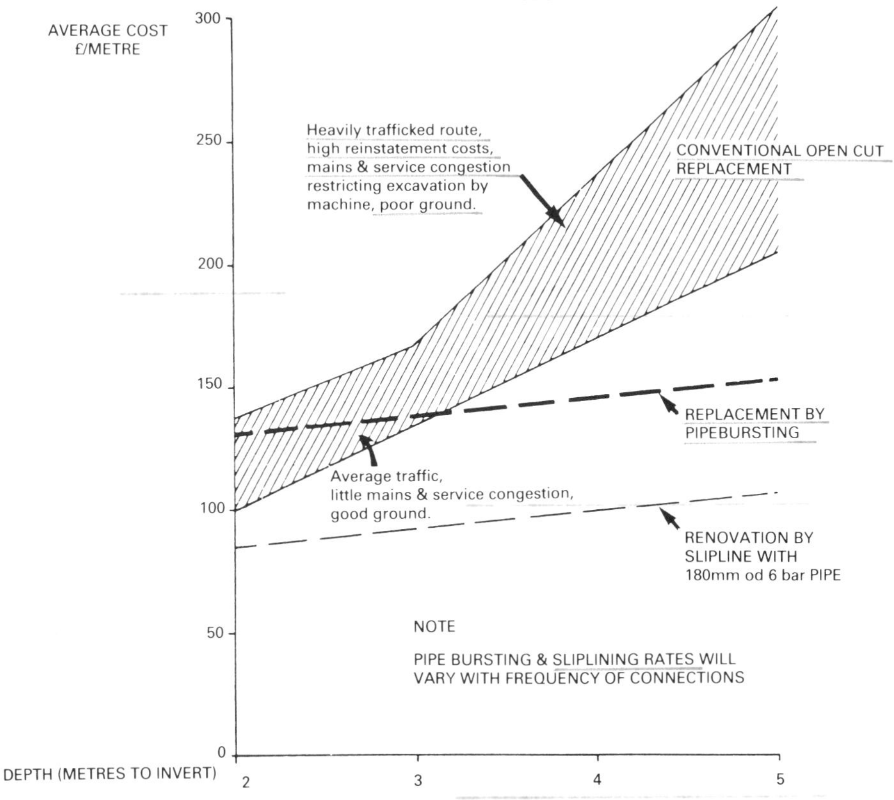

Figure 4.1.6. Cost comparison: renovation and replacement v. open cut for 225mm i.d. pipe, based on 500 1000m contract under highway (see discussion).

4.2 Directionally-controlled drilling for pipelines

John D Hair *Manager Contracts & Engineering, River Crossing Division, Reading & Bates Construction Co.*

Geoffrey E Shiers *Contracts & Engineering Representative, River Crossing Division, Reading & Bates Construction Co.*

The horizontally-drilled, directionally-controlled method of installing pipelines across rivers is an attractive alternative to conventional techniques. The drilled method was developed in the USA where it has now become established practice. Elsewhere interest in the method is growing and a number of river crossings have been completed in Europe, Africa and Asia.

The method involves the drilling of a pilot hole along a designed path, down, under and up to the far side of a river. The pilot hole is reamed to a larger diameter and a pre-assembled pipeline string is pulled, from the far river side, along the reamed path. Once the pipeline is in place it is tested before client handover.

The method is rapid and can also offer other considerable benefits. It is ideally suited to smooth profile pipelines with joints resistant to tensile forces such as the welded steel pipelines used for oil, gas or water transmission and cable conduits.

This paper reviews the background to directionally-drilled pipelines. The method and equipment are explained in some detail. The benefits arising from the method are discussed. Further adaptations of the method are considered along with some pre-contractual design details required before undertaking a directionally-drilled installation.

BACKGROUND

Pipeline river crossings are major engineering works requiring careful design and construction.

Conventional methods are based on either burying the pipeline below the river bed or supporting it across the river on a pipebridge.

The latter pipebridge method is usually reserved for awkward crossing sites with ground conditions such as rock, or natural phenomena such as deep river bed scour. Here the designer is prevented from burying the pipeline safely beyond dangerous phenomena by the limitations of construction equipment, or otherwise high construction costs make river bed burial unacceptable. However the pipebridge is an expensive solution for river crossings, particularly when considering the continual pipeline maintenance costs. Also exposed, working pipelines are vulnerable to damage and nowadays deliberate sabotage is a design constraint along with accidental damage from flying objects and high wind.

The usual river crossing method is burial. Here the designer has been restricted by existing equipment and methods.

Usually a trench is excavated across the river by dredging equipment and the designer has to balance the required safe river bed pipeline cover against the water working depth of dredging equipment. Deep rivers are particular problems for satisfactory cover depth and sometimes even inaccessible to dredging equipment. Normal river traffic is disrupted by river crossings and in some cases authorities are reluctant to issue permits until the construction work has been organized for minimal disruption, which can cause inefficient and expensive construction methods. Prior to pulling the pipeline into the trench, approach ramp excavations are required on each river bank. These are usually major excavations, followed by subsequent backfilling and reinstatement. Not only are the vulnerable river banks disturbed but also the environment, and in sensitive areas environmental disturbance can lead to permit refusal. Additional design restraints of the burial method are dangers from high river currents during pipe positioning and the effect of buoyancy uplift on the in-place pipeline beneath a relatively shallow bed cover depth. The solution is to provide stability by concrete weightcoating. However this adds to the pipe handling and purchase costs without contributing to the pipeline transmission efficiency.

Conventional river crossing construction methods are unsatisfactory and this is confirmed by statistical analysis. A significantly high percentage of river crossing pipeline failures are due to the natural phenomena of bed erosion or bank collapse and accidental damage from dredging, anchors and spuds.

Horizontally-drilled river crossings offer the designer the certain security of deep river bed cover, with minimum environmental damage and no disruption of river traffic. Concrete weightcoating is unnecessary and the crossing is installed in a matter of weeks rather than months.

The method was developed in the early seventies by Titan Construction, Sacramento, USA. The first client installation was for Pacific Gas & Electric Co who installed a 4 inch pipe over 180m across the Panjaro River, Watsonville, California. In 1979 the method was acquired by Reading & Bates Construction Co, Tulsa, USA, a major international pipeline company. Since 1979 the method has progressed to a state-of-the-art where crossing lengths in excess of 1500m have been installed and the largest pipe diameter has been 40 inches.

METHOD

The method involves the drilling of a smooth curved pilot hole approximately 3 inches in diameter from one side of the river to the other. The pilot hole follows the design centreline of the proposed pipeline. The ground entry angle on one side of the river is usually 12°. The pilot hole continues at this 12° angle until it passes through a sagbend to level out horizontally under the river at some 10m deep below the river bed bottom. Once the pilot hole has passed beneath the river it then rises through another sagbend to exit on the far river bank at the predetermined target peg with an angle around 8° (Figure 4.2.1).

The pilot hole is drilled by a specially built rig which pushes the drill rods into the ground. Bentonite drilling mud is pumped down the centre of the drill rods and turns a down-hole motor positioned just behind the drill bit. As the drill bit rotates the bentonite mud acts as a coolant and also washes the cuttings back to a surface reception pit. In line with the down-hole motor is a slightly bent section of drill rod called a bent housing. The eccentricity of the bent housing is accentuated by a steel shoe welded onto one side of the pipe so that the drilled profile can follow a designed path. A curved or straight profile is achieved by steering the drill rods as they are pushed into the ground. The steering is controlled by the position of the bent housing and steel shoe.

The progress of the pilot hole is recorded by a specially designed steering tool system. One part of the system is an instrument package situated in the drillstring, behind the down-hole motor. It records the exact position of the drill bit in the X, Y and Z axes relative to a ground surface origin, along with the relative position of the bent housing. This information is transmitted by radio signals to the other part of the system situated on the ground surface. Here a computer converts the signals into survey data and records them on a CRT display and also a paper printout.

Once the actual position of the pilot hole is known, a comparison is made with the required position on the design path. If there is any deviation, it can be corrected by turning the drill rods and thus moving the bent housing to the correct steering position.

The combination of drill and steering ability along with immediate feedback of the drill bit positions allows the pilot hole to be rapidly drilled along a designed path.

As the pilot hole progresses, soil cuttings mix with the bentonite and increase the stress on the drill rods. To relieve the stress, a 5 inch diameter steel washover pipe is rotated over the drill rods by the drilling rig and the two pipe strings thrust alternately along the design path (Figure 4.2.2), until the drillstring and following washover pipe exit at the far river bank target point.

The complete drillstring is then withdrawn back through the washover pipe, leaving the washover pipe in place to act as a drawstring for the remaining two operations, pre-ream and client pullback. Depending on client pipe diameter, length and soil conditions the two operations are either separate or combined.

Pre-reaming widens the pilot hole to a diameter slightly greater than the client's pipe. A circular cutting tool is attached to the far river bank washover pipe end. The drill rig rotates the washover pipe and cutter, simultaneously pulling them back along the pilot hole. As the cutter is pulled into the ground a further washover pipe string is added behind it.

Bentonite, pumped down the washover pipe, carries soil cuttings to the surface and the combination of bentonite and washover pipe ensure the reamed hole remains open.

Client pipeline pullback is the final operation. Prior to this operation the client's pipeline has been made up in one full length on the far river bank. The pipes have been welded, X-rayed, coated and the full length hydrostatically tested. The pipeline is now ready for ground burial and positioned, on rollers, in line with the drilled hole and adjacent to the target peg. Special tools are then joined between the pipeline pulling head and the washover pipe protruding from the ground (Figure 4.2.3). The first is a fly cutter which widens the drilled hole to its final diameter. Next is a barrel reamer which smooths the hole wall and directs bentonite to the fly cutter and following pipeline. Between the barrel reamer and pipeline is a swivel. The swivel ensures that the rotating action of the cutter and reamer is not transmitted to the pipeline. After all connections are complete the rig rotates and pulls the washover drawstring along the drilled path, while on the far river bank the pipeline moves slowly off the roller stands and into the ground behind the rotating tools and washover pipe (Figure 4.2.4).

During its underground journey the pipeline is cushioned and lubricated by the protective bentonite matrix. Once started, the pullback continues until the pipeline is in place under the river. This operation is usually completed with 24 hours.

All that remains is an in-place hydraulic test and client handover.

EQUIPMENT

A directional drilling rig spread is designed for easy transportation, rapid site erection and the immediate availability of spares and back-up equipment. A typical spread will be equivalent to some twelve 12m trailer loads. It will arrive on site completely equipped to undertake the work. Only water is required and usually this is pumped from the adjacent river.

The major equipment item is the drilling rig. Each rig is a purpose-built item of machinery and can weigh in excess of 30 tonnes. The rig is towed, in folded-down form, to the site working area by a road tractor unit. A site crane is used to unfold the rig, now approximately 25m long × 2.5m wide. One rig end is then lifted, allowing the support legs to be swung into position and the rig is ready to start work. The rig is now essentially a track inclined at 12° to the ground surface, along which a drilling carriage and machinery can move up and down. In this position the rig can drill and extract the pilot hole drillstring and following washover pipe.

Figure 4.2.5 shows a rig drilling the pilot hole. Immediately in front of the rig is a small pit which collects and stores the bentonite returns for disposal. The pit is framed by timber mats for convenience and access. Just beyond the mats is the bentonite mud mixing tank. Here the bentonite is mixed with water from the nearby river to produce a mud gel. Two mud pump units, set on skids, stand in line with the mud tank. One pump unit acts as a standby while the other pumps mud from the tank to the drill rig and into the drilled hole. All major static plant items are set on skids with crane lifting points for easy and rapid handling.

The drill rods and washover pipe are laid out alongside the rig ready for positioning with the adjacent crane.

Also included in the spread are working and standby generators and an office unit.

The on-site back-up support includes spare parts and a mobile workshop, all of which are housed in two large container vans set on their own wheels for road towing.

BENEFITS

A number of important benefits arise from installation by the horizontally-drilled method.

The first benefit is rapid installation. The complete spread is designed for fast movement and handling. It is not unusual for the equipment to arrive at the site on day one and by day two be already set up and drilling the pilot hole. The pilot hole is completed within a few days, depending upon soil conditions and length.

The separate pre-reaming operation, if required, is usually completed within a similarly short time. The following hook-up to the client's pipe and pullback quickly follow, leaving only the in-place hydrostatic test. Installation by conventional methods can be measured in months, while directionally-drilled crossings are measured in weeks.

Deep river bed cover depth is another benefit with the pipeline positioned some 10m below the river bed. Here it is secure from river bed erosion, later canal bed deepening and accidental damage from passing river traffic anchors.

During the installation the river traffic passes unhindered as there is no construction work across the waterway. This benefit is of particular importance to canal and river authorities and can assist with the easy issue of construction permits.

A longer length of pipeline than the immediate bank-to-bank distance is installed as the entry and exit points are some distance back from each river bank. During installation the river banks remain undisturbed and permit issue is again assisted along with the benefit of an easy tie-in for the overland pipeline contractor.

With the pipeline buried some 10m below the river bed, it is safe from the effects of buoyancy uplift. Concrete weightcoating is unnecessary and its elimination offers the benefit of financial savings in pipe purchase and handling.

During a directional-drilled installation environmental disturbance is minimal. This benefit is of growing importance. On one side of the river work is limited to a small site area while on the other side of the river the pipeline is made up in the right of way. There is no need for major excavation of approach ramps, or a dredged channel, no sheet-piling, no backfilling or reinstatement.

ADAPTATIONS OF THE METHOD

Two significant adaptations are pipeline bundles and a beach approach crossing.

A bundle installation is the positioning of more than one pipeline in the same directionally-drilled crossing and is limited to smaller-diameter pipes. The procedure is the same as for a single pipeline crossing. The bundle is made up on the far river bank with the pipelines separated from each other by doughnut spacers positioned along the length of one or more pipelines. The bundle is pulled back as a single unit along the pre-reamed profile.

A beach approach crossing installs the immediate onshore section of a pipeline from an offshore well. The rig is set up on land, back from the beach area. The pilot hole is drilled some 10m under the beach to exit on the sea bed (Figure 4.2.6). Meanwhile the pipeline is made up on the beach in one full string and bottom-towed out to sea. The pipeline tail is connected to the washover pipe and now becomes the pipeline head as it is pulled back along the line of the drilled pilot hole. Using this technique a pipeline beach approach is quickly installed, without disturbing the beach area which is sometimes an environmental or coastal protection strip. There is also no excavation work in the hazardous surf zone and the pipeline is positioned some 10m deep below the beach offering certain security against storm damage and beach erosion.

PRE-CONTRACT INVESTIGATION

A river crossing can be separated from an overall pipeline contract in both technical and contractual terms. There is considerable merit in this separation as difficult river crossings can be installed in advance leaving only simple tie-in connections for the overland contractor. Considering the technical aspects of pre-contractual work the most significant areas are:

(a) **Profile Design.** A section of the pipeline crossing is drawn showing ground, river bed and water level elevations. The pipeline profile is designed to ensure the yield stress induced at the sagbend positions remains within acceptable limits.

The design profile is then superimposed on the section drawing at a nominal 10m river bed cover depth. The profile will show a 12° ground entry angle on the rig side and a 8° ground exit angle on the opposite river bank side. Minor adjustments to these angles can be made.

(b) **Soil Survey.** The profile design can assist the spacing and depth of soil survey boreholes. The standard engineering soil survey should be designed to ensure a continuous soil picture along the underground pipeline route as large gravel deposits and hard rock are avoided. Soil data are required from both below as well as above the proposed pipeline route and generally soil surveys should extend some 15m below river bed level. In addition to soil descriptions, Standard Penetration Tests should be taken along with a sieve analysis, and any rock samples preserved for later visual examination.

(c) **Site Inspection.** A site plan assists the inspection. Access to the rig working area should be available for the specialized equipment which is delivered on standard 12m long road trailers. The standard width rig is also delivered by road, being some 13m long and weighing in excess of 30 tonnes. The rig working areas should be reasonably level, firm and suitable for the movement of rubber-tyred vehicles. Normally it should measure 100m long × 50m wide although a smaller area can be used.

On the other side of the river should be sufficient room for the pipeline to be made up in one full length, on rollers, in line with the crossing axis and ready for ground burial in terms of the client's specification. In certain cases it is possible to make up the pipeline in half or even shorter lengths.

CONCLUSION

Directionally-drilled river crossings offer an acceptable alternative to conventional methods. They also allow difficult river crossing sites to be reconsidered and offer further benefits to the designer and client. The method is now proven and can be supported by an impressive track record of completed installations.

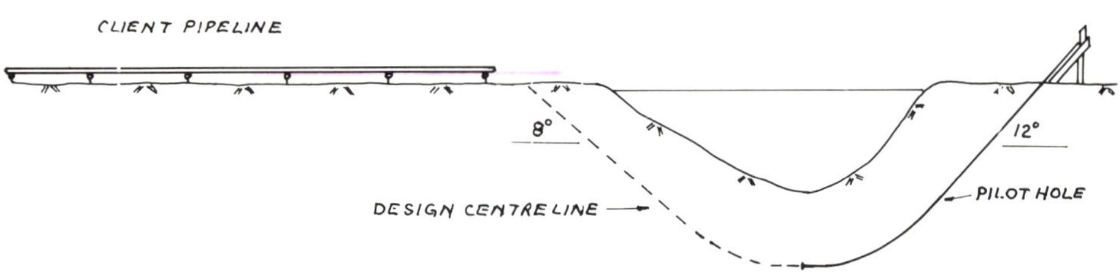

Figure 4.2.1. Pilot hole being drilled along the design centreline. Meanwhile the client pipeline has been made up on the far river bank and set on rollers.

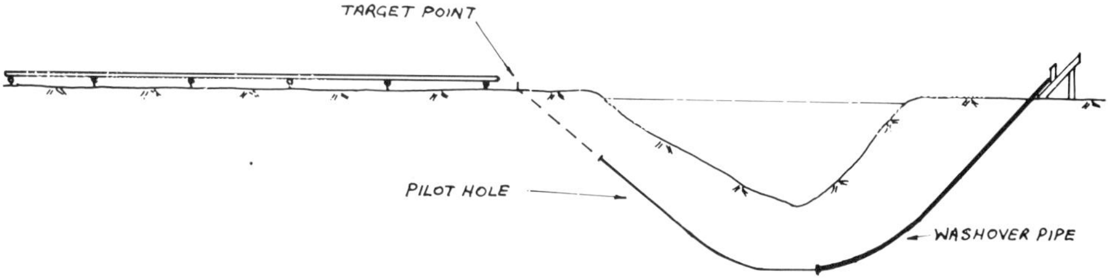

Figure 4.2.2. The washover pipe is rotated over the pilot hole drill pipe and both pipes thrust alternately along the design centreline.

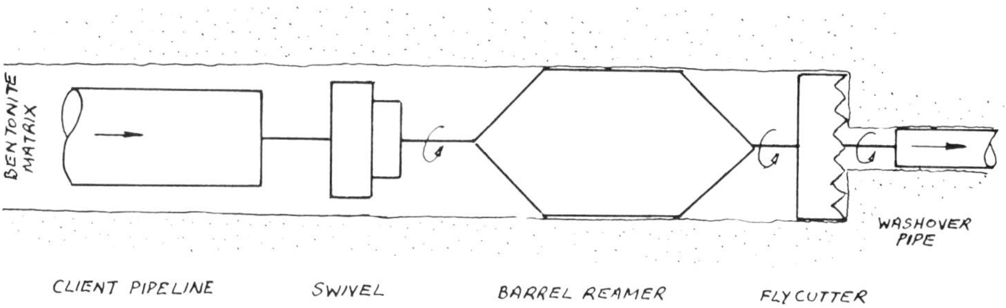

Figure 4.2.3. Washover pipe, fly cutter and barrel reamer being pulled and rotated along the drilled path. The swivel and client pipeline follow behind in a bentonite matrix.

Figure 4.2.4. On the far bank the client pipeline moves slowly off the roller stands and into the ground.

Figure 4.2.5. The rig drilling the pilot hole. The ground entry point is framed by timber mats and support equipment is positioned around the rigs.

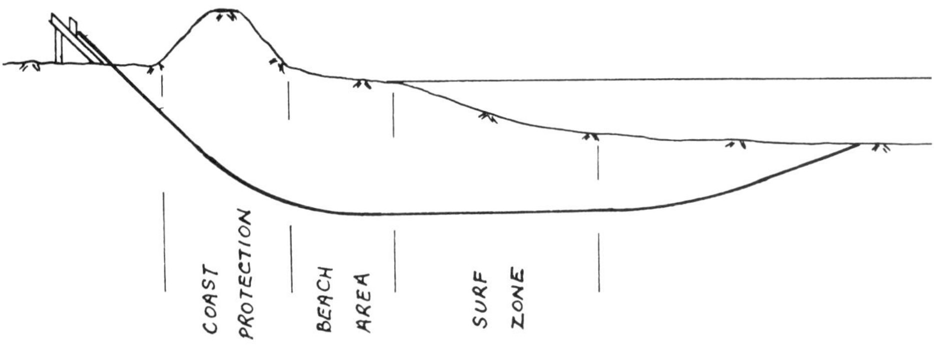

Figure 4.2.6. The pilot hole for a beach approach is drilled deep under the sensitive coastal protection strip, the environmental beach area and the hazardous surf zone.

Authors' introductions

Mr R B Rosbrook: Replacement of small-diameter pipes by pipe bursting

Pipe bursting is the method of inserting a new pipe into an existing pipe of an internal diameter equal to, or smaller than, its replacement by means of fragmenting pipework and forcing the material into the surrounding soil to form access for the following replacement pipe. I emphasize replacement. We really are not talking about renovating pipelines. The structure of the existing pipe is in fact deliberately destroyed.

Figure 4.1.1 in the paper is a drawing of a Merstan Burster unit of the type which has been used for the replacement of clay pipe sewers. The drawing clearly shows the tapered outer casing, the back of the percussive mole, the winch line just going out of the picture to the right, the compressed air line to the left, the fragmented clay pipe and the new polyethylene pipe which is shown attached to the back of the mole.

It is necessary for us to identify the two different pipe-bursting systems: the cast-iron system, that is, the technique of replacing gas, water and sewage pressure pipes, and the sewerage system, the replacement of non-pressure sewage pipes.

The Ryan cast-iron system uses a mole with the added facility of hydraulic breaker arms which are used at more difficult pipe joint locations, collars, etc. These breaker arms can just be seen projecting from the mole in Figure 4.1.2 in the paper. Generally speaking the moling takes place with the breaker arms withdrawn. In the gas/water system a sacrificial protective sleeve is used and inside this the duty pipe is placed. The sleeve is used to comply with the Code of Practice for pressure pipes which demands that no angular fragments are to come into contact with the pipe. Because of the high internal pressures it is thought that damage to the outer surface is particularly critical. The sewerage system differences which are mentioned in the paper are emphasized in Table 4.1.1.

[Mr Rosbrook then showed some slides illustrating the projects described in the paper, and made the following points.]

It is a point very worth while stressing that an early development in the sewerage system was the ability to withdraw the mole from its outer casing, to withdraw it along the inside of the new pipeline and back to the surface for repairs should the occasion arise. A speaker did raise the issue that it was no good having systems if every time they broke down you have to dig a hole in the road. It is obviously more critical with sewers being so deep.

The accurate location of the lateral, the house drain connecting pipe, became an essential part of the development. Minimum excavation techniques are essential to reduce surface disruption and a lorry-mounted auger is used to produce a neat 900mm shaft. In most situations the first metre is dug by hand to avoid those uncharted cables and pipelines.

A rotating-head camera was used by Seer TV Surveys to look sideways and align accurately with the incoming lateral pipe to establish the exact distance along the sewer of the incoming connection. In order that an operative could be lowered into an excavation for depths of 5 metres or more a protective shield system was developed.

Figure 4.1.5 shows the protective shield in position over the lateral connection prior to the pipe bursting sequence. The technique of reconnecting laterals was developed with the aid of an electro-fusion joint to the main sewer and a universal coupling with stainless steel bands provided by Naylor Brothers Limited.

Under the heading, Material Suitability, the paper describes the use of welded polyethylene strings, but points out that for sewers, it is often preferable to use short-section pipes inserted through short trenches or existing access chambers.

Short-section pipes were used on Case History No 1 site in Winchester. These are of polyethylene with snap locking joints and are one metre in length. These particular ones are produced by Stewart & Lloyds.

Demco polypropylene pipes with screw joints and rubber sealing rings were also installed in the trials. As anticipated at the time of writing the paper, some clay pipes have now been produced in the UK for this purpose.

Since writing the paper in 1984 we have carried out on-site monitoring of performance of these pipelines using a light line with a closed-circuit television video camera. The light line is recorded with the sewer camera and stored on video tape. The resulting profile can be digitized onto a computer for accurate assessment.

I would like to draw attention to the cost graph in Figure 4.1.6 in the paper, which can be compared with the graphs in Mr Irvine's paper 2.4. The cross-hatched area, representing open-cut pipelaying, certainly covers a wider range of costs than his model, but that is because of the very varied conditions one meets with in practice. The thicker line for the replacement by pipe bursting shows up quite favourably, and I would like to stress that in the Southern Water Area we have seen direct savings from using these methods. I feel that as the pipe-bursting method becomes more widely used, that line will certainly be shown at a lower point on the graph.

In conclusion, I should like to inform you that work has now started in the Southern Water Area for the oversizing of a pipeline from 225mm to 350mm internal diameter. The purpose of this project is to give a 300 per cent increase in capacity.

Mr G E Shiers: Directionally-controlled drilling for pipelines

The subject of this paper by John Hair and myself is horizontally-drilled directionally-controlled installation of pipelines, a process developed in America and now regarded as an established method of installing pipelines across rivers. It has also been introduced and demonstrated here in Europe.

Using the paper as a basis I would like to explain the process briefly, also refer to the benefits of the process and look at some of the adaptations that are possible.

Consider a typical river crossing as shown in Figure 4.2.1. A pipeline has to be laid across a river. We arrive on site with our specialized drilling equipment and set up the equipment at a position on one river bank. From the river bank position we drill a hole approximately 3 inches in diameter at about a 12-degree entry angle down into the ground along a designed path through a sagbend, levelling out horizontally under the river, passing through another sagbend to rise at a predetermined target point. As the 3-inch hole is drilled, a 5-inch diameter steel pipe is rotated over the 3-inch drill rods to similarly exit at the far river bank target point. The drill rods are then withdrawn.

Meanwhile, on the far river bank the client's pipeline has been made up in one full length, set up on roller stands, the length being equal to the length of the crossing. At this stage of the installation, the drilled pipeline profile represented by the 5-inch diameter steel pipe has to be increased in diameter. The profile can either be separately reamed out to a size larger than the client's pipeline, or if the client's pipeline diameter is small, say something less than 20 inches, the profile reaming is incorporated with the next stage of the process, with the pull-back of the client's pipeline. To the client's pipeline are attached some special tools, a flycutter and a reamer, which widen the hole to a size greater than the client's pipe. By greater I mean an annular space of about 6 inches. The client's pipe is then pulled off the roller stands and into place under the river bed. Next it is hydrostatically tested and the crossing is complete.

The benefits of using this system are considerable. The first is that the pipeline profile beneath the river bed is approximately 10 metres deep. Thus the client has certain deep cover depth security against damage from ships' anchors, barge spuds or later canal deepening.

The largest pipe diameter we have handled so far has been 40 inches, laid under the Houston Shipping Channel. The longest length we have installed from entry to exit point has been approximately 1500 metres, under the Orinoco River in Venezuela.

The second benefit is rapid installation. The method allows a pipeline to be laid across a river in a matter of weeks, sometimes days. We have recently installed a pipeline at the UK coastline where 400 metres of 24-inch diameter pipe was drilled and pulled back in position within six days. Last year in Italy we laid pipelines underneath the Po River, 20 inches in diameter, 300 metres from entry to exit point within five days. On a contract in Bangladesh, the installation time from spudding in to handover to the client took 19 days for a 1000 metre length of 16-inch pipe. Rapid installation is a considerable benefit to the client, particularly to those who equate time with money.

Moving on to the next point — river traffic. During the installation ships pass up and down the river or canal completely unaware that a major civil engineering operation is taking place. Our work is limited to a small work area on one side of the river, while on the far side of the river the pipeline is made up within the right of way. Quite a dramatic illustration was the job we completed under a canal in the Ivory Coast where a pipe bundle, an 8-inch and a 10-inch pipe tied together, was laid deep underneath this busy waterway. We worked on one side and made the pipe up on the other; meanwhile ships and river traffic passed up and down the canal completely unhindered.

Fourthly, because the pipe is deep beneath the river bed — 10 metres — and there is no buoyancy effect, there is no need for weightcoating. The pipes have a cheaper materials purchase, delivery and handling cost. A definite identifiable, financial saving for the client.

The fifth benefit which I think is of considerable importance, particularly here in Europe, is the environment. As we are limited to two small areas, one on one river bank and one on the other river bank, environmental disturbance is minor. There is no need for massive dredging excavations, there is no need for sheet piling, there is no need to excavate approach ramps, there is no need for backfilling. Environmental damage is kept to an absolute minimum, and the installation time ensures that any disturbance is for a short duration.

Turning now to an adaptation of the process which has been used for the onshore section of an offshore oil or gas line. Figure 4.2.6 shows the beach area, the sand dunes or the dykes, the beach and the sea-bed, while somewhere way out to sea is the oil or gas platform. The installation stages are similar to a river crossing. We set our equipment up on the shore work area and again drill a pilot hole down under the dunes, under the beach area, under the surf zone — a difficult area for civil engineering operations — and we exit on the sea-bed. The pipeline itself is made up on the shore and bottom-towed out to sea in one full length so the tail of the pipeline now lies adjacent to the sea-bed exit point and from our rig work area position we pull the pipeline into place, 10, 15 or 20 metres deep below the beach area, deep below the surf zone, deep below any sea defence works. The benefits of the river crossings apply here also. Elimination of weightcoating, deep cover depth protection, no environmental damage and rapid installation time. As an example we installed a beach approach in America, in the Gulf of Mexico, where a profile length of approximately 860 metres was drilled for a pipe of some 10 inches in diameter. From arrival on site to handover to the client of the completed, in-place pipework took four weeks. During those four weeks we had one and a half weeks weather down time. On this particular job the environment was a critical feature as the operator was unable to get a permit for conventional installation methods. By using our unique method the local flora, fauna and the various beach recreation areas were untouched. At the completion of our contract there was absolutely no evidence of our installation work.

I should like to end with the point that this particular method of installing pipelines really is a revelation. Those who have seen it and appreciate it realize that this is the only way to install pipelines across rivers and beaches. I think the engineer who is faced with the problem of crossing rivers should regard our method as an established alternative to the conventional dredging method. We have an impressive track record and can offer a proven technique with considerable financial benefits.

Discussion on papers 4.1 and 4.2

G C Cox (Consultant, M McDowell Co-Partnership): It is stated in the paper by Poole, Rosbrook and Reynolds that the concept of pipe bursting was originally developed in the UK four years ago by British Gas. It was, in fact, developed at least 13 years ago by the then Distribution Engineer of the Metropolitan Water Board, John Pocock, who proposed this as a method of cheaply replacing lead sevice connections by using an expanding mandrel, dragging in a plastic pipe behind it. I am surprised that British Gas claim to have a patent on water and sewerage pipes.

Would the authors explain whether the need to have constant tension winches is because the mole unit needs some additional pull to help bursting, to avoid bounceback, or to act as the drag force for the pipe necklace?

The authors state that the UK specification for closed-circuit television (CCTV) work allows for an error of ± 2 per cent in linear dimension. At the moment there is no such specification. In the current drafting of the first National Specification for CCTV work, in which I am involved as Chairman of the Association of CCTV Surveyors, it is proposed to specify ± 0.3m or 1 per cent, whichever is the greater. Is this accurate enough? How was the level of 0.2 per cent achieved?

CCTV was used to identify laterals not in use, and 28 per cent were found to be dead. In four studies of inner city areas up to 70 per cent of laterals were found to be dead because, although redevelopment takes place on average every 50 to 70 years, the old sewers remain the same. How were the laterals inspected — from the top, or from the main sewer using a flexible endoscope-type unit? On the reconnection of laterals was any thought given to the use of Raychem shrink-on sleeves, as on the Oxford trials?

Was a PUSWA notice issued for this work — indeed, is this common practice for renovation work generally?

On scoring of the new pipe, I have seen very little evidence of damage from the moling process. Indeed I have seen more scoring damage on sliplining pipe, possibly from sloppy lead-in arrangements at the pipe end. Is there any intention to inspect the pipes laid by moling in, say, a year's time, to see what has happened as the ground relaxes back on the pipe?

In the Tonbridge contract, close to house foundations, were there any claims for vibration or other damage? What public relations activity was undertaken beforehand?

The authors say that a job, estimated to take 30 weeks by conventional methods, was programmed to take 12 weeks by moling, but was then completed in nine weeks. Was any penalty or bonus imposed? Is this sort of stick and carrot approach suitable for development work, or is the best bonus a further contract?

Looking to the future, I see an equally important, if not greater, problem to sewer dereliction, and that is the need for greater hydraulic capacity to overcome flooding. There are no firm statistics on flooding, but, extrapolating the major study carried out by Thames Water, in which I was involved, I estimate that 60 000 houses in the UK are flooded with foul sewage on a frequency worse than once in ten years. I see the development of oversize-for-size moling techniques as a most important way of economically solving this problem. Would the authors like to say whether they agree with this, and, if so, what sort of time scale do they envisage?

If they are going for oversize, then the need to know where adjacent services are becomes even more essential. Perhaps this aspect could be included in the Conference message to Professor Horne. Could it be stressed that the industry does not want a wishy-washy compromise which will be out of date by the time it comes to implementation. Will he please be forthright on what is needed for the next, say, 20 years.

May I suggest that by 1995 all services should be recorded in digital map form, and that the caveat on no responsibility being taken for accuracy should be not only withdrawn but turned round, so that if any damage is caused to a service that is not in its recorded position (or within, say, 1.0m), then the owner and not the contractor (or his insurer) bears the cost. I suggest also that no-go areas should be designated for any roads with a traffic count of 5000 vehicles per day or more.

After hearing the fascinating aside this morning from Tomas Brinch, of Norway, on using old telecommunication ducts for new cables, could the authors elaborate on their experience of using any hole in the ground, and could they indicate how they see unit costs for moling developing in comparison with conventional trenching?

Turning to the second paper in this session, by Hair and Shiers, could Mr Shiers enlarge on the limitations for profile design? Is the sagbend dependent on the flexibility or diameter of the final line, the amount the rig can deviate, the friction on pullback, or is it a combination of all three? What is the minimum radius so far achieved?

Could a list of completed installations be supplied, giving, for example, the longest, shortest and largest diameters — or a selection if the list is too long? Were any major snags or failures encountered in the early development of the process, and, if so, how were they overcome?

What sort of cost savings are achieved over conventional methods? Is the method limited to soft ground, and are there any particular soil conditions which the authors would rather not be asked to tackle?

Where do the authors see future development for this fascinating process? Could it perhaps have an application in dealing with an inner city flood alleviation scheme?

D J Irvine (Technical Services Manager, Central Engineering Department, Tarmac Construction Ltd.): I was interested to note that the lower boundary of costs for trench construction quoted by the authors of paper 4.1 (Figure 4.1.6) agreed closely with those presented in my own paper 2.4. However, there is a large difference between the upper and lower cost boundaries (50 per cent at 5m depth). Could the authors give reasons for this, and also give the base date of the comparison?

When new pipes are installed after bursting existing pipes, the new installation must be adequate both hydraulically and structurally. Could the authors comment on the structural design of the new pipe barrel to suit the site conditions of depth, type of ground and live loading? What judgement must be made for the bedding factor provided by the destroyed pipe?

S R Kramer (Project Manager, Gas Research Institute, Chicago): US gas utilities are losing customers because the cost of main extension is too high. Is it possible to adapt the Reading and Bates directional drilling technology to other applications besides under-river crossings? What are the feasibility and likelihood of this occurring?

A L Coe (Distribution Engineer, Thames Water): There appears to be a discrepancy in the upper limit of pipe bursting for replacement. Mr Shiers reports on 3-inch pilot holes being augered out as large as 40 inches, whereas Mr Rosbrook talks about expanding from 225 up to 350mm. In my experience, two specialist contractors have failed to burst out a 250mm pipe to 300mm and had to complete the job in open cut. The length was not very great, and the depth was perhaps 2½ to 3 metres. What are the limiting factors — size, depth, overburden, soil type and density, or a combination of these?

D A Scott (Regional Projects Manager, Thames Water): I am intrigued by what happens when the pipes being burst have mechanical joints with rubber rings. Presumably the rubber ring is stretched, but then springs back and wraps itself around something. Where does it disappear to?

I was involved in a lot of work for the Water Research Centre on stress monitoring and the movement of trenches, and it seemed to me that this was rather glossed over in the paper.

Would Mr Rosbrook tell us more about what stress monitoring was done and what the results were? I am worried about the stresses that might be caused by in-ground methods of enlarging pipelines, and the risks — and costs — of damaging nearby gas or other services. I think this requires more research.

A G Munro (Engineer for Metropolitan Works, Melbourne and Metropolitan Board of Works): Would Mr Shiers please elaborate on the types of ground conditions his system can and cannot handle? Could he also give an explanation of the spoil removal system, both in the original pilot bore and in the later reaming out?

With reference to Mr Coe's question about the limitations on upsizing by impact moling, I would point out that auger boring has no such limitation, because the spoil is removed, not compressed.

Professor T D O'Rourke (Associate Professor, Cornell University): Many gas utility companies in the USA have used metallic joint clamps in their cast-iron systems. The clamps are generally constructed of rugged, ductile irons or steels. Are current pipe-bursting techniques likely to be effective at joints with metallic clamps? What has been the previous experience, and what developments are planned?

Shore crossings, as well as river crossings, often involve changes in groundwater chemistry which can lead to corrosion problems in metallic pipelines. Can Mr Shiers tell us how the horizontally-drilled method of construction affects the protective coatings generally used on high pressure pipelines. Are any special corrosion-resistant practices used in conjunction with the process?

V A Finney (Principal Assistant Engineer, City of Manchester): The re-connection of laterals is described in paper 4.1 and appears to require excavation from the surface at each lateral — inevitably causing disruption. This is a reflection of the fact that there does not appear to be, as yet, an alternative technique to deal effectively with the problem of laterals. Clearly, if numerous connections exist, the overall disturbance to the community can be as great as, if not greater than, would occur if the sewer were replaced in open cut. The paper refers to the innovative challenge which exists in this area. Would Mr Poole in particular care to expand on any work Avon Lippiatt Hobbs are doing in the direction of remote connection techniques?

R T F Clark (Principal Engineer, Arun District Council): As a past and current user of the pipe-bursting system, I remain very impressed with it for various applications. However, there seem to be a few inherent difficulties in certain conditions, and I would appreciate the authors' comments on the following.

1. The application of the method to sewers which are not true to line and level, or are laid on a curve.
2. Long-term deformation of HDPE pipes. Has consideration been given to other materials, particularly clayware?
3. Have any problems been encountered in bursting sewers with rubber sealing rings (as mentioned by Mr Scott)? Have any methods of degrading or corroding these rings before moling been considered?

Mr R B Rosbrook: Mr Cox and Mr Scott are both concerned about oversize replacement and possible damage to other services. We have recently started on a project for enlarging a pipeline from 225 to 350mm. There is a gas main quite close by, and we shall be watching this carefully. But we have close co-operation with British Gas and WRC on this question.

Mr Cox also raised the question of oversizing in relation to flood relief schemes. The project mentioned is for the increase of capacity of 1000 metres of 225mm pipe to an internal diameter of 350mm — a 300 per cent increase in capacity.

The authors would agree with Mr Cox's suggestion that PUSWA notices are still appropriate for this method.

Mr Cox raised the issue of using redundant holes in the ground. Southern Water have recently purchased a gas main from British Gas in Hastings and are renovating it for their own use. Perhaps there are other opportunities for the transfer of holes in the ground which are no longer required for their original purpose, and certainly the process described in the paper could help.

In answer to Mr Coe, we have not had any failures in breaking out earthenware pipes in Southern Water, but there are certain ground conditions — ground that is soft or contains voids, and where the pipes are considerably off line — where the mole may disappear to one side or down below. One has to be aware of this possibility and carry out thorough investigations prior to deciding on the method to be used.

Mr Scott mentioned the problem of rubber joint rings. We have broken out concrete pipes with rubber rings on an oversizing trial. One ring finished up on the nose of the mole, and the others disappeared, presumably somewhere along the pipeline. But we know of a job where rubber rings on pitch-fibre pipes caused trouble to start with, making the mole jump backwards and forwards.

Mr Irvine's question on the structural design of the new pipe barrel is an interesting one. Clearly there have been some unknown factors in comparison to ordinary pipe designs in open cut. It is a problem that the writers have grappled with and have worked with WRC with the object of producing a method statement for the design. Professor Tom O'Rourke is also to be involved in this work, and is working on the production of a mathematical model for the expanding ground theory.

Mr A G Poole: I can assure Mr Cox that the patent situation for pipe-bursting is as stated in the paper.

Mr Coe's reference to two contractors who failed to break out a 225mm pipe to 300mm illustrates the fact that perhaps contractors are being forced to under-price work in sewerage systems, bearing in mind that whereas British Gas have already laid 450km of pipes by this method, only about 10km have been laid for water and sewerage. A tremendous amount of development is required before the method can be recognized as a totally saleable commodity.

To take the question of rubber rings a little further, the new generation of moles have special steel cutters which are there purely to cut the O-ring before it gets caught on the end of the mole.

In answer to Professor O'Rourke's question on metallic clamps, the D J Ryan set of equipment has expanding hydraulic arms on the front of the mole, and these have so far been able to cope with the clamps, though perhaps different kinds of clamp are used in America.

Mr Finney raised the question of remote connection of laterals. It is necessary to disconnect the lateral from the sewer before the mole is passed through. Otherwise the lateral is likely to be disintegrated for a greater length than one would hope.

I think it will be a few years before a complete system for internal re-connection of laterals is perfected. A good deal of development in this direction is going on in the sliplining process, and we at ALH are looking at the problem. However, we developed the system of augering because it is the cheapest form of lateral re-connection we could undertake at the present time.

In reply to Mr Clark, there is a potential line and level problem, because the mole will follow the line of least resistance. I would therefore suggest that the hydraulic capacity should be checked very carefully, so that any deviation off line can be offset by the extra capacity of the new pipe.

Mr J H Reynolds: Mr Cox asked questions about specification for CCTV. We achieved the accuracy we needed on these contracts by tying a tape to the CCTV camera as it went down the sewer, and then measuring back to each manhole. We then measured the whole length of the sewer through the sewer and checked it above ground, so we were 100 per cent certain that we could mark the lateral positions accurately on the road surface.

We deemed that inspection of the laterals was necessary, firstly, to be sure that they were in a sound condition, and, secondly, to know the mode of entry of the lateral into the sewer. This was done by passing a mini CCTV camera down the lateral, and by using a new camera by Sears with a right-angle view to look at the lateral from the main sewer. We noted that many of the laterals on the Edgar Road contract had been capped off at the time of construction — obviously for use in future development.

We received no claims for vibration damage when using the mole, either from other utilities or from nearby buildings.

The question of contract bonuses or penalties was raised by Mr Cox. In the case mentioned, we agreed on a 12-week contract period and on liquidated damages which would be applied after that period. The contractor finished three weeks early and was deemed to have made sufficient profit on his contract, so there was no need to award him any bonus.

Mr Irvine asked about the cost graph. The price base is two months ago (i.e., February 1985), so the figures are quite up to date. The top line on the graph refers to bad ground in a very busy road requiring good quality reinstatement. The lower line applies where there is less traffic, less disruption and a cheaper reinstatement method.

Mr G E Shiers: In reply to Mr Cox, the diameter of the pipe and the radius of curvature of the pipeline are linked. The pipe must be designed to make sure that the outside fibre stress stays within acceptable limits, so both the diameter and the steel grade are relevant.

Friction during pull-back is at an absolute minimum, as the pipe is encased in a matrix of bentonite which acts as a cushion and a lubricant. A practical illustration of minimal friction values occurred when we installed three crossings in Holland last year. Our American superintendent of construction, eager to return home, wrote on the last pipe of the last crossing, in white paint, the words "Baby, I'm coming home". The pipe was then pulled 300m underneath a canal through the bentonite matrix, and emerged on the far side with the words still legible.

We do not specifically design a pipe for a minimum radius, as it in turn is the product of diameter and steel grade. However, in general terms our minimal radii are around 250 metres.

To date we have completed around 130 crossings. The longest crossing length has been 1432m of 10-inch pipeline in Colombia. The largest diameter of 40 inches over 520m was installed in America. The shortest crossing has been 160m. A list of the ten most recent crossings is attached.

We have encountered problems with unknown soil conditions, which were successfully overcome. Rather than discuss detailed technical solutions, I would use these incidents to support the earlier call for adequate pre-contract site investigations. On the subject of failures, I can first say that my company has always completed its contractual obligations. I can also think of two contracts where we expressed serious reservations because of the soil types, but in both cases we went ahead under agreed contract conditions. Both contracts were unsuccessful because of the soil types encountered. One was half successful and was later completed by dredging methods. On the other one, we skidded over and redrilled another hole in the same area to complete the pipeline installation.

The cost savings are considerable, and 130 satisfied clients support this statement. In addition the benefits of rapid installation, with secure deep river bed cover, without disturbing either river traffic or the environment, and the elimination of concrete weightcoating should also be credited in the cost calculation.

Our method can be used in all sedimentary soil types and soft rock. Gravels need particular attention, which returns to my remark concerning adequate soils information.

The principal demand for these installations is in the oil and gas industries. However, as long as the pipe string can carry a tensile load, it can be used to cross under any obstacle.

The answer to the first part of Mr Kramer's question is included in the reply to Mr Cox. The feasibility depends technically upon engineering parameters already discussed, and commercially upon the cost of conventional methods. The likelihood depends upon a client accepting the feasibility of the adaptation.

TEN RECENT CROSSINGS CONSTRUCTED BY READING AND BATES

Date	*Description*	*Waterway*	*Client*
1984	2015 ft of 8-inch	Atchafalaya River, Krotz Springs, Louisiana	Gulf Central Pipeline Co.
1984	4025 ft of 30-inch	Mississippi River, St. Francisville, Louisiana	Transcontinental Gas Pipeline Corp.
1984	1180 ft of 10-inch	River Tevere, Rome, Italy	SNAM
1984	770 ft of 20-inch	River Senio, Castel Bolognese, Italy	SNAM
1984	1310 ft of 24-inch	River Senio, Fusignano, Italy	SNAM
1985	634 ft of 10-inch	Cypress Creek, Houston, Texas	AT & T Communications
1985	700 ft of 10-inch	San Jacinto River, Houston, Texas	American Distribution Co.
1985	460 ft of 10-inch	CIWA Canal, Houston, Texas	American Distribution Co.
1985	1312 ft of 24-inch	Pipeline Landfall, Bacton, England	Hamilton Bros. Oil & Gas
1985	4307 ft of 30-inch	Mississippi River, Baton Rouge, Louisiana	Florida Gas Transmission Co.

I apologize for misleading Mr Coe as we use a soil removal and not a pipe-bursting technique. During the drilling of the 3-inch pilot hole the soil cuttings are washed to the surface, as noted by Mr Munro, by drilling mud, and this technique is similarly used to expand the *in situ* profile hole to its final diameter. A minimum depth of overburden, nominally 10 metres, is necessary to prevent bentonite wastefully breaking out into the river bed. The only limiting factor is the soil type, which is assessed from the soil survey.

Professor O'Rourke is considering the question of long-term corrosion important to all designers. We have handled a variety of pipe coatings without any coating failures. During pull-back the client's pipe is protected against damage by the bentonite surround, which can be confirmed by a visual examination of the in-place pipeline head and a current drainage test. If there was a particular concern about ground water chemistry, one approach could be the use of additional corrosion protection such as a thicker coating, anodes at either end of the crossing or impressed current protection.

4.3 Microtunnelling in Germany

Dr -Ing. E. Kuntze *Erster Baudirektor i. R,*
Präsident Abwassertechnische Vereinigung e. V.

Since the beginning of the 80s the common interest in Germany has turned increasingly towards methods of trenchless pipelaying. At first Japan's experiences and technologies were tested, transferred and developed for German conditions. Today there are more than fifteen different methods on the market and new progress is achieved each month. This paper reports on the tested and reliable techniques.

Since 1972 the tunnelling of larger sewers has made enormous progress in the German Federal Republic. Towards the end of the seventies the general interest turned more and more to processes suitable for the construction of small sewer pipe sections. The reason for this was the ever growing demands made for the protection of the environment with regard to interference with quality of life. Added to this there was the desire to reduce the otherwise unavoidable destruction of expensive road surfaces, the wish to avoid lowering the ground water level and to minimize the risk of moving and/or damaging other pipelines for supply and disposal. The methods which were formerly applied in this field in the Federal Republic of Germany could not meet the demands with regard to the accurate position of sewers. Originally these methods had only been applied for the crossing of roads, railways, and so on, and for very short distances for the placing of cables and water and gas pipes.

Internationally, especially in Japan, progress had developed more rapidly. The particular traffic conditions, the narrow streets and the method of construction of the houses made the subsequent building of sewers by excavation so difficult that the Japanese were forced to apply construction by tunnelling. They built remote-controlled machines which made sufficiently accurate tunnelling of small sewer pipe sections possible, even though the distances were relatively short.

A study trip to Japan revealed that the Japanese processes and experiences could also be applied with success in the German Federal Republic, though they would have to be adapted to German conditions and requirements. The German Ministry for Research and Technology sponsored the adaptation project proposed by a Japanese-German team; this project was carried out in Hamburg at the end of 1980. The result met with general interest. The advantages were so obvious that up to now more than 15 firms have decided to develop their own solutions of the problem. Every development in this field tries to meet the following requirements:

1. During the driving the deflection from the axis must be less than 30mm vertically and 100mm laterally.
2. It must be possible to use the driving machine in different soils (loam, sand, marl, gravel etc.) and under the influence of groundwater pressure. It should reach a driving velocity of at least 12m/8-hour day and be able to produce different pipe cross-sections.
3. The necessary investment for the acquisition of the machine and the devices that belong to it should not exceed one million German marks (about £250,000), to guarantee its economy.
4. The machine must be able not only to locate obstacles underground, but also to remove or destroy them, even larger stones and boulders and foundations.

5. Since the starting and destination pits are very expensive, their number must be kept down by driving lengths of at least 150m.

   In order to further reduce the expenditure for the excavation of the pits the dimensions should be limited for the starting pit to length/breadth = 4.0 × 3.0m and for the destination pit 3.0 × 3.0m.
6. The machine should make the driving of small-radius bends possible.
7. The production of the pipes *in situ* out of concrete or plastic is desirable.
8. House sewer pipes should also be produced underground.

The firms are obviously trying hard to meet those requirements. New and constantly improving methods for the control of the machine, the removal of obstacles, the use of material for the production of pipes, and the transport of soil are being offered and tested. Should this development proceed in the same way, we can expect all the requirements to be fulfilled within the next five years.

An important aspect must not be neglected: the building costs. Unfortunately they are still relatively high, so that new techniques compare favourably with open excavation only at a depth of over 4m. It must however be remembered that indirect costs have not been taken into consideration here; the costs induced by open excavation are high, for instance long-term traffic hindrances and diversions, economic disadvantages to the residents and damage to the environment. The aim is equal costs at a depth of 2m whether by open excavation or tunnelling. This aim should be achieved more easily when the present expensive developments have come to an end.

Let me tell you about some systems which have been tried out recently in West Germany. They have been selected from a large number of systems which are very different as far as methods and results are concerned. Quality was not the main criterion of my choice; I wanted above all to give you an idea of the palette which is being offered.

This is what I have selected:

— the Telemole System and its further development, the Hansemole System;
— the Dywidag articulated steel-pipe shield;
— the pipe-driving unit RUS 100 A of the firm Soltau.

Telemole and Hansemole Systems

The T-M System is a drill-driving system that was developed and tested in Japan. Before it was introduced into Germany it was adapted to German needs. The machine and the pipes are pushed forward hydraulically. The machine itself consists of a steel pipe, 4.6m long, supplied with devices for excavation, control and transport of the soil. It is a remote-controlled shield, which is called an "earth pressure counterbalanced bentonite slurry shield". That means that the face breast is held on the one hand mechanically by a cutting device combined with a full-section scraper disk, and on the other hand, by bentonite.

By adapting itself automatically and maintaining the necessary pressure on the face breast this system makes a safe and efficient excavation possible in practically every kind of soil and also in groundwater. The excavated soil is transported hydraulically. To control the shield four control presses are used, which are operated in a control station from a control panel.

The necessary information is summarized on an indicator panel in the shield and transmitted to a screen by a television camera. A laser beam touches the target plate. With this machine more than 2500m of sewers with diameters of 500 and 600mm have been tunnelled. An excavation length of up to 120m was reached. Experience showed that lubrication of the pipes with bentonite was only necessary in sandy soil under groundwater. According to the number of underground obstacles the advance drive rate was 3 to 18m per 8 hour day; for this four operators were needed. The vertical deflection was less than 30mm; the horizontal deflection about 20mm. The use of this machine is only limited by the size of the stones found in the underground; they must be less than 50mm in diameter. The displacement of larger stones cannot be expected.

In order to drive small sewer pipe sections into soil containing stones of more than 130mm in diameter a new machine based on the Telemole System (Figure 4.3.1) was constructed; the machine is known as the Hansemole. Its characteristics are the changed cutter head with a crunching rotor behind it, which reduces stones of 130mm in diameter to a maximum diameter of 50mm. The cutter head works directly ahead of the shield, so that the pipe that is being driven in finds free room.

Other changes are the reduction of the length of the shield to 3.10m in order to reduce the dimensions of the starting pit, the stronger shield wall, a more efficient motor and better steering jacks. In order to protect the driving machine better, an improved bulkhead was built in.

To be on the safe side, larger openings were made for pressing in the bentonite; this reduces the pipe friction in the case of pipe driving of over 150m. The power of the driving presses was increased to 600 tons (6000kN) so that they could deal with greater driving lengths. The cut soil is transported hydraulically. So far the following materials have been used for jacking pipes:

clay;
plastic reinforced with glass fibre;
reinforced concrete.

The tunnelling was as precise as with the other machine, with the same number of operators. Over 1000m of tunnelling was done in very difficult soil conditions (hard marl with large stones). An average of 9m was tunnelled per 8-hour day. Experience has shown that friction can be reduced by using plain water in heavy soil; in sand or stony soil bentonite should be used.

Dywidag articulated steel-pipe shield

The development of this system, too, was sponsored by the German Ministry of Research and Technology. It can be used in every kind of soft soil, even under groundwater pressure.

The basic idea is that the driving and the installation of the product pipe are two separate operations. First, steel pipes of 2m in length are pushed against the shield. Their inside diameter is 800mm. This makes it possible to reach the machine unit through the pipe in case of a breakdown. If necessary any part of the machinery, including the cutter head, can be taken out and either exchanged or repaired. A kind of very long articulated steel-pipe shield with a constant outer diameter of 860mm is produced, which equals the excavation in length.

Secondly, after this first operation has been concluded, the product pipes chosen are pushed in.

They have the required inner diameter and an outer diameter of 860mm, corresponding to the outer diameter of the steel pipes. The shield and the steel pipes are retrieved in the destination pit. No further mechanical devices are needed inside the product pipes.

The cutter head of the machine excavates every kind of soil, even in groundwater, and overcomes even heavy boulders. Excavation lengths of 300m can be reached with this device. Tunnelling in a curve with a minimum radius of 100m is possible. The advance drive rate is at most 3m/h. The cutter head is built into a scraper head, whose task it is to hold the face breast in combination with bentonite if it is used for transport. The earth which the cutter head has removed is transported hydraulically, either by water or by bentonite. The shield is regulated by four built-in steering jacks, which are also used for curved tunnelling. The shield and the steel pipe are connected by a flexible joint, which makes tunnelling in a curve possible. All the cables and pipes which are needed for the cutter head for supplying and disposal purposes are permanently installed into the steel pipes; they are connected accurately and automatically when a new pipe is added, which saves a lot of time. For the construction of a sewer of 55m including all the corollary work, 40 hours were needed.

For the purposes of controlling and surveying, a reversible CCD-dual-camera is used, which absorbs the light of diodes and shows it on a screen inside the camera. This picture can be shown on a monitor. The data are fed into a computer and transformed into steering commands. Surveying with a laser beam is also possible in straight lengths.

The equipment of the site consists of easily transportable container units. They include a power unit (to ensure independence) and a unit for the separation of bentonite from soil and its retrieval.

Pipe-jacking RUS 100 A of the firm Soltau (Figure 4.3.2)
Contrary to the two methods described so far, this one deliberately limits itself to shorter excavation lengths. The whole system is based on compact machines and component parts and takes into consideration the usual distance between two manholes, which is from 50 to 80m in Germany. It is also taken for granted that the sewer between two manholes follows a straight line. The machine which is constructed for these conditions can be used for pipe cross-sections of 250 to 400mm in diameter. Since about 75 per cent of the sewers which are to be built in the next few years have a cross-section of up to 400mm, this machine can deal with most of the work.

The length of the pipes is limited to 1m to keep the starting and destination pits as small as possible. Suitable pipe-driving units make it possible to dig excavation pits with a round cross-section of 2m in diameter. To construct such excavation pits a concrete pipe of 2m in diameter is sunk into the earth. Since the machines to be described are so consistently compact, a container can hold them; it measures 6.0 × 2.50m, and can be set up directly over the excavation pit, thus making the building site independent of weather and time of day.

An articulated steel head is a part of the mechanical equipment (Figure 4.3.3); it contains a full-section scraper disk which excavates the soil. Three hydraulic control presses are built in for control purposes, as well as an electronic target plate, which catches a laser beam and is used for constant surveying. The results of this surveying process are transmitted to a computer in a control panel, which transforms these data into control orders, which are automatically transmitted to steering jacks. It is also possible to operate the machine by hand.

The earth which has been excavated by the full-section scraper disk is carried by a conveyor worm through a conveyor pipe to the starting pit. The excavation and transport devices can move stones of 40mm in diameter. Experience has shown that larger stones, provided they occur singly, are pushed forward or aside.

A groundwater level between 1m and 1.50m above the bottom of the pipe is no problem. If the groundwater is higher, compressed air must be applied. The necessary device is included in the machine.

The system of pipe driving described was used for the first time in Berlin in the summer of 1982. A sewer of 477m had to be built. The distance between the manholes ranged from 45 to 73m, and the depth of the sewer from 1.70m to 3.70m. Asbestos-cement pipes with a nominal diameter of 250mm (outer diameter 350mm) were used. The pipe friction was at the most $2.98kN/m^2$. The measured deflections from the axis were only a few millimetres. Three men were needed to carry out the work, which progressed at the average speed of 1m per hour.

There is no time to report on other methods of pipe driving. Their number on the one hand, and their constant development on the other, force me to limit my subject. I must add, however that the problem of connecting house sewer pipes by tunnelling has largely been solved.

The simplest method is to sink a concrete manhole of a diameter of 800mm on to the finished sewer. From this manhole several house sewer pipes can be driven. But even today it is already possible to connect house sewer pipes with the main sewer in the street by tunnelling from an excavation pit situated away from the road. Accurate tunnelling and leakproof connections are no longer a problem.

In conclusion we see that the tunnelling of smaller sewers has reached an advanced stage of progress. It is already clear that in the near future the present necessary opening up of roads to place sewers will be an exception.

The future belongs to the tunnelling system.

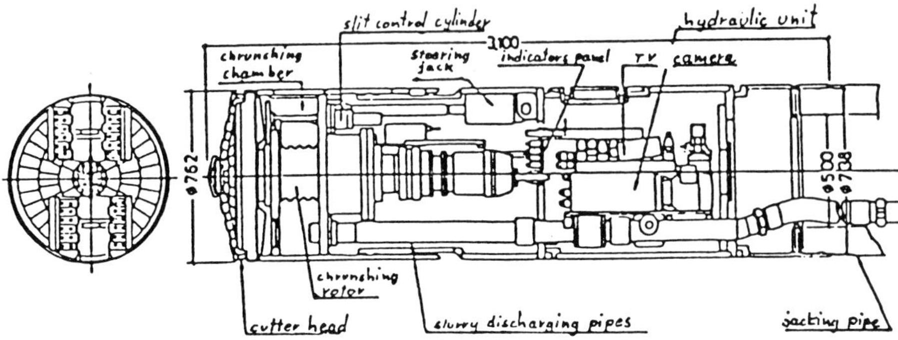

Figure 4.3.1. Longitudinal section of the Hansemole 600.

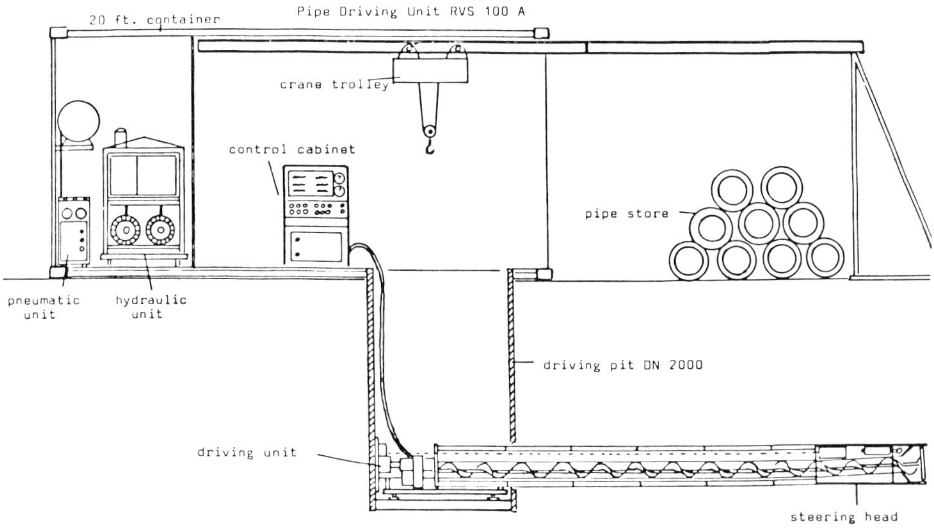

Figure 4.3.2. Pipe-driving unit RUS 100 A with container installation.

Pipe Driving Unit RVS 100 A for automatic driving of pipes of nominal diameters of 250 to 400 mm.
Diameter of driving pit only 2000 mm.
Minimum attendance required.

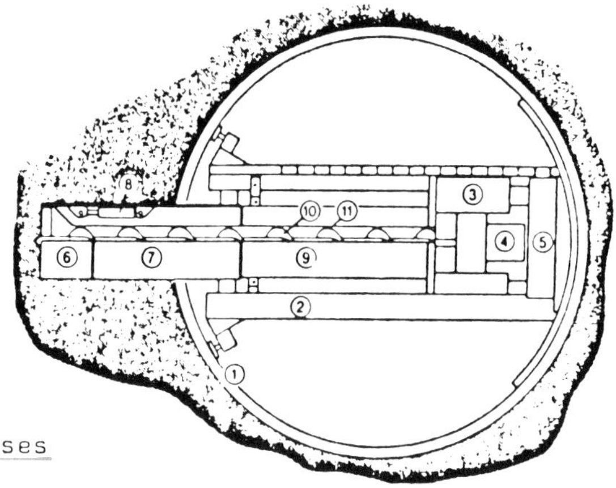

(1) driving pit
(2) driving carriage
(3) press cylinder
(4) worm drive unit
(5) thrust bridge
(6) steering head
(7) follower pipe
(8) direction control presses
(9) machine driver pipe
(10) conveyor worm
(11) conveyor pipe
(12) target plate
(13) laser
(14) laser beam

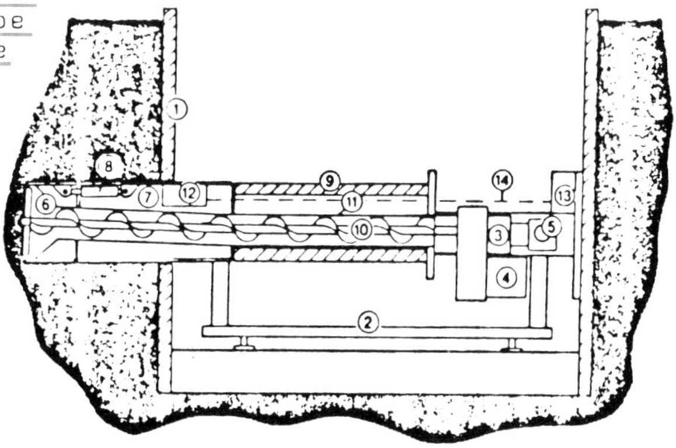

Technical Data Specification

Max. pressing power	1000 kN/500 bars
Max. torque	4000 Nm/200 bars
Press stroke length	200 mm
Driving pit diameter	2000 mm
Total electric power	40 kW/380 V
Hydr. fluid tank capacity	400 liters

Figure 4.3.3. Pipe-driving unit RUS 100 A: schematic arrangement.

4.4 Development of the Iseki Telemole and Telemouse and their application to trenchless pipelaying

Toshio Akesaka — *Director and Chief Engineer, Iseki Kaihatsu Koki Inc*

Alan Rooke — *Director — European Sales, Iseki Poly-Tech, Inc*

This paper covers the background to the development of the Telemole and Telemouse remote-control small-bore slurry tunnelling machines with some details of their design and construction record in Japan and overseas. The paper goes on to describe two of Iseki's most recent developments in the field of small-bore slurry tunnelling machines for installing pipelines to accurate alignments, particularly for gravity sewers in soft and loose ground conditions. Finally, there is a brief review of the potential market for trenchless construction in UK sewerage construction and some proposals for the expansion of this method in the UK.

INTRODUCTION

For the last 25 years the Japanese Government have been pressing ahead with a programme of sewerage and sewage disposal which has accelerated to over £2000 million per annum in recent years. The objective has been to increase the percentage of the 117 million population served by main drainage to 44 per cent by the end of 1985 from a figure of 6 per cent in 1958 and go on to increase this figure to 70 or 80 per cent over the next 15 to 20 years.

Such an objective has no parallel anywhere in the world and required the construction of over 7000 kilometres of sewer in 1981 alone. Earlier in the programme the demand was for large-diameter trunk and main sewers to form the backbone of the sewerage systems but gradually as work proceeded it was necessary to reduce the size of the pipelines to serve small centres of population and individual properties.

Naturally, the larger sewers were laid deep in the ground and a large proportion of these were suitable for construction by tunnelling methods without even considering the implications of social disturbance caused by open-cut excavation. As a result of this demand a design and manufacturing industry grew up to supply suitable boring equipment, and as much of the work had to be carried out on the coastal plains and alluvial valleys in the mountains, where there were large concentrations of population, the equipment designed had to be capable of dealing with these conditions. The introduction of the slurry shield concept was most timely and this idea was developed with enthusiasm, using a range of modern technological hardware to produce safe efficient machines.

It is the Japanese way to study such technological problems in great depth, analysing every piece of data in detail before producing a single nut or bolt. This thoroughness, combined with the enormous volume of work available to evaluate performance, led to the production of a range of efficient soft-ground tunnelling machines and a style of tunnelling quite different from most European methods. A system where the soil and water of the tunnel face are rarely seen and the work is executed by engineers and technicians working in a clean dry environment.

In the early stages smaller-size pipelines were constructed entirely by conventional open-cut methods, but where this involved working in urban areas a demand grew for the work to be executed by less disruptive methods. At this time the technology required for larger-size boring was well developed and it was an obvious step to adapt this for the smaller pipelines.

The pressure for the development of such equipment did not arise until the end of the nineteen-seventies as the market for smaller sewers expanded. The Iseki small-bore machines were not introduced to the market until 1979. In Japan as elsewhere new methods of construction take time to become established, due to pressure from existing interests and the naturally conservative nature of engineers building sewers to last at least a lifetime.

The details of sewers constructed by bored tunnel in Japan in 1981 given in Table 4.4.1 show that a significant proportion were constructed by boring or tunnelling methods. The most common method for sewers larger than 2000mm diameter was segment tunnelling but one-third of sewers in the range 500 to 1850mm i.d. were constructed by pipe-jacking methods. Official figures for later years are not available to this author but from our company records we know that the proportion of sewers in this size constructed by boring equipment and pipe jacking increased considerably in 1982 and 1983 especially in the 500 to 900mm sector where machine pipe jacking is becoming a more economic method in urban situations.

TELEMOLE

A remotely-controlled slurry tunnelling machine for pipe jacking between 730 and 960mm pipe o.d.

The machine has led the field in the growth of small-bore sewer tunnelling in the range of 500 to 900mm i.d. It was developed from the larger manually-operated Mechanical Earth Pressure Counter Balance (MEPCB) shield machine.

Design of the Telemole

The basic MEPCB is a unique machine which incorporates a full-face rotating cutter head which is able to advance and retract under controlled pressure relative to the shield body. This enables the cutter head to exert a constant preset counterbalance pressure on the soil regardless of the rate of excavation and the rate of forward progress of the shield body. The cutter head also contains cutter teeth which adjust automatically to balance the rate of excavation with the rate of advance of the cutter head.

Soil cuttings are transported to the surface by pumping them in suspension in a slurry. The slurry pressure in the tunnel face can be adjusted to balance the natural water level existing in the ground above the tunnelling machine.

It is therefore possible to balance both the soil pressure and the groundwater pressure independently and bore through the soil with a minimum of disturbance to surrounding services and structures. The standard MEPCB however was manually operated by a man located in the shield body and therefore could not be employed on pipes less than 900mm diameter.

Operation of the Telemole

In order to make a machine suitable for boring pipes between 750 and 960mm o.d. it was necessary to make the MEPCB totally remote-controlled. This was achieved by using a closed-circuit television to monitor all functions of the performance on a central display panel in the shield machine together with a remote-control electric/hydraulic operation system to apply any necessary corrections to the shield machine as the boring proceeds. The effectiveness of this method was remarkable; all the functions of the original MEPCB machine were retained but in addition control of the whole operation was now located at one place, enabling the work to be supervised easily by one operator. Furthermore as all information was displayed on a closely-grouped panel the boring could be executed more accurately, as deviations of the shield alignment were immediately apparent and easy to correct. The response of the machine to corrections was also immediately apparent, ensuring that just the right amount of correction could be applied.

Performance of the Telemole

This machine has been used to install a wide variety of pipe materials including concrete, clayware, GRP, asbestos cement and steel. It has been used successfully in a wide range of soft ground conditions including soft clays, silt, fine to coarse sands and gravels as well as harder materials including boulder clay, weakly-cemented sand and coral etc.

The equipment is primarily used where a small pipe has to be installed to a close alignment tolerance especially when the soil conditions are difficult for other methods and where surface disturbance or settlement would be undesirable.

The outside diameter of the installed pipes varies between 730mm and 960mm and the inside diameter is dependent on the pipe wall thickness used and can vary widely. Length of pipe bored in a single operation has usually been limited to 120 metres but recently the maximum range of the 960mm o.d. Telemole has been increased to 200 metres.

TELEMOUSE

A remotely-controlled slurry tunnelling machine for small-bore pipe jacking between 480mm and 650mm o.d. The Telemouse was built in response to the demand for even smaller pipe diameters and to accurately install pipes as small as 350mm i.d. using a boring technique.

Design of the Telemouse

The Telemouse is a development of the Telemole shield machine but the sliding cutter head featured in the Telemole is replaced by a pair of hydraulically-controlled slit doors which perform a similar function by not allowing any soil to enter the machine unless the face pressure exceeds the predetermined figure set on the control panel. The soil support characteristics of the Telemole have therefore been retained.

Operation of the Telemouse

Remote-control operation is also by the well-tried principles of TV monitor and electro/hydraulic controls used in the Telemole, enabling similar standards of performance to be produced.

Performance of the Telemouse

Like the Telemole the Telemouse is at its best in unstable soft loose soil conditions but has also gained a reputation for being a rugged little machine capable of dealing with many of the natural and man-made obstructions found when boring underground in urban areas. By virtue of its high length-to-diameter ratio it is not easily deflected off course by obstructions, and with care has been known to bore through old pipework, concrete and timber. The length of drive in a single operation is up to 100 metres and the pipe may be made of any material provided it can transmit the necessary thrust force required for installation.

CONSTRUCTION RECORD

In the three years, 1981, 1982 and 1983 Iseki have received reports from over one hundred projects where the Telemole has been used to install in excess of 35 kilometres of pipeline and over twenty projects where the Telemouse has been used to install over 5 kilometres of pipeline.

These pipelines have been installed in a wide variety of soil conditions but mostly in the soft-ground range, especially loose sand and soft clay and silt below the water table in the type of soil which causes most difficulty for pipeline construction.

The projects have been located mainly in the Far East, particularly Japan, Singapore and the Republic of China and also West Germany where over 3 kilometres of 600mm i.d. pipe have been installed.

NEW DEVELOPMENTS

The two most significant arguments against installing pipes accurately by trenchless methods are variable subsoil conditions and high costs. Iseki have a continuous programme of research into new methods and techniques and this has resulted in the evolution of two new machines designed to tackle these problems.

Crunchingmole tunnelling machine

The Crunchingmole was developed originally as a large-diameter tunnelling machine to bore through soil containing cobbles and small boulders. It was gradually reduced in size and is now successfully operating as a remotely-controlled machine over the same size range as the Telemole.

In basic operation it is similar to a Telemole except that it incorporates a rotary crusher shown in Figure 4.4.1. This factor, together with a more powerful cutter head, enables it to be used on soil containing large cobble-size stones and in generally harder ranges of soils including soft jointed rock. The small Crunchingmole was originally developed for use in West Germany where it has successfully tackled several projects in extremely difficult soil conditions. It is now the preferred machine for small-bore tunnelling over 600mm i.d. in most soil conditions except sticky clay.

Unclemole super-small tunnelling system

This is the most exciting development in micro-tunnelling machines for many years. It has been designed to tackle both the question of variable subsoil conditions and cost, while still being able to accurately install up to 100 metres of very small-diameter pipeline in a single span to within 30mm of the designed alignment. Sizes presently range from 370mm o.d. to 655mm o.d.

Design of the Unclemole

The basic design of the Unclemole involves harnessing the enormous forces developed within a planetary gearing system to drive a cutter head and a tapered cone crusher as shown in Figure 4.4.2. Iseki have solved the problem of controlling the eccentric rotation of the shaft and use the gear system to reduce the drive speed as well as to apply a massive crushing force to the excavated material.

After the soil is excavated with the spoke-type cutter head it is squeezed into the crushing chamber by the forward movement of the shield where it is gradually reduced to a maximum size of 25mm by the eccentrically-rotating cone crusher, which compacts and crushes the soil perpendicular to the shield axis and so allows it to move with controlled resistance into the slurry chamber, from where it is transported to the separation plant.

Steering and controls are based on a laser datum, TV monitor and electro/hydraulic controls similar to the Telemouse. Pipes are jacked into place behind the Unclemole using a new three-stage jacking system which reduces the length of the drive shaft to 4 metres when using 2-metre long pipes.

Operation of the Unclemole

This machine has been under development and field testing in Japan for the past two years. Experience gained during this period has shown the Unclemole has many advantages over its predecessors. The action of the cutter head and crushing rotor compacts material around the outside of the shield in soft soil providing an improved support for the pipe. Furthermore this compaction reduces the volume of material which has to be transported and separated. The action of the cone crusher is excellent for all types of soils including clay, which tends to be formed into clay balls which are easy to transport and separate and do not block the crusher. Cutter head abrasion is considerably reduced by the eccentric motion of the cutter head.

The counterbalance pressure needed to support the material at the cut face in loose or soft soil is easily adjusted by varying the jacking speed slightly to suit the soil and water conditions at the site.

The crusher will handle cobbles, bricks, pieces of pipe, broken rock, etc. up to one-third the diameter of the shield in size. No bentonite is required for face support or soil transport with this machine and soil separation becomes a simple settlement tank with sludge draw-off as shown in Figure 4.4.3.

Performance of the Unclemole

Field testing of the complete system shows that it is able to install a 250mm i.d. concrete pipe up to 100 metres long in a single operation through a variety of soil types. Control of alignment requires slightly different techniques from previous machines, but once these are mastered the installation of the pipes to within 30mm of design alignment becomes very realistic.

Excavation speed varies with soil type and density but typical rates of progress measured during tests are shown in Table 4.4.2.

Iseki are sure that Unclemole will prove to be a most viable alternative to open trenching for installing pipes up to 500mm i.d. accurately to line and level in difficult or congested areas. The many economies made in the manufacture and operation compared with previous slurry boring machines will reduce the installed cost of the pipes considerably. This machine has great potential for the on-line replacement of existing clay or plain concrete pipe lines, especially when installing oversize pipes in long lengths, possibly to slightly different lines or level and without damage to other services.

SUMMARY

The Telemole has proven to be a pioneer in the field of trenchless pipelaying in Japan, where as shown in Table 4.4.1 over 30 per cent of sewerage pipes between 500 and 900mm i.d. are installed by pipe jacking.

It is our belief that significant inroads can be made into the construction of smaller-diameter pipelines by trenchless methods using new machines like the Unclemole. This is the area where progress must be made if we are able to have any significant effect on reducing the disruption, disturbance and hidden cost of constructing pipelines by open-trench methods in urban areas, because as Table 4.4.1 shows the length of pipe under 500mm i.d. comprised over eighty per cent of the total length of sewerage pipe constructed in Japan. The proportions for sewer construction in each diameter in UK are not significantly different from Japan, although the total quantities are obviously far less.

Therefore there are similar opportunities to make progress in small-diameter trenchless pipe construction here but it will require a very positive attitude on the part of specifying authorities. Careful evaluation of the costs and significance of installing pipelines by both open-trench and boring methods must be carried out before writing the specification. These evaluations should include the costs of disruption to traffic and residents, permanent reinstatement and service diversions as well as the cost of actually executing the work, and the cost of alternative trenchless methods should consider the possibility of shorter routes using trenchless methods which can often give significant savings. When the case for trenchless construction is proved on any particular project it should be unequivocally specified as such.

A method adopted to overcome the initial high capital cost of a boring machine, is for the authority to select the machine type most suitable for their work load and soil conditions and to call tenders on a term contract basis using that type of equipment. This method spreads the cost of the machine over an economic length of pipeline construction. As an alternative to this Iseki are now offering a range of machine sizes for hire in the UK, which also enables authorities to call competitive tenders from contractors whilst only paying a reasonable part of the equipment costs.

Manufacturers of machines and contractors can provide the necessary equipment for trenchless construction but without a real market the experience required to improve the techniques and reduce the costs of the site operations will not be obtained and therefore the potential benefits to the industry will never fully materialize.

Appendix

COST OF NO-DIG OPERATIONS

Large-diameter boring, 1000-3000mm i.d.
Machines for this size of boring have been specifically built as a safe, practical and economical alternative to compressed air tunnelling or soil stabilization using grouting or freezing, when tunnelling in unstable soil conditions. Naturally, such tunnelling costs are expensive and will vary widely depending on individual projects but as a general guide a 500-metre length of 1000mm i.d. pipe should cost between £250,000 and £350,000 (excluding shaft costs) depending on site conditions, which compares favourably with alternative methods of construction currently available.

Small-diameter boring, 300-900mm i.d.
These machines are much newer but we have heard from other speakers that in Japan and Germany there are already many different types of equipment available and manufacturers are all striving to reduce the pipe installation costs by improving the technical performance of their small-bore tunnelling machines.

However, the available machines have been costly to develop and are also expensive to put into operation. The potential for the most significant reduction in pipe installation costs is now in the hands of the pipeline specifiers (particularly the Water Authorities), if they assist by ensuring that this equipment is put to use economically. The case for this is illustrated by Table 4.4.3, which shows the reduction in pipe installation costs which can be achieved simply by better machine utilization, based on a 400mm i.d. Unclemole machine. It must be remembered that such a machine is only useful for installing a single size of pipe, unlike a hydraulic backhoe which can be used for a wide variety of construction activities. Therefore it is important to maintain continuity of work for the machine in order to reduce the installation costs.

From Table 4.4.3 the following conclusions can be drawn.

1. Utilization is paramount in reducing cost. Clients who make a commitment to a piece of equipment can expect to achieve significant savings in total cost/metre.
2. Providing a reasonable utilization is achieved costs are already becoming competitive with trenching methods if the following factors are considered:
   - 2.1. Temporary and permanent reinstatement.
   - 2.2. Damage and reinstatement to other services.
   - 2.3. Traffic disruption costs.
   - 2.4. Commercial disruption costs.
   - 2.5. Social consequences.
   - 2.6. Workers' safety and cost of injuries.
   - 2.7. Potential savings from shorter routes by NO-DIG methods.

   The above factors are most relevant in congested locations and difficult soil conditions.
3. Costs of NO-DIG methods will be significantly reduced by improvements to current equipment and wider acceptance reducing the burden of research and development costs which is presently being borne by a very small amount of real work. The time is right for Water Authorities and utility boards to act for their own benefit.

TABLE 4.4.1 Analysis of sewers constructed by trenchless construction in Japan in 1981

Size pipe i.d. mm	Total length km	Trenchless construction %	Trenchless construction km	Pipejack construction %	Pipejack construction km	Segment tunnel %	Segment tunnel km
Less than 200	1980	0.1	2	0.1	2	0	nil
250- 300	3682	0.6	22	0.6	22	0	nil
350- 450	523	5.0	26	5.0	26	0	nil
500- 900	622	30.2	188	30.0	187	0.2	1
1000-1850	452	51.9	235	32.7	148	19.2	87
2000-2800	141	61.5	87	16.5	23	45.0	64
Over 3000	41	81.6	33	2.2	1	79.4	32
Summary							
Less than 450	6185	0.8	50	0.8	50	0	nil
500-1850	1074	39.3	422	31.1	334<	8.2	88
Over 2000	182	66.2	120	13.3	24	52.9	96
TOTAL	7441	–	592	–	408	–	184

TABLE 4.4.2 Rates of excavation recorded by Unclemole in various soil conditions during field tests

Soil conditions	Max speed mm/min	Average speed mm/min
Loose Natural Gravel	50	25
Med. Dense Natural Sand	80	60
Firm Clay	120	100
Sand and Crushed Stone	40	20

TABLE 4.4.3 Approximate cost analysis for the installation of 400mm i.d. concrete pipes using the Unclemole system

Utilization months/year	Estimated pipe length installed/ year	Present annual operating cost of Unclemole machine	Machine cost/metre	Pipe cost/metre	Support plant/ equipment materials, labour	Total direct site costs	Contractor's site O/H HO O/H profit & risk	Contractor's total selling price/metre
4	560m	£140,000	250	40	65	355	90	£445
6	840m	£150,000	179	40	60	279	70	£349
8	1120m	£160,000	143	40	55	238	60	£298
10	1400m	£170,000	122	40	50	212	55	£267

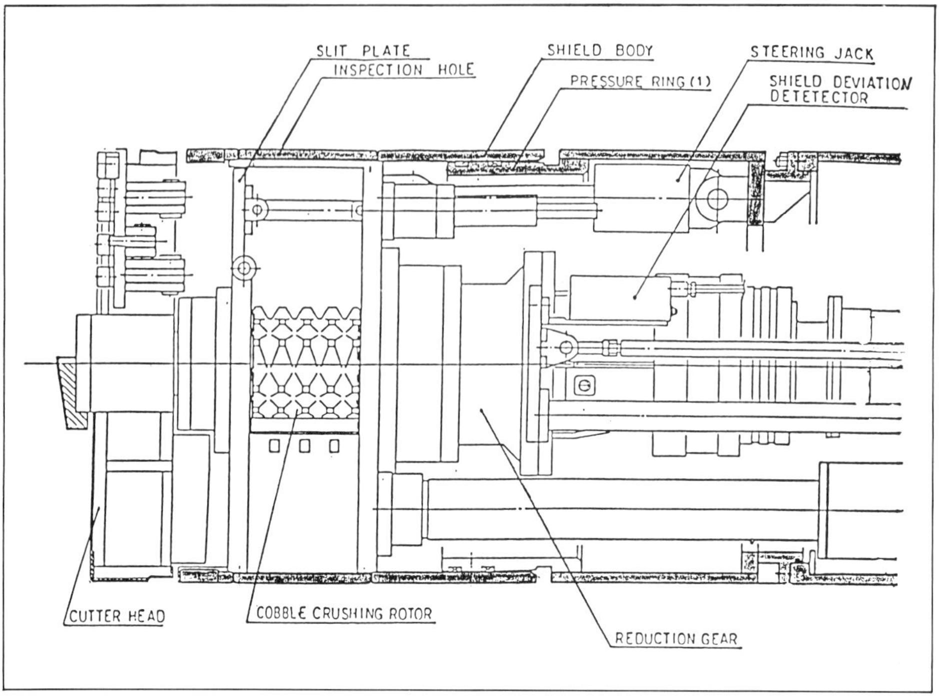

Figure 4.4.1. Detail of cutter head of an 860mm o.d. Crunchingmole showing rotary crusher.

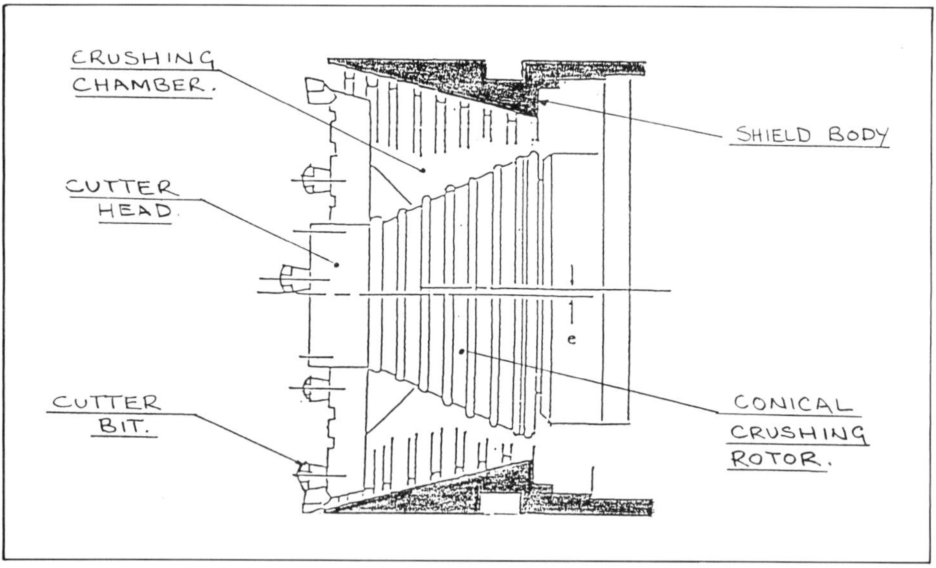

Figure 4.4.2. Detail of cutter head of a 370mm o.d. Unclemole showing eccentric crushing rotor.

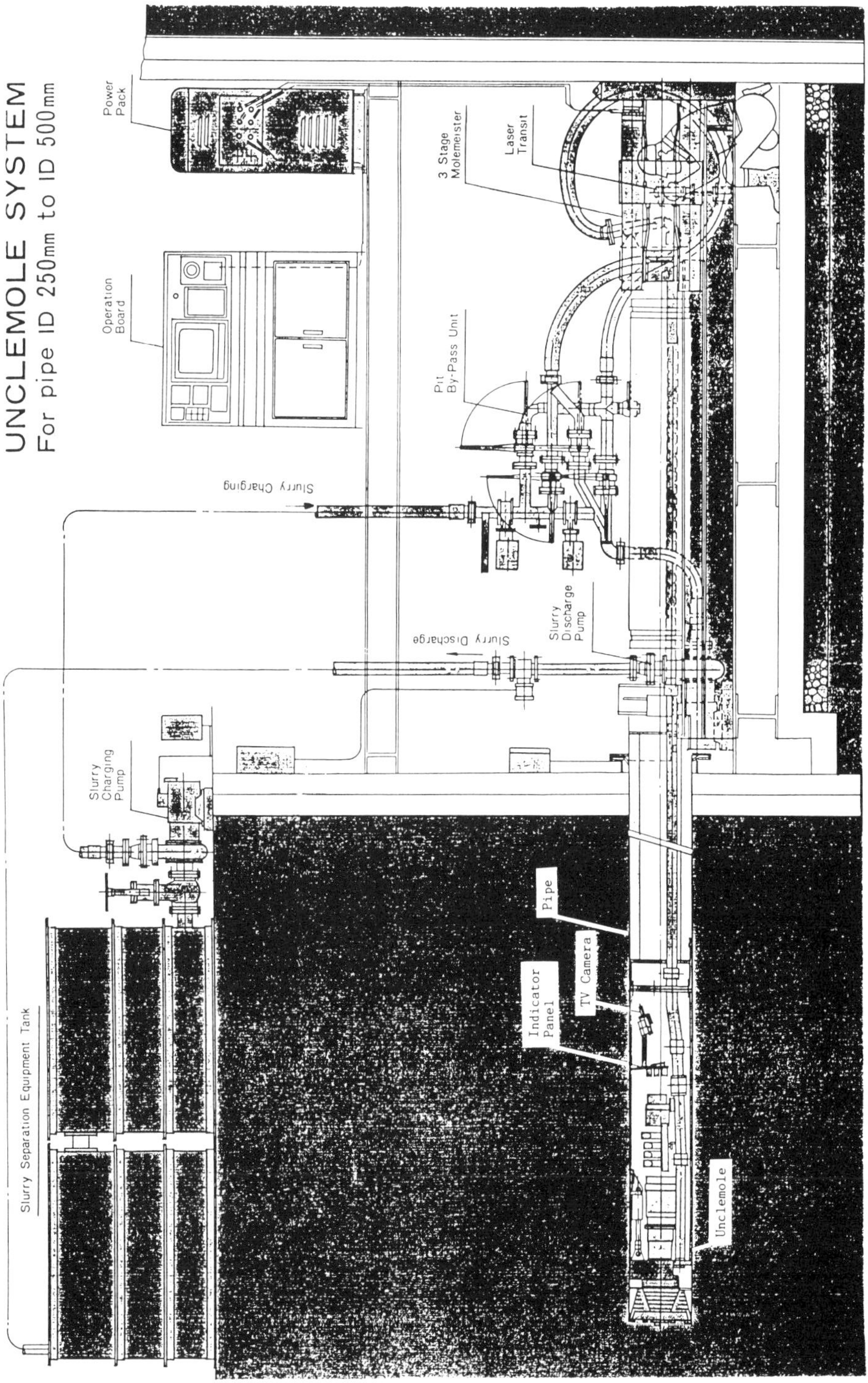

Figure 4.4.3. Operational layout for Unclemole system.

Authors' introductions

Paper 4.3 was introduced by Mr C Schröder, Chief Engineer, Department of Sewer Construction, City of Hamburg, on behalf of Mr E Kuntze, who was unable to attend.

Mr C Schröder: Microtunnelling in Germany

Since 1972 the tunnelling of larger sewers has made enormous progress in the German Federal Republic. Towards the end of the seventies the general interest turned more and more to processes suitable for the construction of small sewer pipe sections. The reason for this was the ever-growing demands made on the protection of the environment and the quality of life. Added to this there was a desire to reduce the otherwise unavoidable disruption of expensive road surfaces, the wish to avoid lowering the ground water level, and to minimize the risk of moving and/or damaging other pipelines for supply and disposal.

From Japan, for this scale, remote-control machines were known which made sufficiently accurate tunnelling of small sewer pipe sections possible, even though the distances were relatively short. The German Ministry for Research and Technology sponsored a project proposed by a Japanese/German team. This project was carried out in Hamburg at the end of 1980, and the result met with the general interest. The advantages were so obvious that up to now more than 15 firms have decided to develop their own solutions of the problem.

Every development in this field tries to meet the following requirements. First, during the driving the deflection from the axis must be less than 30mm in height and 100mm in width. Second, it must be possible to use the driving machine in different soils and under the influence of ground water pressure. It should reach a driving velocity of at least 12 metres per 8-hour day and be able to produce different pipe cross-sections. Third, the necessary investment for the acquisition of the machine and the devices that belong to it should not exceed £250,000. Fourth, the machine must be able not only to locate obstacles underground but also to remove or destroy them. Fifth, since the starting and destination pits are very expensive their number must be kept down by driving lengths of at least 150 metres. In order to further reduce the expenditure for the excavation of the pit the dimensions should be limited for the starting pit to a length and breadth of 4 × 3 metres and for the arriving pits to 3 × 3 metres. Sixth, the machine should make the driving of small-radius bends possible. Seventh, the production of the pipes *in situ* out of concrete or plastic is desirable. Eighth, the house sewer pipes should also be produced underground. The firms are obviously trying hard to meet those requirements.

Last, but not least, the economic aspect is very important. For the moment I can say that the new techniques compare favourably with open excavation only at a depth of over 4 metres. It must however be remembered that indirect costs have not been taken into consideration here.

Let me tell you now about some systems which have been tried out recently in West Germany. First there is the Telemole system, which Mr Rooke will describe in greater detail in the following paper. It is a Japanese drill-driving system adapted to German needs. With this machine more than 2500 metres of sewers with diameters of 500 and 600mm have been tunnelled. An excavation length of up to 180 metres was reached.

In order to drive small pipe sections into soil containing stones of more than 130mm in diameter a new machine based on the Telemole system was constructed. The machine is known as the Hansemole and is shown in Figure 4.3.1 in the paper. Its characteristics are the changed cutter head with a crunching rotor behind it which reduces stones of 130mm in diameter to a maximum diameter of 50mm.

A further development is the Dywidag articulated steel-pipe shield, which was also sponsored by the German Ministry of Research and Technology. The basic idea is that the driving and the installation of the product pipe are two separate operations. First, steel pipes of 2 metres in length and 800mm internal diameter are pushed into the soil. This makes it possible to reach the machine unit through the pipe in the case of a breakdown. If necessary any part of the machinery can be taken out and either exchanged or repaired.

Secondly, after this first operation has been concluded, the product pipes chosen are pushed in. The cutter head of the machine excavates every kind of soil, even in ground water, and overcomes even heavy boulders. Excavation lengths of 300 metres can be reached with this device. Tunnelling in a curve with a minimum radius of 100 metres is possible.

Contrary to the two methods described so far, the pipe-jacking system RUS 100A of the firm Soltau deliberately limits itself to shorter excavation lengths. The very complex system takes into consideration the usual distance between two manholes, which is from 50 to 80 metres in Germany. It can be used for pipe cross-sections of 250 to 400mm in diameter. Since about 75 per cent of the sewers which are to be built in the next few years have a cross-section of up to 400mm this machine can deal with most of the work. The length of the pipes is limited to 1 metre to keep the pits as small as possible.

Figure 4.3.2 shows a starting pit. Soltau pipe-driving units make it possible to dig excavation pits with a round cross-section of 2 metres in diameter. Since this machine is so compact a container measuring 6 × 2.5 metres can hold it. The container can be set up directly over the excavation pit (Figure 4.3.3), thus making the building site independent of weather and time of day.

In conclusion we can see that for Germany tunnelling of smaller sewers has reached an advanced stage of progress. The future belongs to the tunnelling system.

Mr A Rooke: Development of the Iseki Telemole and Telemouse and their application to trenchless pipelaying

Iseki Poly-Tech are manufactures of slurry tunnelling machines from 250mm to 3000mm, but the paper I have presented outlines basically the market pressures and the development of the smaller machines, the so-called Telemole for pipes between 730–960mm outside diameter, and the Telemouse for pipes between 480–650mm o.d. The paper goes on to describe the further development of the Crunchingmole which has been especially designed, as Mr Schröder said, to handle cobbles and boulders by means of an internal crushing mechanism.

What I will concentrate on here today is a new machine. The evolutionary cycle of NO-DIG equipment being what it is, since I was asked to write this paper I regret to tell you the Telemouse is almost extinct; in fact, it has been overcome by a stronger beast, with more powerful teeth, happy to live in almost any soil, sand, gravel, clay, mudstone, cobbles, all sorts of soft rock in fact. It does not need bentonite. It is quite happy to wash down that fodder with a plain glass of water. The new species is known as the Unclemole. It is presently used to install pipes from 250 to 500mm internal diameter, cheaply and accurately, and presently in spans of up to 100 metres. This is because for most sewers of 500mm diameter — and sewers are our main area of concern — manholes are normally at some lesser spacing than that. Details of the cost of operation of the Unclemole are given in the Appendix to the paper.

[Before discussing the Unclemole, Mr Rooke showed some slides of a typical Telemole as used in Hamburg, and the Telemouse constructing a large pedestrian subway in Singapore.]

The latter job was done in very bad water-bearing ground, soft peaty clay with only three or four metres cover below a six-lane highway into the centre of Singapore, and was completed without any settlement or disturbance to the surface whatsoever.

Figure 4.4.3 in the paper is a diagrammatic representation of the Unclemole, showing the component parts, the boring machine on the left, the triple-stage jacking system behind it, the control panel and the power pack.

The triple-stage jacks can expand both backwards and forewards, so that from the very small beginnings they give a large extension, enabling us to operate from a shaft length of about 4 metres, using a 2-metre pipe.

It can be seen from Figure 4.4.2 that the conical crushing rotor is not in the middle of the shield. The reason is that it rotates eccentrically and in the eccentric rotation crushes whatever would enter in there. The result is that the material finishes up as a paste at the back. This is why we do not need bentonite. The material becomes self-compressed until it drops into the slurry chamber at the back of the shield. Only a simple plant is required to separate the crushed material, since the volume of flow is low, and of course the water separation is much easier than bentonite.

We use a simple laser transit for alignment. Because we don't pull the muck out by skips or any method like that we are able to locate the theodolite between the jacks and it is not likely to come to any damage or get knocked or disturbed.

The operator works from a remote control board very similar to the other Iseki machines, with a TV monitor registering the functions of the machine. Along the bottom we have all the operation switches for controlling slurry speed, machine attitude, alignment, and such things, and the monitor relays the result of those operations.

In order to test the capabilities of the Unclemole, it was first placed in an upright position and fed with large cobbles — a successful test which produced fine broken stone without stopping the machine for one minute. The Unclemole was then put to work in a prepared trench using selected backfill material.

For the first trial the whole length was filled with sand and gravel with water and some very large cobbles in order to test the capabilities of the machine in the ground for which it was primarily designed. Subsequent tests were done on the same test site but using a variety of backfill materials including clay and silt. Also, there were some broken bricks and stones and probably the odd bike or two.

The machine enables us to install a robust concrete pipe with a loose steel collar, or a more special pipe with a clay lining. The clay pipe is made first, then the concrete is cast onto the outside to give a good corrosion-resistant pipe with the jacking capability of a concrete outside. In order to monitor the performance of the pipe, strain gauges were used extensively.

After driving the machine for 100 metres in the prepared trench, the resulting pipeline was re-excavated and found to be very accurate in line and level.

I would like to sum up with a few words on costs and would first emphasize that the equipment, with a track record for installing pipes by trenchless means, is now available. The biggest factor in reducing costs will be to use the equipment that is available and in using it develop it further to reduce the cost even more by further sophistication of the methods, much the same as we went from Orville Wright to the 747. If you do not use it you will not improve it.

I was interested to see the representatives from British Telecom taking a very active part in the discussion, but I was disturbed to hear them say that when the costs are right we will use the system. It is down to them for their own benefit to use the system. Use 10 per cent of their 99 per cent open cut, just 10 per cent, by trenchless systems, and you would see that the costs would drop remarkably. It is for their own benefit to proceed with this type of work. Like the old village Post Office, if you do not use it you are likely to lose it.

Discussion on papers 4.3 and 4.4

P Lofthouse (Southern Water Authority): The developments we have heard about this afternoon and in previous papers are really most impressive. There is now a whole range of equipment which enables us to lay, by trenchless method, sewers, water mains and gas mains from the very smallest to the very largest sizes.

One of the areas of discussion which Mr Rooke left with us is the cost, and, at the end of the day, none of the methods will get anywhere unless the costs are right. Some of the processes have been shown to be very competitive with open excavation, and once they are really competitive there is no question which form of construction we as clients and engineers are going to adopt. Social costs are only brought into the equation because of *lack* of competitiveness.

I fully support Mr Rooke's reference to the client's involvement in development. A previous paper showed how Southern Water is involved with the pipe-bursting technique. But there is a limit. We must know that, whatever commitment or investment we as clients are prepared to undertake, we are going to get the right return. In my role as Chairman of the National Sewers and Water Mains Committee, I have just presented a paper to the next meeting of the Water Authority chief executives asking for their support in providing more development sites throughout the country. Not miles and miles of sewers, but reasonable manhole lengths where new and developing systems can be tried out to enable the water industry to develop the sort of confidence in trenchless pipelaying which is necessary, and which ought to grow from this Conference.

I agree with Mr Rooke that the client and engineer must provide the best information about soil conditions, and that the equipment needs to be chosen with the specific ground conditions in mind. But I do not agree with Mr Rooke's suggestion that the client authority should as one option select the machine most suitable for the conditions. If the contractor is given the right information it is up to him to tender on the basis of that information. The client's responsibility is to see that the tender documents convey the correct information to the contractor.

Dr Kuntze in his paper, and Mr Schröder, who presented it, avoided details of house connections to sewers, but, to the water industry client, these are an essential part of the system. I should be glad if Mr Schröder could give us some indication of how they tackle that particular aspect in Germany.

Mr Rooke has referred to the great advantages of Unclemole for on-line replacement and oversizing pipes. Does he think it will have a competitive edge over the apparently much simpler method of pipe bursting, or is it too sophisticated? From the client's point of view, the more competition we have between different systems in the same area, the better.

Dr D Stein (Associate Professor, Ruhr-Universität, Bochum): I can add some more details on the underground construction method for small-diameter lines which we call "microtunnelling" in Germany. In developing this method we had two aims: development of the necessary machinery for remote-control pipe jacking, and finding a suitable pipe material for pipe jacking.

Practically every company in West Germany specializing in boring technology is now in a position to offer the necessary equipment, from the shield to the auger method. Most of this machinery is suitable for use below the groundwater table. If requested, the machinery can be equipped with automatic steering based on an active laser target.

As is well known, the laser system can have problems, particularly on larger jacking lengths. Therefore, in Hamburg we have used, for the first time, a gyro system, including the necessary software, installed in a Telemole machine. The results obtained from a test run over 90 metres were excellent. The method overcame the problem of deviation of the laser beam due to deformation of the point of view of the laser in the starting pit, and, in the end phase of the driving, we were able to run the system even without the laser.

Our aim will be to improve this technology, especially in developing smaller gyro units. This will mainly improve long-distance jacking and automatic steering, as well as driving in vertical or horizontal curves.

With regard to pipes for pipe jacking, we have concentrated from the start on the pipe material being corrosion-resistant. We are proud to have initiated the use of jacking pipes made from GRP and, especially, vitrified clay. According to James Thomson, 70 per cent of sewers laid in Great Britain are below 300mm in diameter, and are mainly vitrified clay. So, even when microtunnelling methods are requested, the normal practice can be maintained.

In spite of the high standard of technology in this field, the development has certainly not yet reached its peak. Further developments will be necessary. First, in improved pipe joints, to reduce the cost of pipes; second, in standardizing the outer diameters of the various pipe materials in order to get uniform specifications for microtunnelling equipment; third, in the matter of pipe lengths. At present, pipe lengths vary between 1 and 3 metres. It is not yet clear whether short pipes with a total jacking length of approximately 80 metres from a rather small starting pit will be more economical than jacking long pipes over distances up to 100 metres, using a large starting pit.

For the promotion of microtunnelling, it is also essential that the relevant authorities change their practice in designing, planning and specifying. Microtunnelling must be allowed as an alternative to open trench, and it must be shown that the total lengths on one site are big enough to justify the rather high investment in the microtunnelling equipment. Furthermore, it is important for codes of practice to take account of these developments, especially in respect of house connections.

All these changes will have a positive effect on inspection, maintenance and rehabilitation. Finally, I come back to the words of Donald Rees, who said, "Certainly there is a NO-DIG future". We feel that in Germany, the NO-DIG future has already started.

J C Thomson (Consultant): Would Mr Rooke say something about the speed of operation of the Unclemole in different kinds of ground? Does it slow down when it is crunching? What happens in higher water-tables with a crunching mole?

I would also be interested to hear some comments on the experiments in Hamburg, where the authorities gave the contractors who had the courage to go into the work with microtunnelling machines a guaranteed work load, at not a very high price (I understand it was equivalent to the open-cut price) over a period of two years. This ensured that the contractors would get some return on their machine, even if they were not going to make their fortunes.

Dr P M Zollman (Managing Director, ZED Instruments): I thought it was an accepted fact that the gyro guidance experiment in Hamburg was not very successful, but if any new information is available I would be interested to learn about the size and cost of the equipment, and the accuracy obtained at the end of the relatively short drive of 90 metres.

B Wilkinson (Principal Engineer (Sewerage), Yorkshire Water Authority): I am very pleased to learn that Iseki Poly-Tech are now making their machines available in this country. There is a great potential, because of the lengths of sewers in the size range 225 to 750mm laid by Water Authorities each year, and because of the savings in reinstatement costs and service diversions. Can Mr Rooke give us more details of how the machines will be made available? Will Iseki themselves provide a contracting service, will they license the process to a particular contractor, or will the machines be available to any established pipelaying and tunnelling contractor?

On first sight the system seems to be vulnerable to failure of the TV camera which views the instrumentation at the face. Has this proved to be a problem?

Mr C Schröder: Mr Lofthouse asked about house sewer connections in Germany. We are very eager to have a system of making house connections underground, and some firms and institutions are working on this problem. In a demonstration in Hamburg we built some underground house connections on a 30-metre length of sewer, quite successfully. First we pushed steel pipes into the soil, and then the product pipes. Three or four systems are being developed in Germany, and I think one will be ready within a year.

Mr Thomson referred to the guaranteed building sites in Hamburg. The Ministry for Research and Technology sponsored a project in which the City of Hamburg guaranteed 2000 metres of sewer construction and paid the equivalent price for conventional construction, the Ministry making up the balance.

Mr A Rooke: I do not entirely agree with Mr Lofthouse's contention that the contractor should be responsible for recommending the machine to be used. I had in mind something similar to the Hamburg example, where the client decided on the machine that would be used and based a contract on a fixed amount of work over a specified period (in the Hamburg case 2000m over 2 years). The effect this has on the cost is fantastic — it can halve the cost of the machine element to have that sort of guarantee. The original Hamburg machine has now completed over 3½km of successful bad ground pipelaying.

Oversize replacement by Unclemole is certainly a possibility, and we are looking at it very seriously. It has some advantages over moling, particularly where the existing line is in poor condition, surrounded by voids or loose material, or is constructed to an incorrect level or line. However, it has some disadvantages — laterals, for example. We have not yet tackled that problem.

In Germany there is a strong move now for *in situ* lining, but I think that, considering the quality of the pipes used for pipe jacking — pre-made and quality-controlled — *in situ* lining is the wrong choice.

In answer to Mr Thomson, a remarkable thing about the Unclemole is that it performs better in clay than in sand or gravel. We have achieved a normal production of 70 to 100mm a minute in clay, but the rate would drop off in sands and heavy gravel to perhaps 50mm a minute. The water level is balanced, as in the old machines, by adjusting the internal water pressure in the machine. The reason we can do away with bentonite is that, as the material feeds into the machine it passes into an inverted cone shape which causes compression and gives the sort of plug effect used in hand-excavated pipe jacking.

At present, the availability of Iseki equipment is by rental, including a competent operator. If a contractor has achieved his own competence, we might consider renting a machine without an operator, but this would be exceptional. We would also consider any realistic contractual arrangement suggested by a client, particularly a prime user, such as a water authority or a telecommunications authority, on the lines of the Hamburg plan, for instance.

It is true, as Mr Wilkinson pointed out, that TV cameras do sometimes go wrong. Iseki has done over 600km of pipe jacking using balanced-head tunnelling machines, many of which have had TV-controlled systems. I am not aware of any failures of a TV camera so far, though it may have happened once or twice.

P J Coller (Iseki Poly-Tech Inc.): Unlike the pipe-bursting technique, Unclemole is not limited to replacing a pipeline by another of the same diameter. It can chew up existing pipes and transport them out of the way, and replace them with much bigger pipes if necessary.

H W Garton (Civil Engineer (Sewerage), Isle of Man Local Government Board): I come from a location where many of our sewers are in narrow streets, with services very close together and pipe sizes typically in the range of 150 to 300mm. We have to protect adjacent property. The real problem is the house connection, and I am heartened by Mr Schröder's assurance that within about twelve months that problem will be solved in Germany. However, I am disturbed to hear Mr Rooke say that his company is not investigating the problem. If any company can present me with a system that will work and will enable me to make house connections economically, I will take it on board tomorrow.

Mr Rooke: The market in Japan is primarily for construction of new sewers rather than on-line replacement. Iseki, and others, are looking at NO-DIG methods of making connections to newly-constructed pipes, but, at the moment, there is not the same need for replacement of existing connections.

SESSION 5. Chairman: R. Tharby, Chairman, National Joint Utilities Group, UK

5.1 Location of underground services by ground-probing radar

D J Daniels

Assistant Division Manager
British Gas Corporation Engineering Research Station

There is a clear financial benefit in using trenchless construction techniques for the installation of buried plant. In order to employ such techniques in the congested underground situations encountered in urban areas it is necessary to 'see' through the ground to delineate volumes clear of obstacles.

The paper describes the work carried out by British Gas Engineering Research Station to develop a ground-probing radar. Although the project is not yet complete the current results are most encouraging. An experimental prototype radar has detected PE pipes buried at a depth of 0.7m in clay, one of the worst-case ground conditions. The same radar is also capable of detecting a 25mm diameter PE pipe at a depth of 0.5m in crushed dolomite backfill material.

The paper describes the principles of ground-probing radar and the engineering problems encountered in the development, and comments on the work necessary to develop the technique to a stage whereby a field equipment can be produced commercially. Although the project has concentrated on the detection of buried pipes and cables, many of the design techniques could be employed to provide information on other parameters of the ground such as relative density, voiding etc.

INTRODUCTION

This paper is a review of the main radar techniques for use in the location or positional recording of buried pipe and cables used for the distribution of gas, water, electricity and telecommunications.

The British Gas Corporation already uses conventional locators to detect the position of buried metallic plant. Due to its replacement programme the proportion of non-metallic mains and services is gradually increasing and a locator to detect all types of buried plant will be useful.

By 1982-83 a total of 77,000km of polyethylene (PE) pipe was in use for the distribution of gas. The total distribution network consists of 213,000km of mains and 171,000km of services. Whilst in 1982 only 13 per cent of the mains distribution network consisted of PE, by 2000 PE will constitute 40 per cent of the mains network. By the end of the century 80 per cent of British Gas customers will be supplied via PE services compared with 29 per cent in 1982. In terms of depth of burial there is no difference in laying practice between PE and metallic pipes.

At present the majority of PE systems do not have any tracer system buried with the pipe and the position of the pipe is recorded in two ways. If the pipe is a service pipe the job record is referenced by the excavation gang. Depending on the standard of the recorded information, the position of the service pipe may be known to within limits of ± 1m. If the pipe is a main the position is recorded by surveying techniques with an inherent accuracy in the order of centimetres. However, as the datum points are in many cases likely to be lost as a result of rebuilding, urban renewal or road building schemes, gradual errors in known position may be accumulated with the passage of time.

It is important that the gas industry has access to an equipment capable of detecting plastic pipe, as considerable economic benefits can be achieved by using such an equipment. The economics of operations concerning gas mains and services are fundamentally based on the costs of excavation and reinstatement.

In the case of PE pipe the justification for incurring costs over and above the base costs for installation lies in the potential for reduction of excessive excavation costs. These costs can be caused by:

(a) Excessive excavation costs associated with small holes.
(b) Excessive excavation costs resulting from inability to use mechanized excavation techniques.
(c) Excessive excavation costs resulting from abandoned trenches and trial holes.

There is a clear financial benefit from using an equipment for locating plastic pipe primarily as an 'eye' for seeing through the ground before mechanized excavation techniques, ie., narrow trenching, thrust boring, etc. are undertaken.

The possibility of being able to 'look' through visually opaque materials has stimulated the efforts of engineers and scientists over many years to design a practical detection system. Some experimental systems have been built and have demonstrated that buried targets can be detected using radar techniques. However, there is no readily available suitable commercial system and for this reason the British Gas Corporation Engineering Research Station (ERS) has started work on developing a ground-probing radar.

The ground-probing radar project has now been in progress at ERS for approximately three years. During this time the project team has concentrated its efforts to gain a basic understanding of the problem and has built several experimental prototype systems.

A major problem has been the lack of adequate data on the electrical characteristics of the soil in the UK. This is an important input to the design parameters. A soil measurement programme has therefore been carried out in order to fill this gap in our knowledge and this has been completed.

It is now planned that the emphasis of the project will progress from a feasibility assessment to the implementation of practical solutions, particularly in relation to recovery of the target signal. It should be noted that an experiment prototype radar system has been shown to be capable of detecting a plastic pipe buried at a depth of 0.6–0.7m whilst the detection of metal pipe appears to be easier.

It is appropriate to consider the principles of a ground-probing radar in order to gain an understanding of the problem.

PRINCIPLES OF RADAR

Several techniques, e.g., acoustic, have been used in the past for the detection of buried objects, but the problems involved have usually been so overwhelming that only the electromagnetic field approach has been shown to be successful.

Essentially the technique is very similar to that of a conventional radar with three very important differences. The first is that whereas a conventional radar operates over ranges of kilometres, a ground-probing radar operates at ranges of a few metres. The second is that the resolution of a conventional radar can be tens of metres whereas a ground-probing radar requires a resolution of some tens of centimetres. The final and most important difference is that a conventional radar transmits through the air, which usually causes little attenuation, whereas a ground-probing radar uses the soil as a transmission medium, the attenuation of which is very high.

The principle of operation of a radar is as follows, and is shown in Figure 5.1.1. A carrier signal in the microwave frequency region is generated and then modulated in order that a timing marker can be applied to 'packets' of transmitted energy. The modulated waveform is then launched into the transmission medium by means of an antenna. Reflected energy from the target is gathered by means of a receiving antenna (usually the same as the transmitting antenna). The receiver processes the information by demodulating, filtering and displaying in a suitable form for the operator. A conventional radar is normally static and the target moves, whereas the ground-probing radar is mobile and the targets stationary.

For an instrument for the detection of buried services it will be essential to provide good resolution in plan and depth.

The plan resolution is defined by the characteristics of the antenna. To achieve an acceptable plan resolution requires a high-gain antenna. This necessitates antenna dimensions greater than the wavelength of the lowest frequency transmitted. This, in turn, means the requirements of portability and plan resolution have to be carefully balanced.

In essence, to achieve small antenna dimensions requires the use of a high carrier frequency which unfortunately does not penetrate the ground to sufficient depth. Hence the most important part of the current development is to correctly determine the exact trade-off between plan resolution, size of antenna and ability to penetrate the ground, which is determined by the carrier frequency.

To give an idea of the frequency of operation the chart shown in Figure 5.1.2 is useful. The wavelength of operation at 500MHz is 0.6m in free space, but whilst travelling in the ground the velocity of propagation decreases and the apparent wavelength decreases proportionally. This helps in that the effective resolution in depth in improved in ground compared with air.

To achieve the required depth resolution it is necessary to employ a timing marker on the carrier signal. The bandwidth of the timing marker defines the resolution in depth and for the ground-probing radar application a modulation bandwidth of greater than 300MHz is needed.

In conventional radar design it is normal to use a carrier frequency that is significantly higher than the maximum frequency of the modulation. This means that the percentage carrier frequency deviation is small. However, in the ground-probing case, soil attenuation in so large at high carrier frequencies that penetration of the ground is limited to a few centimetres.

Hence the design is constrained in that the carrier frequency is in the region of 300MHz and the modulation bandwidth is of the order of 300MHz. These requirements cause design difficulties in that the ground-probing radar requires a percentage bandwidth of approaching 100 per cent compared with a conventional radar percentage bandwidth of 5 per cent.

The most important factor governing the performance of an instrument is depth penetration. The depth penetration of the radar is dependent on the following parameters:

1. Transmitted power;
2. Antenna gain;
3. Ground attenuation;
4. Target-scattered power;
5. Receiver noise;
6. Clutter.

The transmitted power is entirely within the designer's control but safety factors, battery consumption and licensing requirement will have to be considered. In practice an average transmitted power of between 0.1 and 1W would appear realistic.

The antenna gain is a quantity microwave engineers use in providing a reference against an omnidirectional antenna. Ideally, a narrow beam would be used but the physics of the situation require an effective aperture of many wavelengths to achieve a narrow beam. At 0.6m wavelengths, it is obvious that a large aperture antenna is not feasible. However, it is hoped that a gain of 10 may be realizable.

The ground attenuation is perhaps the single most important factor governing depth penetration. The attenuation of an EM wave passing through water-loaded soil is considered to be governed by the complex electrical permittivity of the soil. The attenuation factor of the soil is to a first order defined by the ratio of e''/e', otherwise known as $\tan \delta$. The physical explanation for the attenuation of water is due to the electrical dipole moment of the water molecule. The dipole means that the water molecule is polarized and will orientate itself in an externally applied electric field. In an electromagnetic or radar wave the electric and magnetic fields are time-varying. When this time-varying electric field is imposed upon the water molecule, the water molecule continuously moves to reorientate itself in the changing electric field. The continual reorientation of the water molecule dissipates the energy in the form of heat due to friction. Therefore, any water within the soil serves as a coupling mechanism that converts the energy in the electromagnetic wave into heat in the soil. As the wave travels through the wet soil the wave is severely attenuated due to this water dipole absorbing mechanism.

To provide a perspective into the magnitude of the problem, a comparison of path losses for various situations is given in Table 5.1.1. All losses are quoted for return journeys and ratios are given in brackets.

TABLE 5.1.1 Signal attenuation

Situation	Path loss (dB) (return journey)	Ratio
1m pure water[c]	8	$1{:}10^{0.8}$
1m dry sand[c]	30	$1{:}10^{3}$
1m saturated clay[c]	200	$1{:}10^{20}$
Moon from earth	250	$1{:}10^{25}$
Venus from earth	320	$1{:}10^{32}$
Sun from earth	295	$1{:}10^{29.5}$
1m sea water[c]	6530	$1{:}10^{653}$

[c] = calculated value at 1 GHz

The fundamental limitation to signal recovery is the dynamic range of the complete equipment. The dynamic range can be defined as the ratio of the transmitted signal to the minimum detectable signal. The latter is fundamentally defined by the noise level at the input of the receiver caused by the movement of electrons.

In practice about 120dB would appear practical and the effect on actual range performance can be seen in Figure 5.1.3.

From Figure 5.1.3 it can be seen that there are fixed system losses which cannot be reduced. These losses are composed of the effective target scattering area, the antenna gain, the transmission losses from the air into the ground and vice versa, the reflection loss from the target and radar system loss. Added to these losses are the calculated ground attenuation losses for two cases, dry sand and wet clay. It can be quickly seen that saturated wet clay could seriously limit the effect range of a ground-probing radar. However, the calculated value may be pessimistic as practical results have shown that buried non-metallic pipes can be detected in clay subsoils.

DISCUSSION OF RADAR DESIGN

The basic design relationships involved in a ground-probing radar are shown in Figure 5.1.4.

The most fundamental element is the target characteristic; pipes and cables are linear features, hence this can be exploited in both the design of the antenna and the signal processing. The interrelationship of these elements is complex and beyond the scope of this paper but it has occupied considerable effort to fully understand the design trade-offs.

The major problems associated with the design of the radar are shown in Figure 5.1.5 and the work carried out by ERS has concentrated on solving these.

A major consideration is whether it is desirable to design an antenna which is ground-contacting. From the radar design point of view, this approach has advantages in that it reduces the significance of the air/ground interface and maximizes the transfer of energy into the ground. However, this would place a severe limitation on operational speed and therefore a non-contacting antenna with its attendant difficulties has been designed and constructed.

In addition to the design work carried out on antennas, microwave systems have also been built for signal generation and receiving.

Signal-processing techniques have been, and are, under investigation and software has been developed to implement some of the techniques. This is an important area of work since system-generated clutter is a major limitation on radar performance, and signal-processing techniques which reduce its effect are necessary. At an early stage in the project it was appreciated that powerful signal-processing techniques would be needed to provide a clear presentation of the area of ground of interest.

The signal processing falls into two categories:

(a) Improvement of the basic return signal;
(b) Image-processing of a set of returns.

The first category involves the use of advanced mathematical processes associated with deconvolution of the wanted signal from system response. At present these are implemented on mainframe computers, but in a production radar it is envisaged that dedicated processors will carry out these functions.

The basic signal return from a plastic pipe is as shown in Figure 5.1.6.

Figure 5.1.6 represents a measurement by the radar at one particular plan position. An array of results over an area can be combined to produce a map giving the intensity of signal strength, as shown in Figure 5.1.7.

Further work is planned to improve the clarity of the images, possibly using colour displays.

OPERATIONAL USE

With conventional inductive locators, which can detect metal pipes only, the issue is a simple one. These instruments generally cost around £500 per unit and are thus cheap enough to be made available to all persons who require them. They are also easily hand-portable since there are physical restrictions on their size.

However, with radar equipment the position is different. Microwave systems involve precision engineering and the unit cost of a system will be at least an order of magnitude greater than the inductive type instruments.

The current generation of ground-probing radars suffer from several operational weaknesses.

(1) Inadequate resolution for pipe and cable location;
(2) Inadequate performance in very lossy materials, i.e. wet clay;
(3) Unwieldy antenna operation;
(4) Difficult interpretation of data.

Since we do not think that it will be possible to judge whether or not a pipe is present on the evidence of a single measurement, a mapping system will be necessary and may be a distinct advantage. The implication of this is that a more sophisticated data-gathering and signal-processing system will be required, preferably with the capacity to provide results on site. Such a system would probably be mounted in a vehicle with perhaps only the antenna together with microwave generator and first-stage signal-processing being portable. This would naturally lead to a link with digital mapping techniques.

It will be assumed that a non-contacting antenna system can be designed and that therefore scanning the antenna over the ground is a matter of designing the appropriate mechanical system. The choice here would probably be either a trolley-mounted system or a boom-mounted system operated from a vehicle. In either case, a vehicle would be necessary and the information gathered by the antenna and microwave system would be processed by computing equipment installed in the vehicle.

The rate of mapping is governed by the maximum speed at which the microwave signal can be converted to audio frequencies and changed into digital form suitable for computer storage. In addition, before a map can be produced these raw data will be manipulated by the signal-processing algorithms to remove the system-generated clutter. It is the latter manipulation of data which will probably take the longest time but this need not prevent the collection of data.

With the limited knowledge of the capabilities of the final system available at present any estimate of mapping rates is, at best, an educated guess. However, the following figure is based upon the time taken to convert the microwave signal and change it into digital form and store it. For one particular measurement this process will probably take approximately 500ms. (This includes the time taken to store the data on hard disc).

A further consideration to be taken into account is the spatial density of the samples. Since the spatial resolution of the radar will be of the order of 20cm, there is no point in taking samples which are spaced closer. Therefore the sampling density is $25/m^2$. Therefore the estimated time for data gathering would be $12.5\ s/m^2$.

The time required to process the data gathered to form a readable map is more difficult to estimate since it is dependent upon the capability of the computing hardware. If a map was required on site in a reasonable space of time, say 10 min, and the area surveyed was say $10 \times 25m$, then the total number of measurements would be 6250. This would require the computer to perform approximately 64×10^6 multiplications, or one every 10^{-5}s. The present generation of 8-bit microcomputers are at least an order of magnitude slower than this, but this speed might be possible using 32-bit machines and special purpose hardware.

For a ground-mapping system, the radar system, including computing equipment for signal processing, would probably cost in excess of £20,000. Also, if a ground-mapping system is chosen then some reliable means of establishing datum positions on the ground would need to be established.

CONCLUSIONS

The design of a ground-probing radar equipment can be seen to be a specialized adaptation of a conventional radar design. To scan successfully a volume of ground and reject false targets will require significant data storage capacity. Obviously when such an instrument is developed it would have to be fully utilized to achieve economic use.

Figure 5.1.8 shows an artist's impression of one realization of a ground probing radar. The present state of development is shown in Figure 5.1.9.

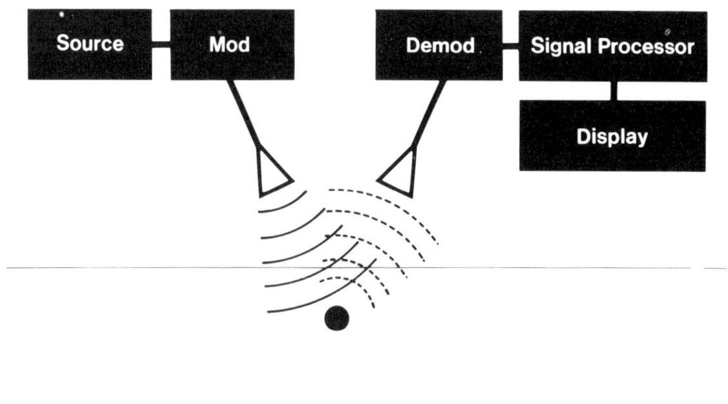

Figure 5.1.1. Generalized block diagram

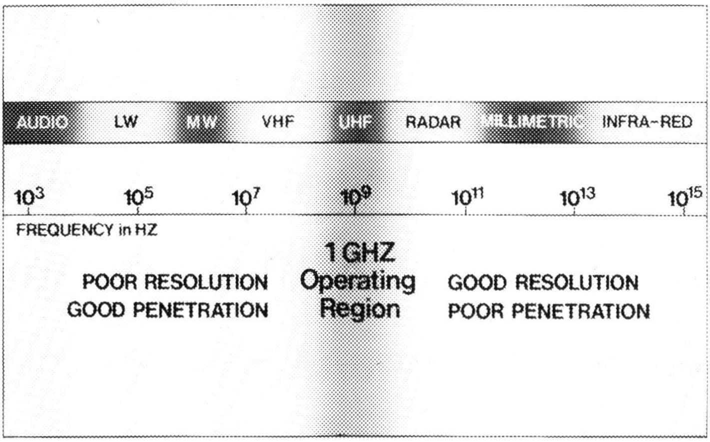

Figure 5.1.2. Operating frequency

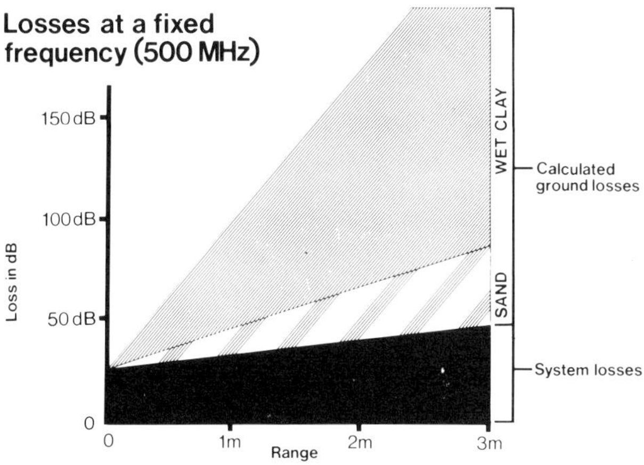

Figure 5.1.3. Losses at a fixed frequency

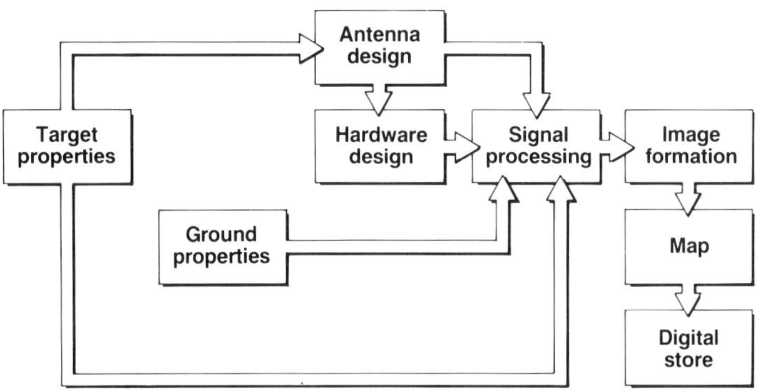

Figure 5.1.4. Design relationships

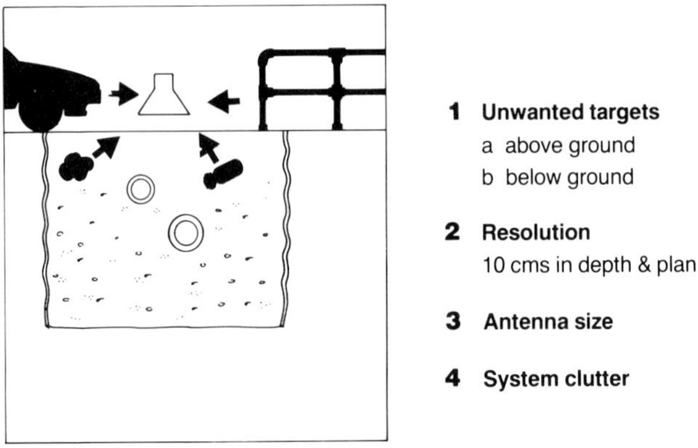

Figure 5.1.5. Design headaches

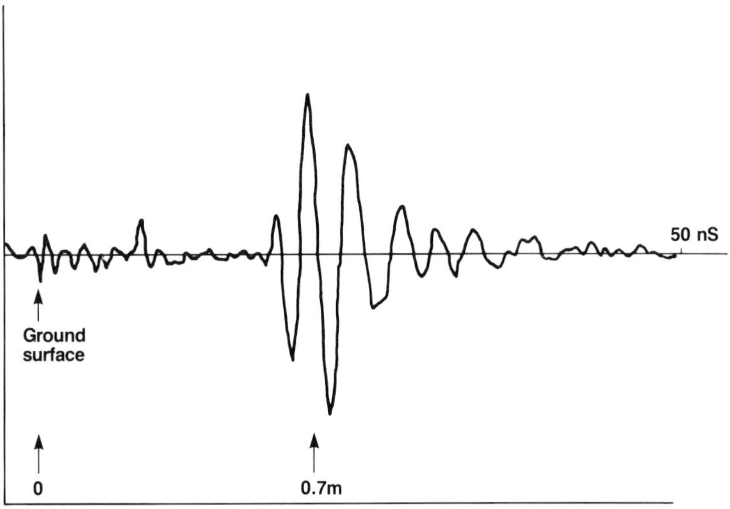

Figure 5.1.6. Signal return from plastic pipe

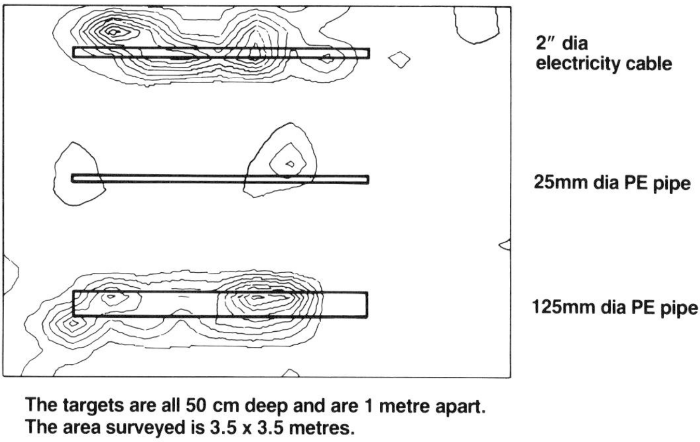

Figure 5.1.7. Plan image of three buried objects

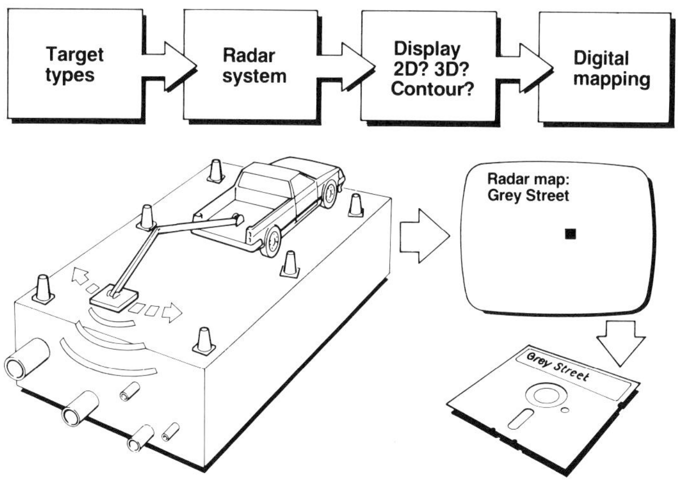

Figure 5.1.8. Artist's impression of radar

Figure 5.1.9. ERS prototype radar

5.2 Location of underground services by alternating magnetic field technology

A F Garnett *Managing Director, Electrolocation Ltd*

Knowledge is the key to effective operation in any field. Using NO-DIG techniques to repair or extend buried utility lines or digging down to them is no exception. And knowledge is difficult to acquire because of the metre or more of intervening soil.

The same techniques, and so to some extent the same equipment, used for locating buried conductors before digging, or for defining what lies in the path of new underground works, can also be used for monitoring and guiding many of the new generation of underground instruments, tools and machines.

Techniques for locating buried cables or conductors using alternating magnetic field technology have been known and used for the last seventy-five years (Figure 5.2.1). In 1931 Dr Gerhart Fisher founded a laboratory in California to research and exploit alternating magnetic field location. In 1937 he produced his Metallscope or M-scope and the first popular pipe locator was launched. The successor of this instrument is available today from the same firm.

There are a number of firms, particularly in this country, in Germany and in the United States, who specialize in developing this technology. My firm, Electrolocation Ltd. in Bristol, is one of them and we consider ourselves to be among the leaders in the field.

Three or four years ago we felt that we and other firms like us had exploited the technology to its limit. We thought we should explore other technologies for our company to prosper.

Today we know different. We are no longer just in the business of making pipe and cable locators, though that is still the most important part of our business. We are in the information business. Our technology measures, monitors, guides, locates and can even command through soil or rock.

Locating

Locating buried pipes and cables with a hand-held instrument from the surface is how the technology began. At the moment there is considerable interest in locating, its contribution to safe digging and to reducing excavation costs. There is also widespread misunderstanding about locating.

Reasons for locating

There are two reasons for locating buried lines. The first is because there is a need to dig and it is necessary to find out if there are any buried pipes or cables in the way. If buried lines are located the position of the trench may be changed or it will be necesary to dig with special care to avoid damage. Knowledge of the identity of the buried lines is not required; it is enough to know that something, some buried conductor is present and must be avoided. This is 'locating to avoid'.

Owners of buried plant have another reason for locating. They need to identify their line at a particular point. This means going to a point where the line can be identified, a valve box, hydrant or manhole, applying a captive signal to it and then tracing the signal to the point where the information is needed. This is 'locating to find' or mains tracing.

A utility locating to find a buried line and needing to dig to it in order to repair or extend it then needs to 'locate to avoid' so that the trench down to the line is placed to avoid other nearby lines or to give warning of the presence of other lines that cannot be avoided.

Most locating, even including a complete subsurface survey of an area, is a development of locating to find.

Different types of magnetic field locator

There are two different types of locator. The first is known as a *passive locator*. It locates lines by detecting existing or 'natural' signals on them. The best known signal is 50/60Hz energy on a loaded power cable. This energy transfers easily to other nearby buried lines, and in practice most of the conductors under the pavement can be located by detecting power energy signals. There are other 'natural' signals present on buried conductors. In particular, VLF radio transmitters use high energy ground waves that induce strong currents that seek a return path through ground where they encounter buried conductors that act as aerials (Figure 5.2.2) and re-radiate the energy. Some modern locators are designed to locate buried cables and metal pipes by detecting this re-radiated radio energy signal. At the moment the radio signal only exists in detectable strength on lines with sufficient length and cross-section. But this R-detection mode is a useful supplement to the 50/60Hz mode because it easily locates buried lines such as metal pipes, main telecom cables and buried HV lines which, because of the balanced load on their conductors, often do not emit a detectable 50/60Hz signal.

The other type of locator is an *active locator*. It comprises a transmitter which applies a signal to a pipe or cable by connection or induction. The receiver is tuned to the same frequency and is used to trace the pipe or cable. An active locator can also be used to search an area; the transmitter is carried across the area and beams its signal onto any buried conductors which are in turn located with the receiver carried a few paces away.

The uses of active or passive locators can be summarized:

Active: Trace and identify. Thorough search. Time-consuming to use and requires a definite skill.

Passive: Quick and effective sweep. Not suitable to trace or identify. Simple and quick to obtain results.

Passive locators are suitable for 'locating to avoid' and active locators are necessary for 'locating to find'.

Some manufacturers combine active and passive modes in a single set of equipment: the receiver by itself is used to 'locate to avoid' and the transmitter is added for 'locating to find'.

Locators for everyday locating and locators for the specialist

There are many problems in deciding on the design of a modern high-performance locator. The most difficult is to decide the balance of sophistication, performance and complexity against simplicity of use. My own company solves this problem by producing two separate ranges of equipment. One range, the CAT and Genny (Figure 5.2.3.) (originally called CAT for Cable Avoiding Tool and Genny for Signal Generator) is for the non-specialist and is designed for use by workmen. The other range, the Radiodetection 600 (Figure 5.2.4) is for the engineer or the specialist.

Both sets of equipment use the twin aerial system. They pick up the signal at two separate points on the antenna and use the circuitry to reject interference and provide a narrow response to the user. There is a different characteristic response from the circuits in the two instruments. The CAT adjusts itself automatically to its electromagnetic environment and is then silent until close to the signal from the buried line when it switches on and then switches off after crossing the signal; the user only fine tunes the gain control to obtain a peak response to pinpoint the objective. The RD600 leaves total control of the response to the user and also gives him three different ways of using the instrument's aerial configuration.

There are many other differences between the workman's locating tool and the technician's instrument, but perhaps the most important is the signal frequency. It is generally recognized that low frequencies are difficult to apply to a line, particularly by induction, but once applied they run for a long distance and tend not to couple to nearby unwanted lines. High frequencies are easily applied to lines, even high-resistance lines such as iron pipes with rubber sleeves, but attenuate quickly and couple all too easily to other nearby lines. With the CAT & Genny we compromised with a medium frequency of 22kHz and this has been found to be very suitable for locating in the street. With the problem-solving RD600, whose performance would often be stretched to the limit, it was clear there was no single frequency that would be suitable for the wide range of locating, searching, tracing, identification and fault finding. So the transmitter/receiver was designed with four different frequencies ranging from 1kHz at the bottom end to 130kHz at the top. The transmitter can not only be switched to apply each frequency separately, but it can also apply all the signal frequencies at the same time. This is an advance in the technology that will make locating in the most difficult areas more certain and quicker.

When our company's Design Team were discussing the performance specification of the RD600 there were several devil's advocates who said that the extent of the versatility and sensitivity of the RD would seldom be required and that customers would therefore not pay the price of including all the features. This view was overruled on the basis that once the specialist is called out he must be equipped as comprehensively as possible to solve any locating problem. Failure to locate usually means digging, the most expensive of all the alternatives.

How does the user choose the right locator for his needs?

There are several steps to take when choosing a locator. Define the work it will have to do; locating to avoid? locating to find? finding and avoiding? Decide the type of employee who will use it. Then ask the various suppliers to lend a set for a fifteen-day evaluation by the people who will be using it. At the end of that period make a choice on the basis of performance, ruggedness, suitability, simplicity of use and finally, and most important, its usefulness to your organization.

Locating non-metallic drains

The first step to making small self-contained signal transmitters able to transmit signals through soil from below the surface came with the introduction of instruments for tracing non-metallic sewers (Figure 5.2.6) to pinpoint blockages or collapses.

The technique is very simple and is now in use with local authorities throughout the country and municipalities around the world. The transmitter is a self-contained watertight container with a transmitting coil, circuit and dry-cell battery. It transmits at a frequency that passes through all types of soil but that does not couple to intervening pipes or cables or concrete reinforcing mesh. Several manufacturers produce these transmitters, known as sondes. They range from weighing as little as a few ounces with a range of 1.5m to large sondes detectable as deep as 25m. The receivers used on the surface to pinpoint the location of a sonde are often fitted with a pushbutton to give an estimated depth reading on the meter with accuracy at least down to 5m. The daily duty of being pushed down a drain on the end of rods or banging through the drain attached to a high-pressure water jet has meant that considerable effort has been spent making the sonde housing robust and the circuitry ultra-reliable for this inhospitable environment.

Other Applications of Sondes

Today a range of sondes of various shapes and sizes is available. And today firms and organizations are putting them to work in situations often far removed from sewers.

Example: A cable crew had been having problems winching cable into old ducts. They needed to know precisely where to dig to free the blockage. Solution; insert a small sonde inside the pulling sock giving instant information at the surface.

Example: Some years ago a Yorkshire Electricity supervisor found that soil-piercing tools or moles sometimes went astray with possible disastrous results. Solution; he attached a sonde to the feed hose just behind the tool; he then knew the position of the mole. He had the knowledge to retrieve it if it deviated or he knew exactly where to dig if it got stuck.

Example: PTT in suburban Paris check ducts for blockages by blasting a small sonde through with compressed air.

Example: There was an accident in a Breton granite quarry because the borehole holding the explosive charge instead of being parallel to the quarry face was bored at an angle towards it. The firing resulted in stone falling over the nearby village and highway. Solution; lower a sonde down the bore hole and measure the distance to it from the quarry face. A solution now adopted by the French Nobel Explosifs Company and offered to users of their explosives.

Sondes are becoming increasingly used for pinpointing water leaks in plastic water mains. The combination of plastic pipe buried in certain types of soil makes acoustic methods of leak detection ineffective. In the past exploratory digging was the only way to find leaks. Several water companies, notably the Sunderland and South Shields Water Company, used and developed the technique of inserting plastic foam swabs fitted with a sonde.

The swab acts as a piston in the line and by closing or just cracking the downstream valve the piston is propelled to the point of the leak (Figure 5.2.7) where it stops or where its rate of progress substantially decreases. The progress of the piston is followed from the surface and the position of the leak pinpointed.

This technique is used by one of the British Gas regions during inspection of new plastic pipe to detect kinks or other internal problems.

These are examples of using existing equipment to solve problems that would otherwise become expensive and troublesome.

Building the sonde into underground tools and systems

The next step has been entirely predictable. In the UK, the Pearpoint Company designed a completely new inspection camera for drains. It incorporates a number of new features; amongst them is the ability to pinpoint the position and depth of the camera from the surface while looking at the image it is producing. Pearpoint wanted their camera to be an integral unit and so the sonde was incorporated inside the housing; the circuit components have been added to an existing printed circuit board and the sonde element is powered from the same supply as the camera. The Pearpoint camera gives the user important additional information at only very marginal additional cost. This information will soon be desired from other types of underground inspection or monitoring equipment.

Soil-piercing tools or moles are getting increasingly enthusiastic acceptance. Their position can be pinpointed by attaching a sonde to the hose behind the tool (Figure 5.2.8). But clearly this is only a first step to the ideal. The second step of incorporating a sonde element in the tool has already been taken. Trials with a mole have shown that the sonde can be engineered to provide a spot signal source rather than using the length of the tool as an imprecise signal source and have shown that not only can the position of the sonde be pinpointed from the surface, but its direction and depth can be defined. The next step is to increase the sophistication of the receiver to give the inclination of the mole in addition to the other information.

Manufacturers of moles are only a short step from making a mole that can be steered from the surface. Magnetic field techniques will almost certainly be used to detect the position and depth of the obstacles in the path of the mole and will then provide the necessary information about the position and direction of the mole so that it can be steered safely through the obstacles. There are possible variations to this theme: a cable energized with a low frequency signal can be laid on the surface and the mole, perhaps two metres underground, triggered to follow the path of the cable. Something similar is already in operation and undergoing field trials: a small tractor (Figure 5.2.9) with a laser beam reflector follows the line of a cable to guide a tractor spraying fertilizer or pesticides over large areas of agricultural land.

The ability to transmit a signal from a precise point through soil has many possibilities that have not yet been exploited. For instance the technique can almost certainly be used to pinpoint laterals while lining sewers and it can be used to register spot points during tunnelling operations.

Other aspects of the technology also solve problems

Although it hardly falls within the scope of a NO-DIG meeting the technology of locating can be harnessed directly to machines that do dig. As far as we know this has only been done on an experimental basis on a large drainage ditching machine in Holland: a locator by the cutting wheel of the machine sounds an alarm just before crossing a buried cable. More work on directional antennas is required but there appears no reason why any machine digging, cutting or piercing soil should not be fitted with a device which will detect any conductor crossing its path.

Already magnetic field techniques are increasingly being used for monitoring faults in wrapping in transmission pipelines.

Magnetic pipeline wrap monitoring varies in many ways from the traditional Pearson system. It highlights low resistance faults, that is to say, faults where electrochemical reaction is taking place. It only needs one man to do the walk with the receiver, and ground penetration is not required. The survey can therefore be made over paved areas and over rocky soil.

Already some pipeline operators are drawing magnetic profiles of parts of their systems. In the future only spot checks at discrete points need be made to check any deterioration of the line wrap.

Electromagnetic technology has also been used and is likely to be developed to produce more positive results in detecting the presence of boulders and anomalies in clay and other soils.

Other location techniques

Other techniques already exist or are contemplated for extracting information from below the ground. High frequency radar systems, seismic reflection techniques, variations in static magnetic fields and of course dowsing, a technique that is known to have produced results for more than four millennia. Beside them alternating magnetic field techniques have one great disadvantage. They cannot locate or trace plastic pipes or other non-conductors unless a transmitter is inserted in the pipe and propelled along it. This is a greater disadvantage in the UK than in other developed countries, where tracer wires buried with plastic lines are often required by law and are almost universal.

However, alternating magnetic field location technology does combine a number of qualities that together constitute a technology worth exploiting, and that are not available in alternative location technologies.

— The technique can be used regardless of soil conditions.
— It does not interfere with existing communications or other systems.
— A recognizable signal can be applied to a particular line and that line identified up to several miles away.
— It is possible to house a transmitter into small self-contained units that can be fitted to underground devices.
— Depth to a signal source can be estimated simply, often by pushing a button on the receiver.
— The equipment is portable and can be hand-held.
— If required transmitters/receivers can be developed to transmit information.

And the future? For the moment devices used for underground work, inspection or monitoring are powered from the surface. This enables the power line from the surface to convey information and commands. But when these devices become self-contained and we do not require an umbilical to the surface, it is probable that magnetic location technology will be used to get information to the surface and also to transmit commands to the device.

The technology that is used today to enable hand-held locators to detect buried conductors is being developed and is now available and waiting to transmit information from the subsurface to permit an enormous increase in the scope and possibilities of NO-DIG techniques for locating and for underground work, measuring and monitoring.

Figure 5.2.1. An early cable locator, probably Germany about 1905. The search coil can be seen on one corner of the roof truss.

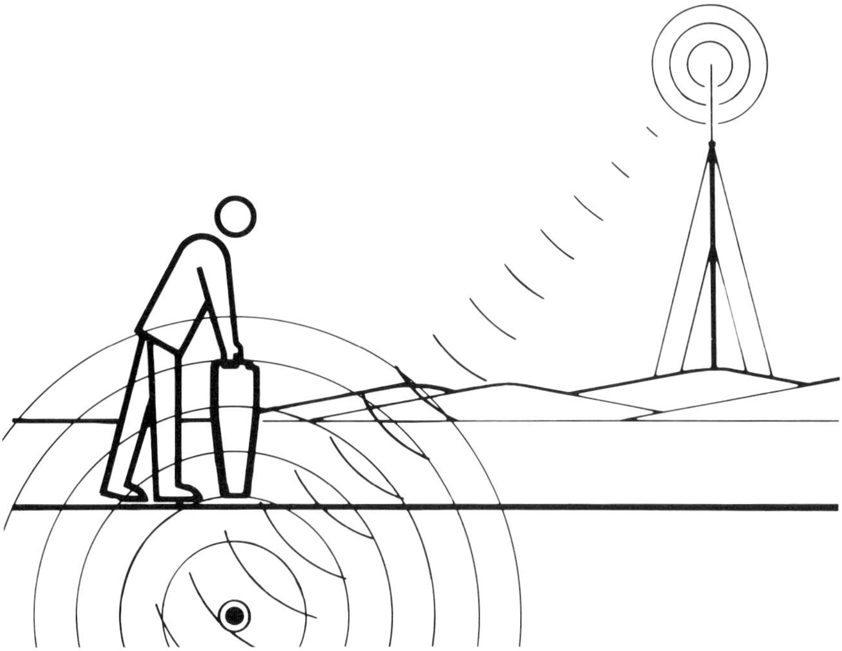

Figure 5.2.2. *VLF radio energy penetrates soil and is then re-radiated by buried conductors which can then be located from the surface.*

Figure 5.2.3. *A locating tool designed for used by workmen on site.*

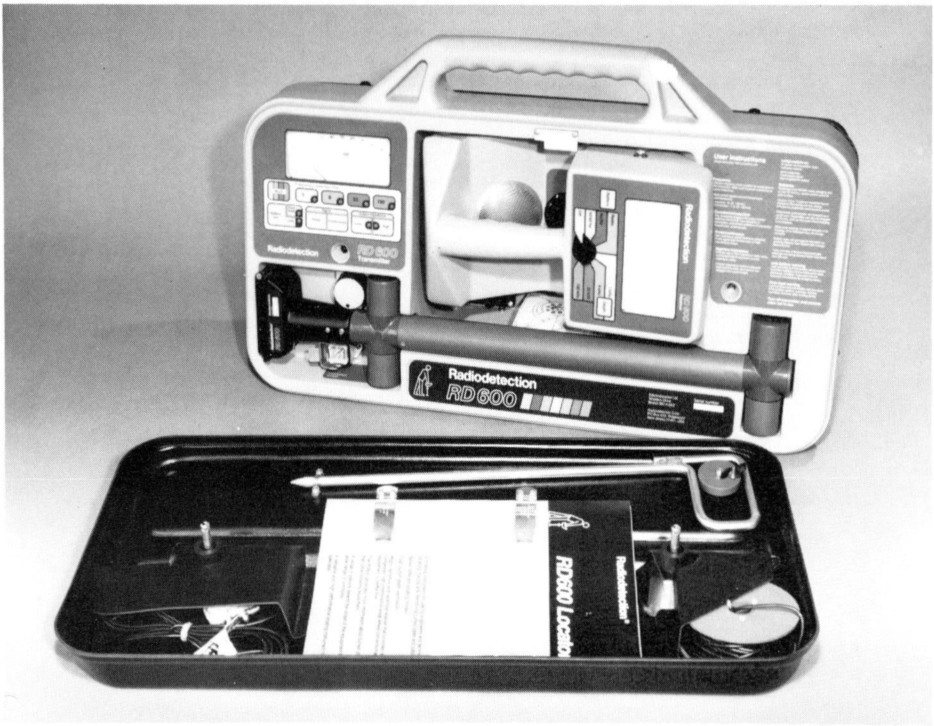

Figure 5.2.4. A specialist's instrument with seven modes for locating. The user controls all aspects of the application and reception of the signal.

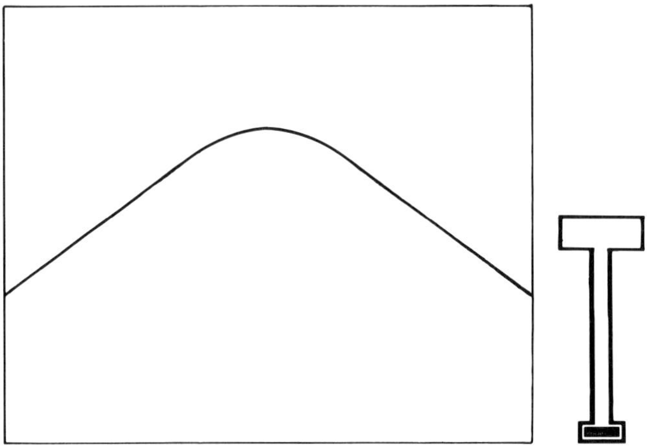

Figure 5.2.5. *Signals.*
(a) Response from a conventional single aerial locator.

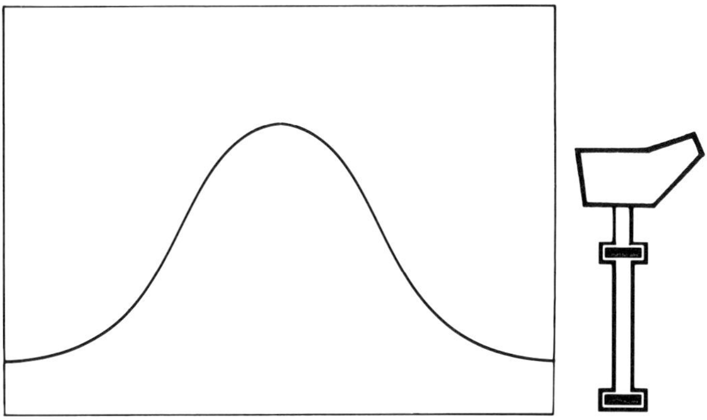

(b) Response from twin aerial antenna.

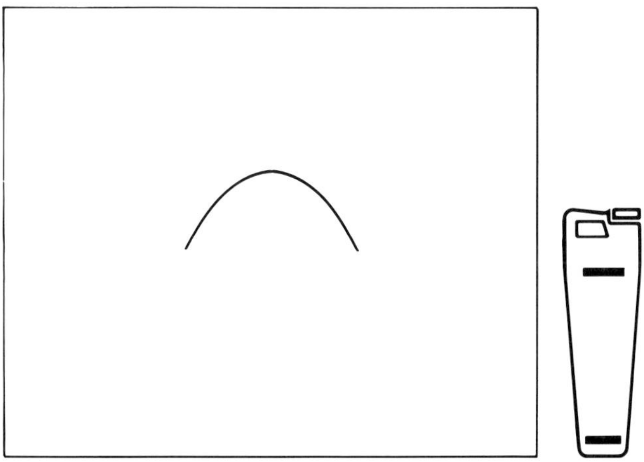

(c) Response on the CAT locating tool using twin aerials to provide sharp narrow response over the line.

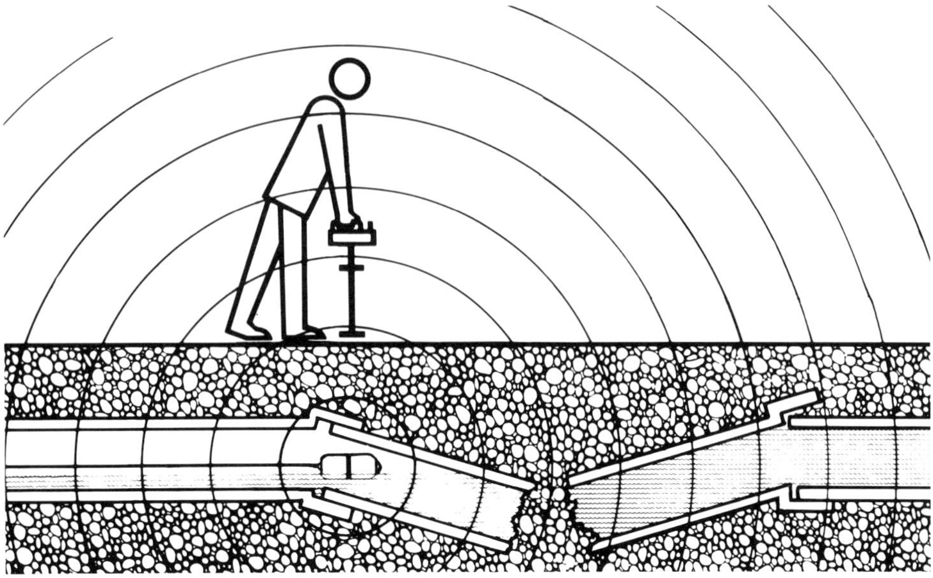

Figure 5.2.6. *Pinpointing a collapse in a sewer by locating the position of a sonde from the surface.*

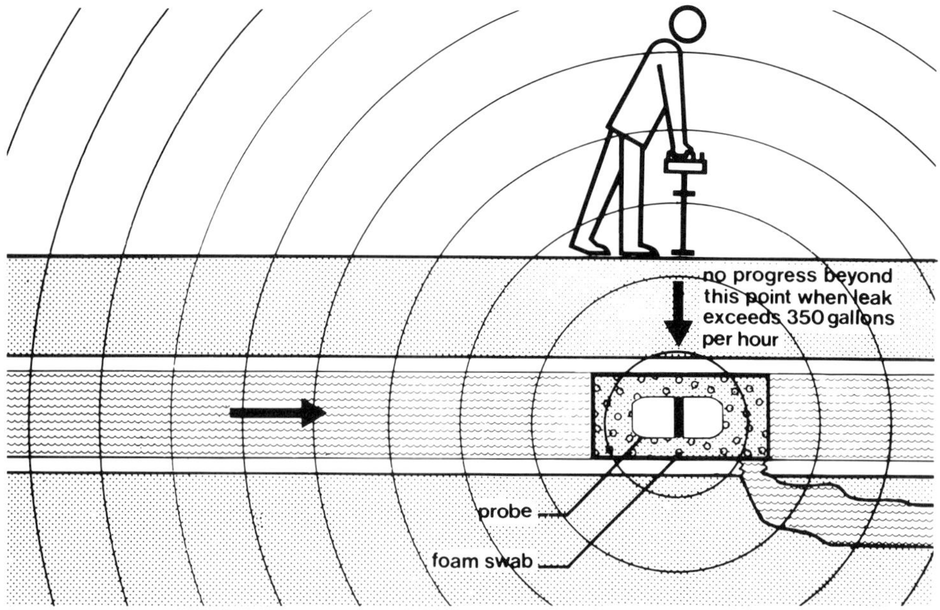

Figure 5.2.7. *Using a sonde to locate water leaks in plastic pipe.*

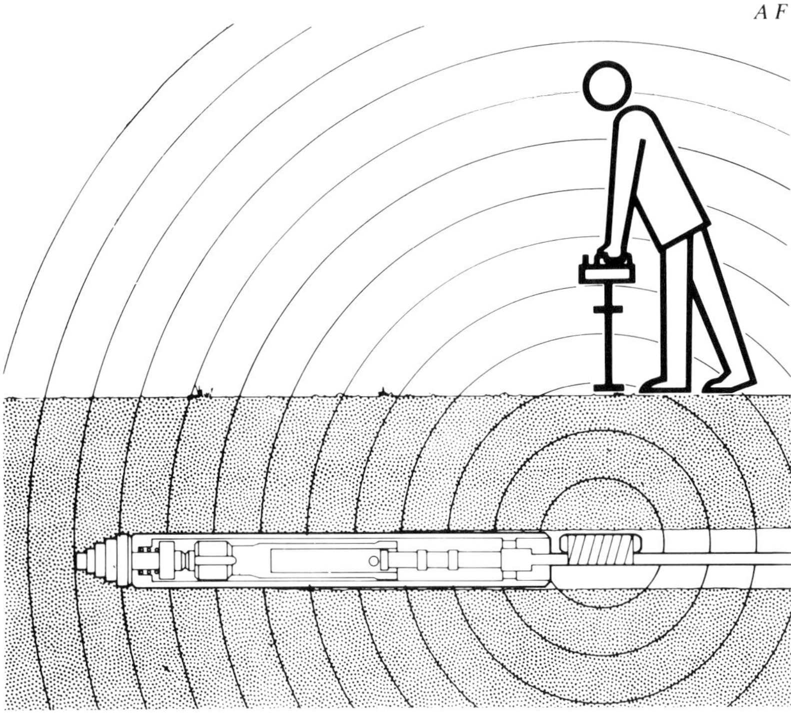

Figure 5.2.8. *Soil-piercing tool or mole with transmitting sonde. The position of the sonde can be pinpointed from the surface.*

Figure 5.2.9. *Prototype tractor following the line of a buried cable across agricultural land.*

5.3 UK Development in underground mapping

M J Slipper *Water Research Centre*

J L Wilson *National Joint Utilities Group*

N J Pearce *Water Research Centre*

The role and activities of the National Joint Utilities Group are described.

An overview is given of both conventional map record-keeping systems and of digital mapping systems. Particular attention is given to potential problems which may prevent the exchange of digital map information between the mapping systems of different utilities.

Examples are drawn from recent experiences in the United Kingdom where the exchange of digital map information has been tried.

The National Joint Utilities Group (NJUG) was established in 1977. It is a voluntary association formed by British Gas, British Telecom, the Water Industry and the Electricity Supply Industry, and has a permanent secretariat jointly funded by the member utilities.

The Group was formed to promote co-operation between the public utilities on distribution engineering matters. Its prime objective is the improvement of safety among operatives and the public when work is being carried out. Equally important is the potential for savings in time and money resulting from a co-operative approach to long-term development and short-term maintenance and repair work.

The NJUG has issued a number of publications concerned with safety, underground plant location and plant identification. It has also issued in co-operation with the National Federation of Building Trade Employers and the House-Builders Federation a *Model Guidelines for the Planning and Installation of Utility Supplies to New Building Developments* to improve the efficiency of utility installation and building work.

The United Kingdom is a densely populated country with a strong public concern for the visual environment. The trend over many years has been for the public utilities to place their plant underground in urban areas. This means under the streets such that in 1983 the total length of utility apparatus amounted to 1,650,000 kilometres with a replacement cost of £117,000 million. The current workload of the utilities includes over 18,000km of new and replacement mains, 1,664,000 new and replacement services and over 2 million small openings per annum.

Traditionally, all utilities maintain records of the location of their underground assets by drawing the position of pipes and cables onto large-scale printed maps. Indeed, in the United Kingdom, it is usually a legal requirement that such records are held and subsequently maintained to indicate where changes have occurred. It is common to find that copies of the same map records are held by several departments within an organization. For example, as well as a master set of maps held in a drawing office, a water undertaking will provide copies of these maps to local inspectors; headquarters; network planners; the estates department and others who need to know about the location of the water mains. Problems arise when one of these copies is amended to reflect a change in the network since the other copies in circulation will remain unaltered.

The United Kingdom is fortunate to have a government-sponsored national mapping agency, the Ordnance Survey (OS), responsible for producing and maintaining map coverage of the entire country at a variety of scales. The OS have an international reputation for the quality of their work and their larger-scale series of map sheets at 1:1250 and 1:2500 scales are used as the basis for almost all utility plant records. Some utilities photographically enlarge OS maps to a scale of 1:500 before plant details are drawn on to them, in an attempt to improve clarity.

The location of pipes and cables are either drawn directly on to the OS map sheet using pen and ink or, more recently, plant positions have been drawn on to transparent overlays which are viewed in conjunction with the printed map sheet. The overlay technique drastically reduces the redrawing effort required when the base map is revised to reflect changes to road outlines or building development, but has the disadvantage that it is cumbersome to view both the plant location overlay and the base map sheet together.

Whether the overlay system or direct drawing on the base map are used, it remains the case that making alterations to the records of underground plant location is a tedious process. In practice records are often not modified even when inaccuracies are found. Certainly, one of the reasons that service connections to premises are often not shown on records is because of the difficulty of collecting, drawing and maintaining such large quantities of information.

An important use that is made of existing record systems is in complying with the requirements of the Public Utilities Street Works Act 1950 (PUSWA). Under this Act, any utility needing to make an excavation to reveal its underground equipment must, whenever practicable, first notify the other utilities and the highway authority of its intention. Typically, the notification will include a copy of a map on which the plant to be excavated is shown. In response to this notification, the other utilities will often provide maps showing their own equipment in the area which might be affected by the digging. Although the use of maps is not essential to the working of the PUSWA requirements, maps are frequently used as a simple way of passing the information.

Many excavations are made under emergency conditions to carry out repairs and it is often not possible to carry out the PUSWA pre-notification procedure. In these circumstances, extra caution and full use of pipe and cable locators helps reduce accidental damage to other equipment. In an attempt to improve safety during emergency repairs there are now several regions within the United Kingdom where certain utilities routinely exchange full sets of microfilmed plant records. These microfilm copies of the existing paper map records can be read at the scene of the excavation by use of portable viewers.

In addition to the exchange of plant records in an attempt to reduce accidental damage during excavations, records are also exchanged between utilities when new pipe or cable laying is planned. In this way routes for equipment can be found which cause the minimum of disruption to existing underground plant. A problem when considering inter-utility exchange of maps of underground plant location is the accuracy of the records themselves. Sources of error are numerous but some of the more common are listed below.

- records are drawn to show the positions of pipes or cables as planned, not as actually constructed;
- records are 'improved' by the draughtsman to smooth out small deviations and irregularities in the linework to improve the aesthetic appearance of the map;
- where a large number of pipes or cables are laid in close proximity, the draughtsman actually draws a diagrammatic representation to show the plant layout, rather than showing each item of plant in its correct geographical position,
- when records are redrawn onto revised map sheets transcription errors often occur;
- records are simply not kept up to date and changes which have occurred since they were first drawn are not shown. Even when faults are found in existing records there are often no mechanisms to ensure the records are updated.

It is very unusual for utilities to state, or even know, the accuracy to which their records are held. In practice, whenever copies of records are passed to other organizations, under schemes such as PUSWA, they are accompanied by a cautionary note regarding their accuracy.

Traditional record systems are slow, cumbersome and expensive to maintain, while the records themselves are often inaccurate or missing. Such record systems are frequently viewed as an unavoidable burden on the organization rather than as an important asset to efficient operation. It is the view of the authors that by adopting digital mapping as a replacement for existing techniques many of the failings of the conventional records systems can be overcome. A digital mapping system is a computer with the necessary hardware and software to enable the lines which make up a conventional printed map to be stored and subsequently displayed and manipulated on suitable computer graphics terminals. In addition to being able to store the base map in digital form, the digital mapping system will have facilities for the user to store details of his own plant locations so that they may be displayed along with the map background.

There are several different technologies used in digital mapping systems, but by far the most common are systems which hold the map information in a vectorized form. In this type of system, each line on the map is held by the computer as a series of easting and northing coordinate pairs. A simple straight line needs only one pair of coordinates to define its starting point and a second pair to define its end point. Curved lines or angular lines can be described in a similar way, only this time additional pairs of coordinates are used to indicate positions of intermediate points along the lines. Thus, all lines are composed of a series of one or more short line segments, or vectors. In practice, most mapping systems will have special facilities which can be used to define curves, circles, polygons and other features, but this extra complexity need not concern us here.

Each line on the map will also be assigned a feature code (or 'level'), a colour and a line style. The feature code indicates what feature in the real world the line respresents, for example a road edge, a property boundary, a river bank, a cable or a pipe. The colour and line style determine the appearance of the line when the computer draws it onto a graphics display terminal or a plotter. By asking the computer to display only lines with particular feature codes it is possible to produce specialized maps on which only relevant features are shown. For example, a map showing only road outlines and gas pipes over 100mm in diameter could be displayed.

Finally, the computer will be able to store non-graphical information in an attribute database and to associate this information with the individual lines forming the digital map. A line representing a water main could have an associated entry in a database to hold the pipe diameter, material, age and length. The variations in the database facilities offered are very wide, but the important point is that there will be the ability to store and retrieve information associated with the map as well as to view the map itself.

A digital mapping system will allow maps to be displayed at a wide range of scales. Typically, the coordinate systems used will allow map information to be held with a resolution on the ground as high as 1mm or more. With this degree of resolution, and the ability to selectively display different feature codes, it is possible to rapidly display or plot maps at scales from 1:1,000,000 up to 1:25 or larger. Unlike conventional maps which are produced only at certain defined scales, digital maps can be viewed at any scale convenient to the purpose for which it is required.

A major potential benefit of digital mapping systems is that digital plant records may easily be transferred between the mapping system of one utility and the mapping systems of other utilities in the same area. In this way it will be possible to routinely access plant records of other organizations for operational purposes and for planning new mainlaying. If such exchanges are to be made then great care must be taken to ensure that the records held in the various systems are capable of exchange. Following installation of a mapping system there will be a major exercise to transfer information from the existing manual records to the computer system. It is during this process that mistakes can be made which will seriously impair subsequent information exchanges.

The first stage in setting up a digital mapping system is to purchase and load copies of the Ordnance Survey base maps in digital form. At present, only about 15 per cent of the large-scale maps of the United Kingdom are available in this form from the OS, with the remainder due for completion as soon as resources allow. If available, the OS-produced maps are the ideal basis for a mapping system but currently few utilities are fortunate to have wide coverage of their areas in digital form. Alternative sources of base maps include private companies using similar production techniques to OS, private companies employing automatic document scanning and vectorizing techniques and in-house production of base maps. Each of the alternatives has its advantages and disadvantages and the solution chosen will depend very much on specific local requirements.

Once the base maps are loaded it is necessary to convert the existing underground plant records to digital form by a process called digitizing. The important goal in digitizing is to ensure that the computer-held records conform as exactly as possible with the existing manual records. This is most easily achieved by displaying the base map at a large scale on a computer graphics terminal and positioning the lines of cables and pipes onto the base map using a cross-hair cursor or a light-pen. The actual positioning of the lines is done by eye, the operator having a copy of the existing records available for constant reference during the digitizing process. This method is very quick and produces a computer record well within the accuracy at which the original records were drawn. Of particular importance, the lines of plant appear in their correct position relative to base map features even when there are positional inconsistencies between the base map used for the original records and the digital base map now being used. Positional inconsistencies as great as 3 metres on the ground are known to occur between successive revisions of large-scale OS maps. If lines of plant are transferred directly from the existing records to the computer system without regard to these inconsistencies, then a pipe or cable shown on existing records as lying below the footpath may appear on the computer record as being laid below the roadway.

If the records are to be exchanged with other utilities, it is vital that the positions of plant relative to map features in one mapping system remain unchanged when the records are transferred to another system. If the Electricity Board mapping system displays a cable as being 2 metres from the kerb-line then it must appear at this location when transferred and displayed on the mapping systems of the other utilities.

If these exchanges are to occur successfully then the following conditions must be met:

- all mapping systems must be loaded with identical base maps on to which the plant records are digitized. If one utility buys digital maps direct from OS and another digitizes the same maps in-house, the chances of proper registration of records from one system when displayed on the other are slim;
- any geometric transformations used on one system to manipulate the base map must be applied in each of the other systems. Examples might include inducing distortions to edge-match adjacent map sheets; simplifying the base map structure by coordinate data reduction; applying elastic body transformations to remove distortions observed in the base maps; etc etc. If base maps in one system have undergone geometric transformations, then records held on that system are unlikely to register well with maps on other systems unless a reverse transformation is applied before exchange;
- a standard format for the interchange must be agreed between both the issuing and receiving utilities;
- a computer medium (eg magnetic tape or floppy disk) able to be written and read by the computers of both the issuing and receiving utilities must be chosen or a telecommunications link established for direct transfers.

Already NJUG have drafted a preliminary standard format for records interchange between utilities. The standard covers ways of encoding positional information together with conventions for passing non-graphical attribute data. The first version of the exchange format is based on extensions to the Ordnance Survey Transfer Format (OSTF) currently used by OS to transfer digital map data to customers. Later versions of the NJUG interchange format may be based on international graphics standards such as IGES or GKS if these prove suitable.

With a view to investigating the problems associated with the exchange of information held in a digital mapping system, NJUG established a trial at Dudley in the West Midlands starting in 1982. The Dudley Trial is described fully in a paper by G J Yarrow[1], but some of the practical problems of map exchange already found are worth considering.

Unusually, at Dudley a single computer system is shared, for trial purposes, by Dudley Metropolitan Borough Council, West Midlands Gas Board, Midlands Electricity Board, British Telecom and the South Staffordshire Waterworks Company. Each participant has now transferred its records to the mapping system and checked their accuracy against existing records.

Staff of each utility are able to view and update their own records and are also able to view selected information about the location of the plant of the other utilities.

It has always been the intention to issue detailed maps showing locations of all underground plant to work gangs making excavations. Almost immediately it was found that the single A4-sized monochrome map plot issued was very difficult to understand when gas pipes, water pipes, electricity cables, telecommunications cables and sewers were all marked on the same sheet. Furthermore, when the records from the five participants are superimposed and then enlarged to a scale of 1:200 or greater, it is tempting to assume that the relative positions of the various items of plant in the ground will be the same as shown on the printed map. In fact, because of the errors associated with the original records it is not possible to infer relative positions when separations of lines represent no more than 1 metre separation on the ground.

To reduce the complexity of the maps issued and to lessen the temptation to infer false positional information an alternative scheme is being tried where several maps of the same area are issued, each showing plant of a different utility.

Once an excavation is made, it is possible to establish whether the records on the maps issued are accurate or not. Potentially, inspectors can not only verify the records of their own utility but also those of the other utilities whose plant was uncovered during the excavation. It remains unclear whether such feedback will be acceptable!

During 1985 an investigation will be made at Dudley into ways the PUSWA notification system can be made to operate more effectively when all participants routinely have access to each other's map records.

As the Trial continues much other useful information will emerge about the practicalities of record exchange. Similar collaborations between groups of utilities in the South and South West of England will again be a useful source of experience upon which nationwide record exchange schemes can be defined. In the future, one of the many benefits of digital mapping systems will be the ease with which information about underground equipment can be exchanged between the various utilities.

REFERENCE

[1]Yarrow G J, *Digital Records Trial — NJUG Dudley Distribution Development*, March 1983

Authors' introductions

Mr D J Daniels: Location of underground services by ground-probing radar

My talk is on ground-probing radar but I think most people are more familiar with the type of radar which uses a very large dish antenna to look at a moving target. In the case of the ground-probing radar the situation is exactly the reverse. You are looking into the ground at a fixed target and the size of the system obviously has to be fairly small so that it can be moved through the streets. Those are the major differences between the ground-probing radar and conventional radar.

The objective of the work that British Gas is carrying out is to develop an equipment to locate metallic and non-metallic plant, and I think is an important point to make — that ground-probing radar will detect all types of buried plant. It will also detect things like half bricks and tin cans, and that is one of the engineering problems — but we leave that for the moment. What we would like to do is to link the radar to digital mapping technology, basically as an eye to see through the ground. Locally we intend to have field trials of a prototype system later on this year.

First, let us look at the type of target the radar is going to detect. If we consider the disposition of mains and services in the footpath in the UK, we will see that most of the pipes and cables are in very close proximity. This means that any detection system has got to have the ability to resolve, that is to distinguish separately, those pipes and cables both in depth and in plan. That is one of the real problems of designing a radar. It is quite easy to build a radar to detect objects a long way in the ground; what is not easy is to design a radar that will give a separate signal from each of those targets. Because of that there are certain constraints upon the frequency in which the radar operates, as shown in Figure 5.1.2 in the paper. The frequency of operation is round about 1 gigahertz or 10^9 hertz, and it lies just about at the beginning of the radar frequencies and the end of the UHF frequencies. If you go higher in frequency the radar starts to have a much better resolution, but unfortunately you can no longer get the penetration into the ground, so the range of the radar is reduced. If you go lower in frequency, the range of the radar increases, but its ability to resolve, that is to separately distinguish adjacent targets, falls very dramatically. So in essence that is the radar problem. It all sounds a little bit gloomy particularly when you do a few back-of-the-envelope calculations as to the kind of losses which occur.

If we imagine a radar looking down at a plastic pipe in the ground, as in Figure 5.1.1, the ratio of the power that is received to the power that is transmitted is 10^{-20}. So the signal level is unbelievably low. This would be for two metres of absolutely waterlogged UK clay.

There are also constraints upon the radar designer. In the UK we cannot put out a great deal of power. The power that we can transmit is in the order of 50 milliwatts, and of course we cannot have a very large antenna — about 1 metre is the maximum — so there are some fairly significant engineering problems. Finally the signal will take roughly 10^{-9} of a second to go out and come back again so we do not have a very long time to do the signal processing. With all that it is an achievement to get returns back from plastic or metallic pipes, but the kind of signal we do get back is shown in Figure 5.1.6. That is from a pipe 125mm diameter, buried at about 32cm in sand. This is a result that we obtained two or three years ago. Since then we have improved the performance of our prototype radar quite significantly. What we envisage in the future is that we will have a radar which will have some kind of antenna which will be scanned over the ground and will produce a two-dimensional plan of what is buried underground, and we can store that on floppy disk or in some other way.

To illustrate further some of the problems that such a scanning radar would have, let us assume we have a PE gas pipe which has been sloped to a depth of 0.8 metres and at each point over the pipe we get a signal return. As we go deeper into the ground the signal gets lower and our problem is to try and present that kind of signal in a format that enables a non-skilled operator to interpret the results clearly. One of the weaknesses of the current generation of radars is that you cannot interpret the results that easily. In a typical example, we know there is a target there, but it would need some persuasion to convince other people.

The radar itself has been a fairly difficult design problem but what we have identified are the target properties. Pipes are unique in that they are linear features. We have now developed an antenna to quite a reasonable scale which can be scanned over the ground, without having to contact the ground. This is one of the points we identified operationally — that most radar antennas have to have fairly close contact with the ground. This one does not, but we can still get a fairly sensible return.

What we did was to take a test site with a crushed dolomite backfilled material in which we buried a number of plastic pipes. The radar was scanned over the pipes and the signals were recorded in the form of a computer diagram. On this we could see that the strongest signals corresponded approximately to the position of the pipe. By the time we had done the full scan we had a very crude image of the signal returned from the pipe. We know that we can tidy that up to make it a little bit more visible, so that we are now beginning to get to the stage of producing some kind of image of buried plastic pipe. We do not bother to demonstrate signals back from metal pipes because they are much easier to detect. The plastic pipe is a lot more difficult.

We could see that in one corner of the diagram there was a small signal that we believe was coming from the edges of the trench. This is one of the problems and we will have to develop software to discriminate against that kind of return.

At the moment we have reached the situation where we have a van which is equipped with the radar system and a small antenna which we can trundle over the ground (Figure 5.1.9). We hope in the next two or three years to refine this system and to be able to produce proper two- and three-dimensional maps of what is buried underground.

Mr A F Garnett: Location of underground services by alternating magnetic field technology

Information is a most important ingredient in any undertaking. It is especially important in civil engineering, whether for a multi-million project or for an afternoon's work. The difficulty of obtaining information before any DIG or NO-DIG work is that it's not visible: a suitable technique has to be used to obtain the information from underground.

Several techniques are available. The most obvious one is looking at maps and documents. Then there are the possibilities of seismic reflection, a technique that has never been exploited. We believe there are also possibilities in measuring the differential of static magnetic fields but there is the difficulty of the very strong magnetic field of the earth. Downward looking radar: other papers in this meeting have brought you up to date on the progress of this technology. Witchcraft: that usually means dowsing and I expect we have all seen more surprising manifestations of it. There are people who are prepared to undertake dowsing by payment on results rather than payment for presence. And then finally there is my field, alternating magnetic location technology.

I must start by telling you that this magnetic field location technology has one massive disadvantage compared to some of the other techniques mentioned. It cannot locate or trace non-metallic lines. I will come back to that later. For the moment let us consider the advantages of alternating magnetic field location. For the sake of simplicity I refer to a buried cable or a metal pipe or a conductor capable of carrying an electric current as a line.

To enumerate the advantages:

1. It can search an area from the surface for all buried lines;
2. It can trace and identify a particular line, the one that you need;
3. It can trace and identify sewers or other non-metallic lines to which there is access at some point;
4. It can measure depth from the surface;
5. It can monitor the progress of NO-DIG tools and instruments and equipment;
6. It can provide a long-lasting signal source for retrieving NO-DIG equipment that may be lost underground;
7. Very important, the equipment is portable;
8. The equipment is simple to use. In many cases it is already being used by workmen;
9. The equipment works in all soil conditions;
10. The component parts of the technology are cheap. The equipment is cheap enough to be purchased by small contractors and is cheap enough to be issued on a large scale by regional or national organizations;
11. We have found that the technology can also be used to steer equipment or tools both above the ground and below the ground;
12. It finds cable faults and it can monitor pipelines for external corrosion faults;
13. Last and not least, the technology exists today.

These advantages are substantial. None of the other technologies that I mentioned can feature more than two of these advantages. Now add to these advantages the fact that almost all the functions I have mentioned can be performed with just three building blocks of equipment. I want to give you two examples that I have selected at random.

Example 1: Information before and during moling (soil piercing) along a street.
Equipment required: a receiver, a signal transmitter and a transmitting sonde.

The technique? Use the receiver to do a quick and effective locating sweep over the area to be excavated for the mole launch. Use the receiver to locate the buried lines in the projected path of the mole. Use the receiver and transmitter to estimate the depth of these lines. Find out the depth of any sewer or sewer lateral by rodding a transmitting sonde up it and using the receiver to estimate its depth. Choose the optimum depth for launching the mole and then monitor the progress and position of the mole by attaching the sonde to it.

Last, use the receiver to do a locating sweep to make sure there are no buried lines over the position of the retrieval pit.

Example 2 is in a quite different field: pinpointing leaks in plastic water pipes after acoustic and other conventional methods have failed.
Equipment required: a receiver, a transmitting sonde and a plastic foam swab.

If the leak is in a known section of plastic pipe, isolate the line by closing the valves at each end. Then choose a foam swab of suitable size so that it acts as a piston in the line. Put the transmitting sonde in the swab. Insert the swab into the line either by cutting into it or pumping it down through a hydrant. Control the progress of the detectable piston by closing the downstream valve and throttling the upstream valve. Then use the locator to trace the detectable piston, which will stop or change speed at the point of the leak.

I could give many other examples of the flexibility of the technology for obtaining information from underground, but I would like you to notice that the same equipment is used for both those two very disparate examples.

Now let us return to the problem of non-metallic lines. In most utilities outside this country it has been a discipline to lay a tracer wire or tape with plastic pipes. The best and probably the cheapest system is an insulated wire bonded to meters, valves, or property boundaries. In this country some of the leading national utilities have chosen, as a matter of policy, not to lay a tracer of any description. Instead they are spending large sums of money, probably more than the cost of tracer wires for the whole of their system, to develop devices to locate plastic pipes. But the point is that they miss the point. The man who needs to know the position of a buried pipe in a hurry is the man who deals with an emergency. It is the contractor installing a new line and needing to thread his way through existing buried lines. It is the fencing contractor or perhaps the local authority workman installing street signs in ten different sites in a day. They do not have the resources or the skills to make use of sophisticated devices. But they can, and in many cases already do, make use of alternating magnetic location tools. I believe it should be mandatory to bury a tracer wire with every non-metallic buried line, so that its position can be determined using equipment already in use.

Alternating magnetic field location technology can make a contribution to making plastic lines easily locatable by installing unattended sealed signal generators at intervals on an insulated tracer wire. These would make the plastic lines not only locatable but identifiable with even the crudest hum-detecting locating tools. The system has already been demonstrated in two or three installations overseas on important fibre optic communication cables.

Paper 5.3 was introduced by Mr. N. Cullen (Water Research Centre)

Mr N Cullen: UK development in underground mapping

This paper is about digital mapping and the work of the National Joint Utilities Group. I am not going to talk about the National Joint Utilities Group because John Wilson, one of the joint authors, is here to answer questions on that and on the Dudley digital mapping trial. Another reason is that there is an excellent paper by Dick Nitze of NJUG in the current (April 1985) journal of the Institution of Public Health Engineers and I would commend it to you.

I am going to address the problem that all utilities are interested in. How do you keep track of all those pipes in the ground? I would like to show you some slides illustrating what digital mapping can do, and then mention three major issues that we in the UK need to address now if digital mapping is to be utilized fully and effectively.

[Mr Cullen showed a number of slides illustrating a system which the Water Research Centre was using. It consisted of two television monitors, one to show maps in colour, and the other to show text, controlled by a menu with a cursor to select the particular function required. Mr Cullen demonstrated with slides how it was possible to view an area at any scale desired, using colour for different layers of information, eg, buildings, water mains, electricity cables, etc. All of this information was held digitally in computer storage and could be shown on the screen in various formats at will.]

The water industry is interested in digital mapping database as a potentially powerful management tool. The utilities however, as a group, are interested in the interchange of information about plant, pipe and cable location, which has already been mentioned this morning. This is where I come on to the three problems that I think have to be addressed. They are mentioned at the end of the paper. First of all the accuracy of records; secondly data exchange standards; and thirdly the need for common base maps, ie, the Ordnance Survey maps.

Let us look at accuracy first. We have assumed that records are generally not very precise, but in order to get operational you put on your existing records. Every time you have a water main burst and you dig a hole, or someone else excavates a trial pit and discovers a water main, you can put a star on the map to indicate that location. Obviously, once you have collected enough stars, you can "join up the dots" and put in the main where it was actually found. Also the information is kept on the map so that anyone who gets a copy of that map knows what has gone on before. Each star can have a database entry saying, for instance, who found it, when, and details of what else was found in the hole. Over a period it will then be possible to build up a more accurate picture of where everything is and you will be able to overlay a whole series of utilities.

We could also interchange between different utilities, but we do need to agree now on the format and coding of data so that we can transfer maps electronically or on magnetic tape between the computers held by different utilities. NJUG is making good progress towards a national agreement on this problem.

Finally of course the plant always has to be referenced to the base maps. You are not interested in the precise location with regard to the National Grid, but much more "how far is it from the kerb?" "how far is it from the building?" or "how far is it from the water main or the electricity cable?" We need to reference through a common grid. You will probably say, "we have got a National Grid, won't that do?" Unfortunately the answer is no, because each edition of an Ordnance Survey map may be positioned slightly differently on the National Grid. Different utilities use different editions of the base maps and the plant shown on the road in one may be beneath the pavement or a garden on another. Therefore the utilities need to agree locally, now, in each area of the country, so that all the utilities use the same edition of the base maps.

In conclusion, firstly digital mapping is expensive and it requires considerable skilled effort in data acquisition and management. However, it has very great potential for asset management. Secondly, data interchange between utilities needs national agreement on data codes and formats and media, and local agreements at least on the editions of base maps which are going to be used.

Discussion on papers 5.1, 5.2 and 5.3

R H Morgan (Distribution Manager, Rickmansworth Water Company): While the remainder of the Conference is concerned largely with the techniques of trenchless construction, these cannot be sensibly undertaken without a knowledge of what exists underground. This is always important, but is a particular problem in urban areas where streets are congested with utility apparatus. The accuracy of utility records tends to be a bit of a joke, and I doubt whether there is a single utility representative present who is fully satisfied with the accuracy of his records. The findings of the PUSWA Review are likely to criticize the state of utilities' records.

However, much of the utilities' apparatus has been buried for up to 150 years, and records can be improved only very slowly on the basis of actual sightings when excavating. Perhaps the work described by Mr Daniels and Mr Garnett will allow overall surveys to be made in the future, but the technology is not yet sufficiently developed.

There are two aspects to improving records: the location of the apparatus, and the method of recording. Both aspects have been well covered in these three papers.

Mr Daniels indicated that the depth of penetration of the radar instrument was dependent, among other things, on the ground attenuation, which in turn was governed by the water content of the soil. This may have been an obstacle from the Gas Board's point of view, but it might perhaps be of advantage to a water undertaking looking for hidden leakages. It might even lead to a reduction in the number and size of excavations.

I must take exception to the statement in Paper 5.3 that "the first stage in setting up a digital mapping system is to purchase and load copies of the Ordnance Survey base maps in digital form". If that is prerequisite, then few, if any, utilities would be able to start thinking of digital records this century. The current forecast by the Ordnance Survey for coverage of the whole of the country is the year 2015, assuming an accelerated programme.

Ordnance Survey digital information should certainly be used as soon as it becomes available, but overlay techniques can provide a perfectly satisfactory interim solution. Rickmansworth Water Company will within the next few months have digitized all its records information, making it one of the first utilities to have done so. This has been done with the intention of predominantly using the overlay principle.

The paper suggests that the preferable method of digitizing is by use of the graphic screen. In my experience this is not necessarily the case, and in most instances it is slower than using a digitizing table. The Ordnance Survey, who have done extensive research into digitizing techniques, favour blind digitizing, i.e., they do not use a screen.

I am particularly pleased that the paper on digital mapping has been included in the Conference. In my opinion, all utilities will have to adopt digital records systems sooner or later, and the sooner the better.

P Mills (Distribution Engineer, Wessex Water Authority): The ground-probing radar described by Mr Daniels can detect PE gas mains down to a depth of 0.8m. Will the equipment be effective in locating PE and uPVC water mains at the more usual depth of 1.0m, or possibly 1.5m? Since saturated ground can seriously limit the range of the method, what would be the effect of the water within a non-metallic pipe?

There is often a requirement to trace a single non-metallic main, for instance for new service estimates or possible diversions, rather than to map a complete carriageway. Is there a possibility of a portable second-generation machine which could be used by individual inspectors and gangs, like the magnetic field instruments used at present for locating metallic pipes? I would be interested to know of any similar work being done in the water industry to locate non-metallic pipes.

Mr Garnett made a pertinent remark that if tracer tape had been laid with all non-metallic mains we might not have needed to develop this new and expensive location technology. I should be interested to hear the views of other utilities on this point.

A G Munro (Engineer for Metropolitan Works, Melbourne and Metropolitan Board of Works): In my experience with pipe and cable detectors, some years ago, it was found that the system also picked up boulders and ridges of rock. Could Mr Daniels and Mr Garnett comment on how their systems handle this situation?

Mr A Garnett: Most pipe and cable locators are based on alternating magnetic field principles. By definition, they are unable to locate any feature which does not conduct an electric current. In practical terms, I have never heard of boulders or rock being picked up by this type of locator. Perhaps Mr Munro was using some other system.

Mr D J Daniels: British Gas, British Telecom and the electricity authorities have a joint collaboration on research into radar, so by this means we have an eye on development costs. When we looked at the feasibility of a number of techniques for detecting pipes and cables down to a depth of 1½ metres, it became obvious that the radar approach was the only way in which a non-contacting system could be developed, since with seismic methods the sensors must be right on the ground, which slows the rate of scanning very considerably.

Radar can also give greater accuracy. With four or five pipes in the ground, the ability to locate them accurately and get separate indications from each is highest in principle with radar. The low-frequency electro-magnetic field technique is good with one or two pipes, but not in a cluttered situation. It is also labour-intensive.

The idea of using tracer wires with non-metallic pipes would be fine if they did not corrode over a 50-year period and have to be replaced at great expense. There are stories of workmen pulling tracer wires out of the ground, not knowing what they were there for.

In reply to Mr Mills, radar will detect pipes down to 1½ metres, provided they are a larger size. Water in the pipe makes detection easier.

The radar we have designed in intended *not* to detect things like rocks, tin cans and other junk. It would be possible to design a radar that would pick up such objects, but we are concentrating on locating pipes and cables, and exclude these false targets.

Mr J L Wilson: It must not be thought that when records are digitized we will automatically have an accurate set of maps. Since much of the buried plant has been there for up to 150 years, we have to rely on plans which have been passed on from one authority to another, and probably copied more than once, with accumulating inaccuracies. The only way of improving the accuracy would be to dig holes — the very thing we are under attack for doing. If we dug a hole every 10 metres on each main line we would still not know what happened in between. With an estimated 1 650 000km of mains plant alone, and pricing each hole at £50, the total cost would be £8250 million. The resultant increase in unit rates would hardly endear us to our customers.

Mr N Cullen: As Mr Morgan pointed out, Ordnance Survey digital maps will not be available for most of the country for many years. To overcome the difficulty, WRC has entered into a commercial agreement with a company called Syscan to have maps scanned in three levels so as to separate the streets from the buildings and the text. In comparison with an OS map, which retails at about £30, a scan map costs about £105, even allowing for the fact that the necessary three hours of computer processing time is provided at a knockdown price on the WRC computers, using overnight processing. Now, if all the utilities in an area got together on this, the price could be brought down to something approaching the cost of an OS map. I would encourage the utilities to consider this option.

The question arises of whether one can operate with ordinary underlay OS maps with a computerized overlay. This is fine in the short run, but there is still the question of how to update an underlay map. Remember also that a Regional Water Authority has many functions, and maps are required for all of them. Why not update a map once so that everybody has access to it? Similarly, why not have the map updated once on behalf of all the utilities? The advantage of having a scan map is that, if a digital map becomes available later, digitized into 130 different levels, you can overlay one on the other, and throw away the old scan map. But it is important to agree with the Ordnance Survey which edition of their map is going to be digitized, so that that is the one chosen for scanning.

On the question of digitizing on a table versus digitizing by eye, our considered opinion at WRC is that it is right to go ahead on screen digitizing by eye. However, there will be continuing discussion on this, and the purpose of the trial in South West Water Authority is to study the various options.

I agree totally that digital mapping will not improve the accuracy of records. Nor will it simplify management. The cost of administering the data is still a lot more than the cost of the hardware and the software. But, at the same time, I feel that, if one is spending a lot of money on collecting information, it would be unwise not to use computers now to handle it, because of their power in searching, collating and managing.

Mr Garnett: I feel strongly about the question of laying tracer wires with non-metallic buried lines to which Mr Daniels refers. Most of the world's great utility companies insist on tracer lines being laid and also being tested.

Like any other product the life of a tracer wire is finite, but its life could be greatly extended if only a small part of the expenditure into researching new methods of finding non-metallic lines had been allocated to the simple matter of improving the specification and life of tracer wires.

Speaking as a practical man I believe that tracer wires are the only practical way of finding the position of non-metallic pipes and cables. The great advantage is that the location work can be done with a piece of equipment already owned by contractors and the utilities.

D Wilson (Assistant County Surveyor, Berkshire County Council, and representing the Institution of Highways and Transportation): I am very pleased to hear of the progress being made in locating and recording buried equipment, but I would like to ask whether, every time a utility opens a road, it makes a record of what it finds — of its own apparatus at least, and possibly of others. If not, it does not matter how much digital work you do, the information is only as good as what your eyes see. I regard it as absolutely essential that everything that is found is recorded. This, of course, applies equally to highway engineers with regard to their gully connections, highway drains and signal cables.

D B Field (Director of Engineering, Water Research Centre): Mr Morgan asked earlier in the discussion whether ground-probing radar might be useful in looking for leakages. We at WRC are not aware of any attempts at leak detection by radar. In principle, a large volume of water should be detectable, but there are several difficulties which discourage this application.

1. Radar is most sensitive to abrupt changes in subsurface conditions, such as rock strata. A change in moisture content produced by a leak may be too gradual to give sufficient signal. Changes in the soil type could be confused with changes in moisture content. A high local groundwater level would probably obscure the change produced by a leak.
2. Suitable radar equipment would be bulky and probably vehicle-mounted. A matrix of measurements would be needed which would take time, and a skilled operator would be required to interpret the results.
3. There is no commercialy available equipment of sufficient sensitivity to undertake trials at present. Development costs for such equipment are high, and the unit cost of a mass-produced radar would still be at least £10 000 to £15 000.

While existing techniques such as noise correlation can be applied, radar does not appear to be worth pursuing. If British Gas succeed with their plastic pipe detection system, it would be worth arranging a leak detection trial with this equipment, but it is probable that it would only work where the surrounding ground conditions were favourable.

G Cook (Head of Renovation Unit, Halcrow Water): I gather from the three papers that we are not only moving towards methods of detecting all types of existing pipes, but also are able to track the route of the mole or auger as it passes through the ground. Do the authors consider that, by combining their various techniques, it is yet possible to produce a computer three-dimensional plot of all services in the area of an intended new tunnel in order to choose the best route for the construction, and then to guide the machine along it using computer control?

If this is not yet possible, how far off is such an ideal, and do they consider that those of us who are involved in writing contract specifications could do anything to accelerate progress? If we did include a specification for a survey of services before finalizing the route in a contract today, how feasible would it be for the contractor to comply, with realistic results?

Mr Garnett: We are still a long way from having the technical capability of making a computer three-dimensional plot of buried services. We are certainly still further from it becoming an economic possibility.

Any contractor should be able to pinpoint almost all buried lines. Organizations such as the gas industry laying plastic lines claim to keep records of the precise position of their buried lines. Combining this information plus information from the other utilities, and confirming as much of this information as possible with a pipe and cable locator and a drain locator, should give sufficient information to finalize the optimum path for a new buried service.

Mr Daniels: I do not believe that at the moment it is practicable with any of the technologies that are available to give an accurate three-dimensional view. We may get to that stage in four or five years' time, but with the current range of tools I do not think anybody could fix positions in three dimensions to within, say, five centimetres. So it might not be sensible at present to require that level of accuracy from contractors because it would put up the surveying costs.

[The following written contributions were submitted after closure of the oral discussion.]

J H Reynolds (Senior Engineer, Southern Water Authority): Southern Water has, for the past two years, been closely involved with the development of small-diameter sewer renovation and replacement techniques. As described in Paper 4.1, location of underground services is an essential facet of sewer renovation systems. The three papers on location techniques have emphasized location of services in plan, but for sewer renovation work, accurate depth measurements are also required. I would therefore ask the authors' views on the following.

1. The ability of the systems described to measure depth from road surface to service pipe.
2. The effect on instrument accuracy of (a) increasing depth, (b) traffic vibration, and (c) water on the road surface.
3. The ability of the location systems to identify voids around and above service pipes.

Mr Daniels: In response to Mr Reynolds' questions:

1. The low radio frequency instruments commonly used for metallic pipe and cable detection will only indicate depth accurately for single pipes. Multiple pipes will cause serious errors.

   Radar systems can, in principle, measure depth accurately, but will depend on the homogeneity of the ground.
2. Increasing depth will degrade the accuracy of depth measurement for radar systems. Traffic vibration should have no effect, but water on the road surface may be a problem as yet unquantified.
3. Radar can, in principle, detect voids, but our system is designed to detect plastic pipes and discriminate against other types of target.

Mr Garnett:

1. Techniques to provide a push-button depth estimation from road surface to a buried metal line exist and have been commercially available for at least five years.
2. Neither traffic vibration nor water on the surface or in soil affects the capability of today's alternating magnetic field type locating instruments. The limit of the depth of a service that can be effectively located depends on the strength of the signal that can be applied to the line. If a strong signal can be applied or is already present, lines down to 4 metres or 17 feet can be located from the surface.

V A Finney (Principal Assistant Engineer, City of Manchester): In the City of Manchester we have tried all the recognized detection techniques with varying degrees of success and failure. Most of them suffer from a lack of reliability in the typical urban situation of cluttered services.

In practical terms, however, the majority of ducts, sewers, water mains, etc. are in reality purpose-built longitudinal voids. Why, therefore, do we concentrate on locating individual types of pipe rather than setting the *void* as the main target?

Mr Daniels: The radar system actually detects the void in non-metallic pipes, and Mr Finney's understanding of the principle of the radar is quite correct.

Mr Garnett: The technology for locating voids already exists, but I do not believe the necessary applications engineering has been done to make it commercially available.

Despite complexity and congestion in heavily built-up areas it is probably more accurate and more simple to obtain location results using existing techniques.

THURSDAY 18 APRIL

SESSION 6. Chairman: R. Tharby, Chairman, National Joint Utilities Group, UK (for Paper 6.1)

Professor P. F. Stott CBE, King's College, University of London, UK (for Papers 6.2 to 6.5)

6.1 Water-jetting techniques: existing and potential uses for underground services

R M Fairhurst — *Higher Research Engineer, BHRA The Fluid Engineering Centre*

D H Saunders — *Principal Engineer, BHRA The Fluid Engineering Centre*

Water jets are used for a wide range of cutting and cleaning activities, both in their own right and in association with mechanical techniques. Existing underground applications include mining and tunnelling, drilling, pipe and sewer cleaning and renovation, and pile driving. These application areas will be outlined, concentrating on recent developments.

Water jet-cutting trials on a wide range of materials will be assessed with reference to conditions to be expected underground. This will include the use of abrasive-entraining water jets to cut hard target materials, including metals. The suitability of water as a transport medium to remove soil, rock and other debris from the cutting areas will also be referred to in the paper.

INTRODUCTION

Probably the first significant application of water jets under pressure was in the field of hydraulic mining. As early as 1852 gravel banks were excavated using water jets which issued from simple nozzles. Similar equipment was later used in the open gold fields of California and Alaska.

As water-jetting equipment has developed and pressures increased, many applications have been found in cutting and cleaning. It is the ability of high-pressure water jets to focus large amounts of energy and to remove a maximum amount of material which has made them so attractive in many industrial cutting and cleaning applications. Uses range from cleaning pressure vessels, heat exchangers, tanks and pipework in chemcial and process plants, to removing marine growth in the ship and offshore industry. Sewer and drain cleaning has benefited greatly from the use of high-pressure jetting to the extent that it has become a major user of high-pressure water pumps. Production uses include the cutting of foam rubber, plastics and packaging materials. Mining and tunnelling machines have also been developed that use water jet assistance to economic advantage.

The use of abrasives with high-pressure water jets is an even more recent addition to this field which has not only improved cleaning and cutting rates but has also allowed many materials to be cut at pressures which can be reliably generated, and in environments which are potentially hazardous.

WATER JET DRILLING

High-pressure water jets have two attributes that make them attractive for underground drilling and boring. Firstly they have been shown capable of cutting stones and rocks, using a method of hydrofracture. They have also been successfully applied to opencast mining of soft coal and china clay as well as soil excavation where the water is an excellent transport medium

Plain water jet drilling

To drill or bore a hole underground using water jets usually requires the jet or jets to be rotated. Much work has been done to design practical drilling machines since there are both hydraulic and mechanical problems. It is only with the relatively recent development of reliable rotating seals that such a technique has become a commercial possibility.

Plain water jet drilling can be done using twin rotating jets and a typical design is shown in Figure 6.1.1 (taken from reference 1). Such rotating jets have been shown capable of drilling through sandstones and granites at advance rates in the order of metres per minute.

Water jet-assisted mechanical drilling

Water jets may be combined with mechanical means to improve performance and to enable hard and abrasive rocks to be drilled. The water acts as a lubricant and coolant to the cutting tools as well as hydraulically fracturing ahead of the tool, thus reducing drilling forces. A drilling head with water jets is shown in figure 6.1.2, which is taken from reference 2. Nozzles may be placed between cutting tools or may be incorporated within the cutting tip. Such heads have been successfully applied to blast-hole drilling and roof bolting.

Underground cable installation and replacement

One recent application of water jets to trenchless methods may be seen in the development of soil-boring machines for installing or replacing buried electrical cables[3]. The method could also be applied to any similarly sized pipe.

Several impacting-type soil-boring tools are commercially available but are restricted to fairly short runs (around 25m) in uniform soil. Water jet-assisted impactors can improve penetration rates two or even four fold and can be used in a wider range of soils and clays. Two approaches have been used; one is to develop a steerable boring tool for installing new cables, and the other is to replace an existing cable using the redundant one as a guide for the boring tool.

A steerable water jet-assisted impactor is shown in Figures 6.1.3 and 6.1.4. The penetration rate may be increased without increasing the power delivered to the tool since the water jet pilots the hole and reduces friction. Since the design is steerable it can also avoid obstacles and can operate in rocky soil. The high-pressure water jet aids significantly in the steering ability of the device by jetting in the desired direction of travel. This particular prototype operated with a single 0.64mm diameter jet at a pressure of 544 bar, could advance at 25mm/sec, and was capable of a 2° course correction.

A cable-following tool is shown in Figure 6.1.5 which is a pure water-jetting machine using an array of five nozzles to excavate soil. Using water pressures of 340 bar, the double-acting gripper system has reportedly obtained penetration rates of up to 10mm/sec in field trials on US primary distribution cables. The replacement cable can be installed either by direct attachment to the tool or by connecting it to the failed cable and pulling it into place as the redundant cable is removed. In this type of design the water plays an important role in the transportation of waste slurry from the hole.

Cornering water-jet drill

Another example of a water-jet drilling machine may be seen in applications to methane drainage in coal mines and the linking of vertical process wells for underground coal gasification. Such techniques are equally applicable to urban trenchless construction. A development project [4] discusses work which started in 1975 and recently culminated in field tests of a cornering water jet drill. This was designed to drill horizontally through a coal seam from a single borehole and is shown in Figure 6.1.6. The technique is first to drill a 16-inch diameter cased borehole to the required depth. The bottom of this hole is then underreamed to a 6ft diameter using multiple rotating water jets. The drill string can then be inserted down the borehole and the shoe moved into the drilling mode. The water jets are rotated to provide an 8 to 12-inch diameter hole as the drill head is mechanically advanced using a gear drive. A submersible pump is used to remove the water which accumulates in the sump from the drilling jets and flushing jets which are situated 20 inches behind the drill head. Figure

6.1.7 shows the drill head assembly with twin rotating jets, rotary seal and hydraulic drive motor. At a water pressure of 690 bar and a flow of 25gpm, advance rates of 5ft per hour have been obtained in proof of concept tests at an underground coal seam. The cutting head path may be monitored continuously using accelerometers and magnetometer generators within the drill head, thus making possible a vertical deviation of only ±5 inches over a 50ft-long hole.

WATER JET-ASSISTED TUNNELLING

Although there are possibilities for pure water-jet tunnelling machines, almost all the applications to date have been in combination with mechanical cutters. However, there seems every reason to expect that development of techniques, similar to those used for drilling, will eventually result in a pure water-jet tunnelling machine.

Pure water-jet tunnelling

A production machine that loosely fits into this category is the 'Hydrojet' shield [5,6]. This machine has been used in pipe jacking down to 1.5m diameter and uses three oscillating jets that sweep across the cutting face as shown diagrammatically in Figure 6.1.8. Bentonite slurry is pumped through the nozzles at pressures up to 20 bar and although this can hardly be described as high-pressure jetting, it has been capable of excavating soft soils and sands. The bentonite is also an ideal medium for transporting the waste which may then be separated at the surface before the bentonite is recirculated to the nozzles. The application of high-pressure water jets to this technique will be discussed later.

Tunnel profiling

Water jets have provided advantages in conventional drill-and-blast methods of tunnel excavation. Their role has been in the smooth profiling of the tunnel by adding some direction to the blasting in medium and hard rock. The jets are used to cut slits in each of the outer shot-holes allowing the blast to break the rock along the lines of the slits to form the boundary of the tunnel[7]. The time required to perform the jetting operation is negligible compared with the excavation time, and the procedure reduces overbreak of the rock.

Multiple rotating water jets have also been found capable of road profile cutting in rock[8], although the main application is likely to be assisting hard metal tools to increase their range of application.

Water jet-assisted disc cutters

The use of high-pressure water jets to optimize the performance of mechanical cutters is an established technique, especially in hard rock excavation where advance rates are slow and cutter wear and costs are high. Full-face heading machines have been built[9], and display some of the possible economic advantages. A 2.6m diameter disc tunnel borer is shown in Figure 6.1.9 and showed a reduction in feed force of as much as 55 per cent by combining an effective water-jet cutting tool with a disc breaking tool. The jets cut grooves either side of the disc, putting free surfaces into the rock, thus reducing the shearing force required by the disc. The multiple jets required 100-200 litre/min to operate at 4000 bar and the pumping equipment is available. The water jets offer other advantages since the reduction in feed force potentially reduces the size and weight of equipment, reduces tool wear, suppresses dust and sparks, and also provides effective flushing of the tunnel floor to remove the waste. Design must obviously ensure that the nozzles are well protected from impact and abrasion by the tunnel face.

Water jet-assisted boom-type roadheaders

Water-jet assistance has also been applied to mechanical picks used with boom-type roadheaders[10] as shown in Figure 6.1.10. The cutting head uses 15 nozzles positioned on the forward part of the head where wear is usually the greatest. A 45-litre/min, 700-bar intensifier pump is mounted on the boom and supplies the tungsten carbide nozzles through a rotary seal. Trials on sandstone showed no reduction in power, but on harder limestone, forces and machine vibration were reduced and tip life increased. Normally the hard limestone would be considered uneconomical for excavation by this method, and thus the water jet technique looks very promising. Rotating seal problems have been effectively solved by using either a layshaft gearbox or a water passage within the boom. At present there are four modified boom headers undergoing field tests in the UK and all the benefits expected from the use of the technology have been achieved to some degree.

Water-jet assistance has yet to be applied to mechanized pipe-jacking machines but there seems no reason why similar advantages could not be realized.

TRENCHLESS SEWER REHABILITATION

Water-jetting methods have been extensively used for sewer-cleaning operations for many years. The technology has proven very effective at removing obstructions ranging from build-up of silt and scale to intrusion by tree roots, bricks, lateral connections and even steel reinforcement. Such operations normally require the use of multiple-jet cleaning heads which are fed underground on flexible hoses and supplied by surface-mounted high pressure jetting pumps.

The introduction of trenchless sewer rehabilitiation methods, especially for non-man-entry sizes, has spurred the development of a new breed of water-jetting machine. There are a number of techniques for restoring such sewers and the majority of these require any obstacles to be removed from the pipe prior to restoration. Although many obstacles can be cleared using conventional water-jet cleaning nozzles, there are some which require a controlled and precise high-pressure water jet. This requirement has led to the development of special motorized carriages which traverse the jet, or jets, around the inner surface of the pipe at a controlled speed. At pressures of 700 bar, lateral pipe connections and most other intrusions can be cut. The carriage will usually be winched into place and extensive use is made of CCTV. Some hybrid mechanical/water-jet machines have also been developed that provide similar functions.

After the sewer is cleared, it may be relined or a new pipe of smaller diameter than the original can be inserted. If a pipe of the same or larger diameter than the original is required, an expander is used to break the old pipe and consolidate the soil, allowing insertion of the new pipe behind. The problem is that the new pipe covers the lateral connections which must now be reopened. Although the connections can be cut from the lateral it can be easier to do this from the main sewer, even though the position of the lateral is obscured. Rotating water jets or mechanical cutters can be used for this task and accurate positioning of the carriage may be accomplished from previous measurement or electronic sensing devices. Figure 6.1.11[11] shows a twin rotating water-jet cutting device for reopening lateral connections in polyester and PVC sewer linings.

PILE DRIVING

The application of water jets to underground construction methods may be extended to the driving of sheet piles or extraction of RSJs in soft rock. Work has shown [12] that fan jets can fracture the ground ahead of piles, reducing the resistance and thus increasing efficiency two to five times over conventional methods. Field experiments have also shown that if water jets are used to cut 'mudstone' around a pile, it can then be withdrawn using a vibro-hammer. This would not have been possible using conventional means without excavation.

The techniques can be applied to clay, sand and soft rocks. When extracting piles, pilot holes may be drilled before the jet is inserted, which is then manipulated to loosen the surrounding earth. For driving, the jet can be attached to the extreme edge of the sheet pile. In addition to reduced resistance, damage to the piles is minimized and also vibration greatly reduced, which can be important if there is a nearby construction.

ABRASIVE WATER JETS

In general, as the pressure of a jet increases so does its potential to do damage. In the early days of water-jet cleaning it was soon realized that higher pressures were required as the deposits became harder. Jetting equipment was thus designed for higher and higher pressures until a reliable limit for conventional piston pumps of around 1000 bar was reached. However, even this pressure is not sufficient to remove some stubborn deposits or to clean metal to a 'white' metal finish which is ideal for recoating. The very high-pressure intensifiers which have been developed to operate above 1000 bar can cut a range of hard materials although the flow is relatively low and thus the time required to clean large areas can be prohibitive. Also the necessity to use rigid small-bore pipes has rendered such high pressures impracticable for many site operations. The pressures required to cut steel with pure water are beyond those obtainable with conventional intensifier equipment. Another limitation is that brittle materials tend to break instead of cut, and pressure build-up within fissures causes spalling and unpredictable fracture. In sewer cleaning applications, excess pressure or flow can cause considerable damage to the sewer and surrounding soil.

Adding abrasives to cleaning jets at pressures up to 700 bar increases the scope of materials which can be removed, increases cleaning rates, and provides a 'white' metal finish. The focusing of such abrasive jets into a cutting tool has enabled many hard materials including steel to be cut, while using reliable pumping equipment and flexible hoses.

Equipment

A complete abrasive-jet cutting system comprises a high-pressure water pump and delivery hose together with an independent abrasive feed system connected to the cutting head containing the high-pressure nozzle. Several systems are currently available, although they are usually tailored to suit particular applications.

In a typical abrasive-feed system (Figure 6.1.12) the supply hopper is loaded manually and the abrasive is mixed into a water supply at a metered rate. This is then pumped in a slurry form at a relatively low pressure (2 to 10 bar) to the remote cutting head. It is also possible to dry feed abrasive to the cutting head but this increases the risk of sparks during cutting. The most important element in an abrasive-jet cutting system is the actual cutting head, a schematic of which is shown in Figure 6.1.13. High-pressure water is accelerated to high velocities through a nozzle. The jet which issues from the nozzle passes through a mixing chamber to an outlet tube. The action of the jet causes a reduction in pressure within the chamber allowing the abrasive to be entrained and accelerated by the water.

The chamber and outlet tube are designed to ensure thorough mixing and even distribution of abrasive particles, which also avoids high rates of wear. The components within the cutting head that are exposed to abrasives are designed for ease of replacement and are made from wear-resistant materials such as tungsten carbide.

The choice of abrasive depends upon the target material and can range from expendable sand and copper slag, to expensive abrasives such as silicon carbide and aluminium oxide. High-cost abrasives would normally be recycled, making them more suitable for fixed installations. In mining applications, one material that is generally available in large quantities is the stope material itself. In crushed and graded form this could possibly be utilized as the abrasive in the cutting process, and has the advantage of being virtually self-generating.

Cutting trials

Cuts have been performed on materials as diverse in nature as cast iron, sandstone and quartzite rock. However, a high proportion of the work has been concerned with cutting steel and concrete sections (both plain and reinforced) [13, 14]. As an illustration of the results, a reinforced concrete block 700mm deep (Figure 6.1.14) was cut through using a multi-pass technique, whilst 13mm mild steel plate was cut in a single pass at 200mm/min. Figure 6.1.15 demonstrates that good-quality cut surfaces are obtained on a wide variety of metal sections. In order to investigate the behaviour of abrasive water jets in explosive environments, cutting trials have been conducted on steel and sandstone in flammable atmospheres. A total of 79 cuts were carried out (equivalent to approximately 26 minutes of continuous cutting) in explosive mixtures of various compositions without causing ignition. The cut surface is continuously cooled by the water and the jet creates a damp working atmosphere which suppresses dust.

THE WAY FORWARD

Much of the equipment discussed above is only at prototype stage, although it is hoped that after further development and testing, more of the techniques will exist as standard practice.

Pure water-jet machines

Possibly the most promising new area of application for pure water jets in trenchless construction is in pipe jacking and small-bore tunnelling. The hydrojet shield described earlier has demonstrated the use of jets at very low pressure. The technology exists to design a shield that would excavate and bore using water jets, permitting a considerable reduction in jacking forces. If the advantages seen in other applications, such as pile driving, can be applied here, then the impact could be considerable. The addition of water jets would have the potential of increasing jacking lengths, or reducing the need for intermediate jacks, and hence reducing cost. Hypothetical designs of water-jet shields as shown in Figures 6.1.16 and 6.1.17 would employ water jets to fracture the earth ahead of the shield and also as a means of excavation. For soils and sands, fixed fan jets may be all that is required, although it is suspected that on harder earths and soft stone, an array of rotating straight jets (as in Figure 6.1.17) or oscillating straight jets would be necessary. The excavating jets could be positioned perpendicular to the work face or could be directed forward slightly to aid the perimeter jets. An advantage of the technique is that the jets could effectively ream a tunnel of larger diameter than the shield thus reducing the jacking force. The jets should also increase break-up of the waste and hopefully reduce the number of stoppages necessary to clear any obstructions. The remaining of the workface to a larger diameter ahead of the shield will increase steerability and this could be most beneficial driving pipes on both horizontal and vertical curves. Since curved jacking also increases resistance, water jets may keep the forces within practical limits.

The pressure and flow requirements of the machine will depend upon the conditions underground although in general, low pressure and high flows would suit soft soil and clay while soft rock would require high pressure and lower flows. The water supplied to the workface would also allow flushing of the waste and transportation of this waste to the surface. If necessary the water can be recirculated to the nozzles after suitable filtration.

Another area of application would be to transfer the technology of water jet-assisted mechanical cutters to mechanical shields. Such jet assistance has already been proven to have economic advantages in mining and tunnelling field trials.

Abrasive water jet applications

Abrasive water jets could also be used in combination with mechanical cutters or explosives in a similar manner to conventional hydraulic cutters. However, the cutting capability of an abrasive water jet greatly exceeds that of a plain water jet. Therefore existing equipment for combined cutting operation is unlikely to be suitable for this purpose. The main application for the technique is likely to be in mining or tunnelling of hard and abrasive rocks which is at present uneconomical using conventional means. Abrasive jet cutting is particularly advantageous in environments where risk of explosion exists, such as the flammable or dust-laden atmospheres frequently encountered underground. The actual cutting head is dimensionally small and can be used in confined or otherwise awkward areas where access for conventional equipment is not possible. Also the reaction force from the head is comparatively low and is constant and independent of both target material and stand-off. The cutting head is thus highly suited to automated control. A further advantage is that the power unit and abrasive-feed system can be located remotely from the cutting head. At present distances up to 240m are possible between the cutting head and abrasive feed.

One application of the abrasive water jet in tunnelling could be profiling of the workface ahead of mechanical cutters or explosives. If some depth could be achieved by using an abrasive jet to traverse the perimeter of the tunnel, the free surface available would reduce the cutting force and overbreak. This method may well extend the range of mechanical cutters and could provide improved profiles.

Abrasive water jets can also be advantageous in sewer cleaning and renovation. The increased performance at lower pressures, compared with pure water jets, means that lighter and more flexible hoses can be used. There are thus some operational advantages to be gained from the use of this technique. Also, as discussed previously, some cleaning and rehabilitation methods require controlled and precise cutting. The abrasive jet considerably extends the range of materials that can be cut, including steel, and can improve the quality of cut on brittle materials. This, combined with the lower pressures, has the potential to reduce damage to the sewer and surrounding soil. BHRA, WRC and some contractors have demonstrated the advantages of abrasive water jets in such applications and the results are very encouraging. Sand would typically be used as the abrasive in such operations and this can be trapped and removed at the nearest convenient downstream manhole.

CONCLUSIONS

Pure water jets have been found capable of increasing the applications and economics of many underground construction methods. Some solely water-jet machines have been developed to prototype stage and have been most effective in soils, clays and soft coals. The ability of water to efficiently excavate, provide lubrication and aid transportation of waste materials is foremost. However, water jets can have operational advantages by increasing steerability and hence the accuracy of tunnels.

The relatively recent development of abrasive water-jet cutting methods has spawned many new applications to underground mining, construction and renovation, and this is bound to be a most exciting area for development. The precise machining and cold-cutting ability of such abrasive jets are very useful qualities of the process.

[1]Vijay, M M, Grattan-Bellew, P E and Brierley, W H, An experimental investigation of drilling and deep slotting of hard rocks with rotating high-pressure water jets, *Paper H2, 7th International Symposium on Jet Cutting Technology,* BHRA Fluid Engineering, Cranfield, June 1984.

[2]Lefin, Y and Hurel, A, Rotary drilling assisted by water jets and other recent developments of high-pressure water jets for the cutting of rocks in France, *Paper H3, 7th International Symposium on Jet Cutting Technology,* BHRA Fluid Engineering, Cranfield, June 1984.

[3]Reichman, J M, and Kelley, D P, Development of water-jet tools for installation and replacement of buried utilities, *Paper H3, 6th International Symposium on Jet Cutting Technology,* BHRA Fluid Engineering, Cranfield, April 1982.

[4]Shirey, D L and Engler, B P, Development of a cornering water-jet drill, *Paper H4, 7th International Symposium on Jet Cutting Technology,* BHRA Fluid Engineering, Cranfield, June 1984.

[5]Thomson, J C, Sewer rehabilitation in Hamburg Part 1: Pipe jacking and Hydroshield tunnelling, *Tunnels and Tunnelling, 15,* 10, p25, October 1983.

[6]Anheuser, L, Experiences with the Hydrojet shield, *Small-bore Tunnelling and Pipe Jacking: Supplement to Tunnels and Tunnelling,* p.13, October 1984.

[7]Honda, H, Okuda, K, Igarashi T, Kanzawa, M and Hoshino, K, Application of water-jet cutting to economic tunnel excavation, *Paper P7, 7th International Symposium on Jet Cutting Technology,* BHRA Fluid Engineering, Cranfield, June 1984.

[8]Baumann, L and Koppers, M, State of investigations on high-pressure water jet-assisted road profiling, *Paper G2, 6th International Symposium on Jet Cutting Technology,* BHRA Fluid Engineering, Cranfield, April 1982.

[9]Baumann, L and Heneke, J, Attempt of technical-economic optimization of high-pressure jet assistance for tunnelling machines, *Paper C4, 5th International Symposium on Jet Cutting Technology,* BHRA Fluid Engineering, Cranfield, June 1980.

[10]Plumpton, N A and Tomlin, M G, The development of a water-jet system to improve the performance of a boom-type roadheader, *Paper G1, 6th International Symposium on Jet Cutting Technology,* BHRA Fluid Engineering, Cranfield, April 1982.

[11]Emery, J A, New techniques in non-man-entry sewer renovation, *Paper D1, International*

Conference on the Planning, Construction, Maintenance and Operation of Sewerage Systems, BHRA Fluid Engineering, Cranfield, September 1984.

[12]Yahiro, T, Yoshida, H, and Nishi, K, Sheet pile driving and H-steel pulling by high-speed water jets, *Paper D4, 5th International Symposium on Jet Cutting Technology,* BHRA Fluid Engineering, Cranfield, June 1980.

[13]Saunders, D H, A safe method of cutting steel and rock, *Paper K5, 6th International Symposium on Jet Gutting Technology,* BHRA Fluid Engineering, Cranfield, April 1982.

[14]Barton, R E P and Saunders, D H, Water/abrasive jet cutting of concrete and reinforced concrete, *Paper K4, 6th International Symposium on Jet Cutting Technology,* BHRA Fluid Engineering, Cranfield, April 1982.

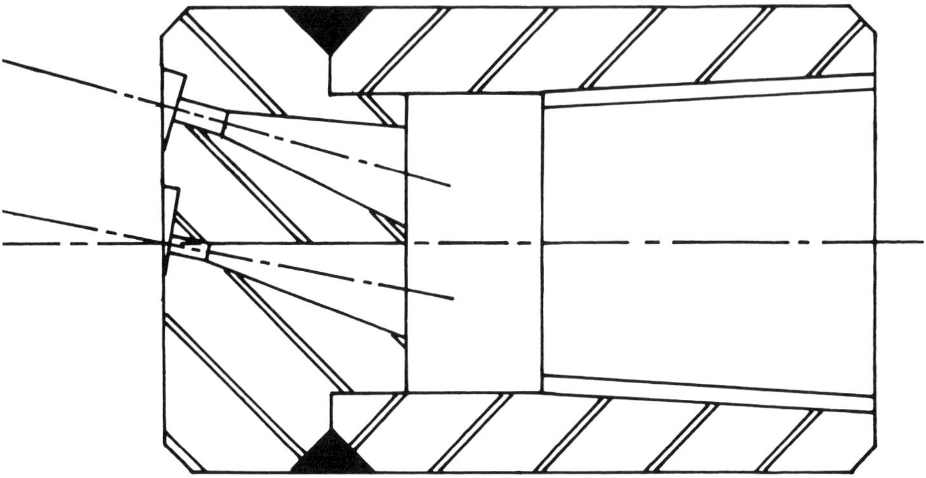

Figure 6.1.1. Nozzle body incorporating twin conical entry orifices.

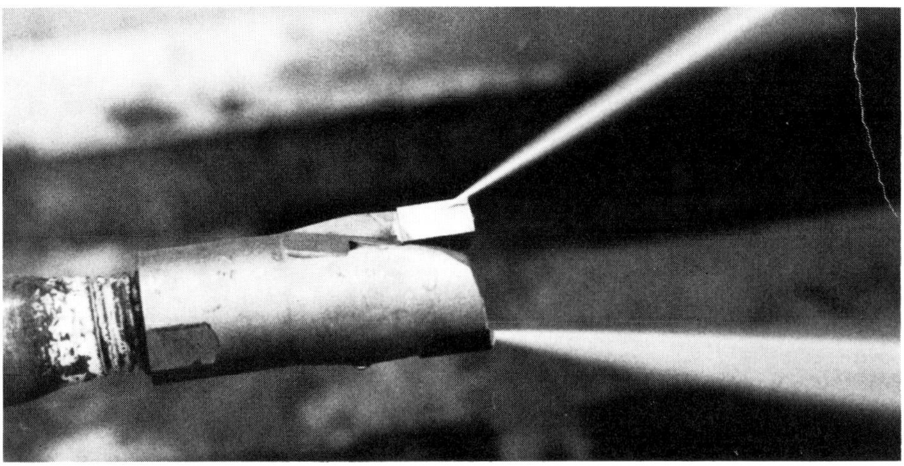

Figure 6.1.2. Assisted ring bit with nozzles in the tungsten carbide inserts.

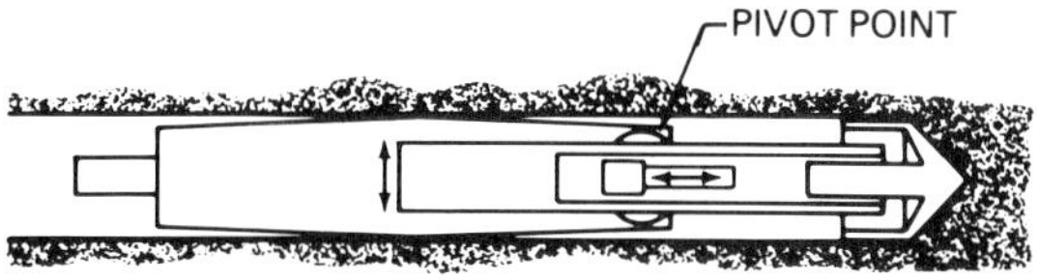

(a) Straight-line advance.

(b) Maximum correction.

Figure 6.1.3. Steerable soil-boring tool.

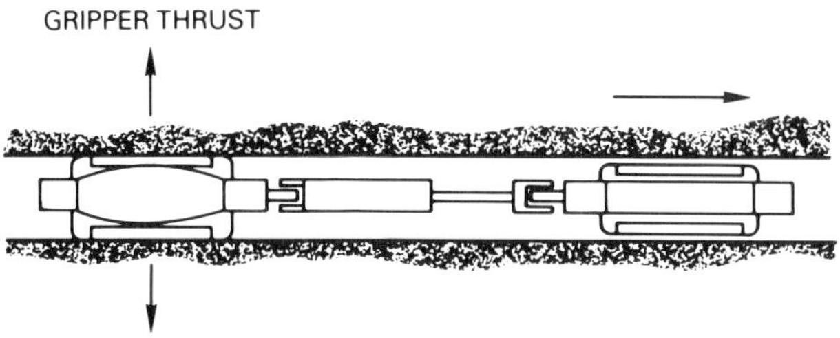

(a) Rear gripper activated (cylinder advancing stroke).

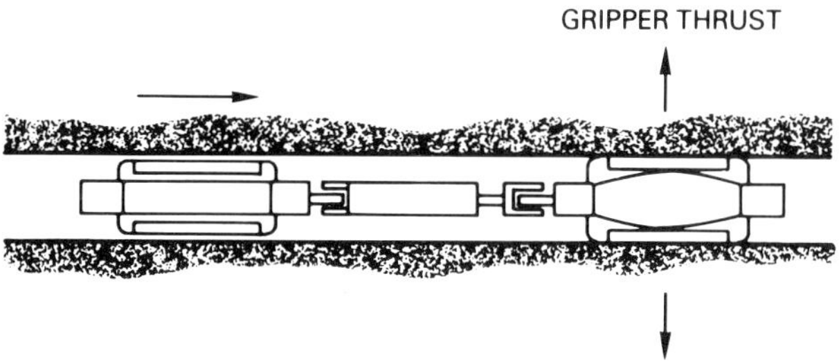

(b) Front gripper activated (cylinder retraction stroke).

Figure 6.1.4. In-hole advancing section.

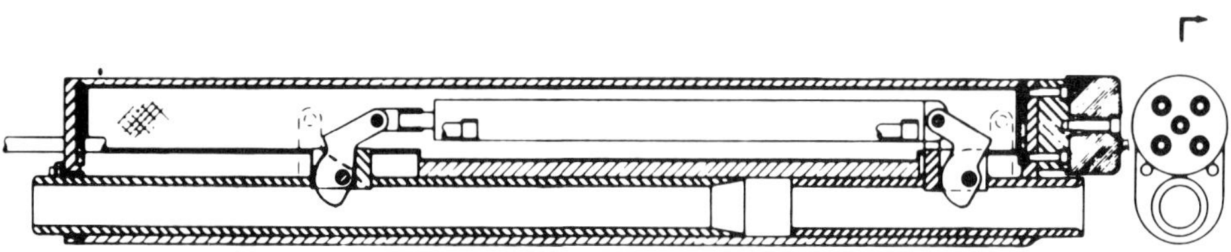

Figure 6.1.5. Cable-following tool design.

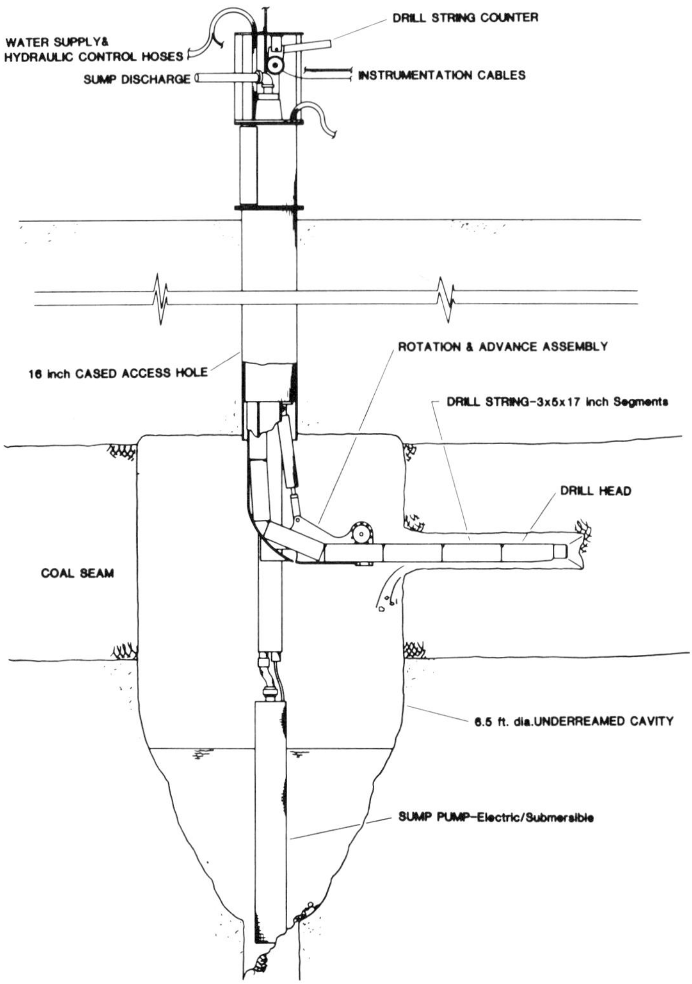

Figure 6.1.6. Water-jet drilling system layout.

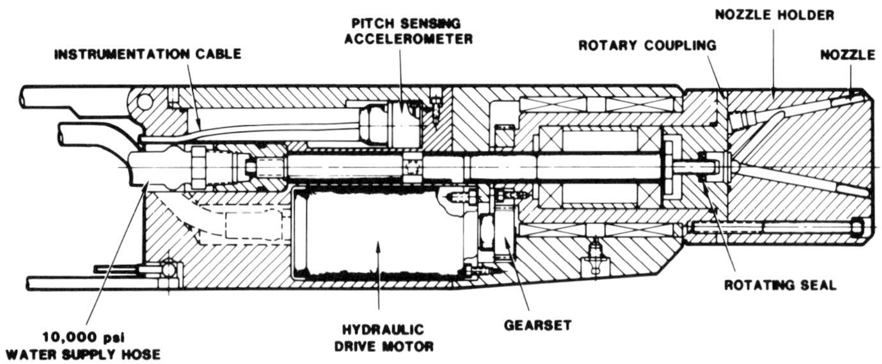

Figure 6.1.7. Drill head assembly.

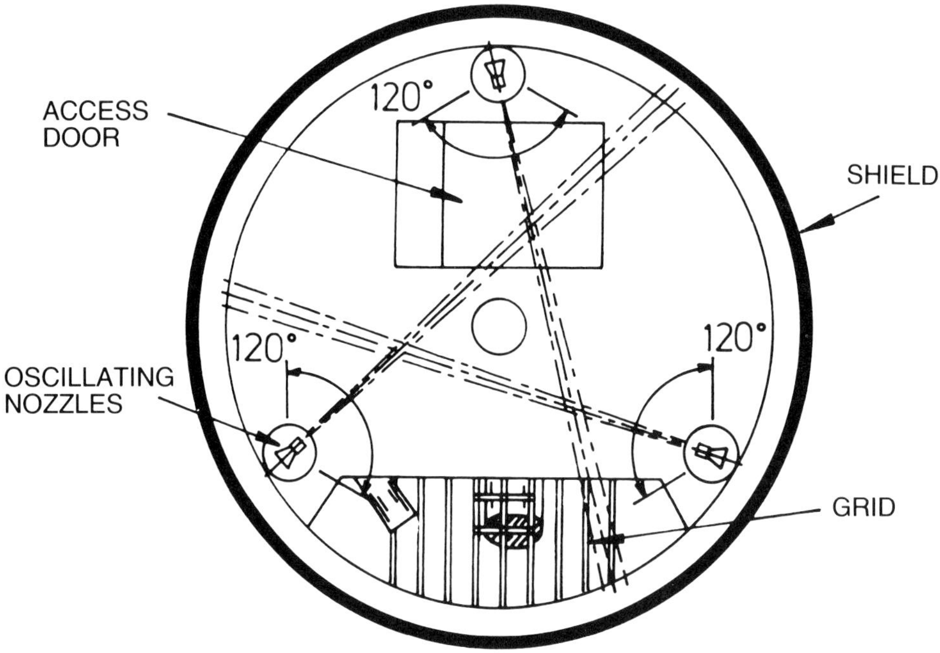

Figure 6.1.8. Principle of Hydrojet shield.

Figure 6.1.9. 2.6m-diameter water jet-assisted tunnelling machine.

Figure 6.1.10. Roadheader prior to cutting in limestone.

Figure 6.1.11. Water-jet cutter for reopening laterals.

Figure 6.1.12 Abrasive-feed system

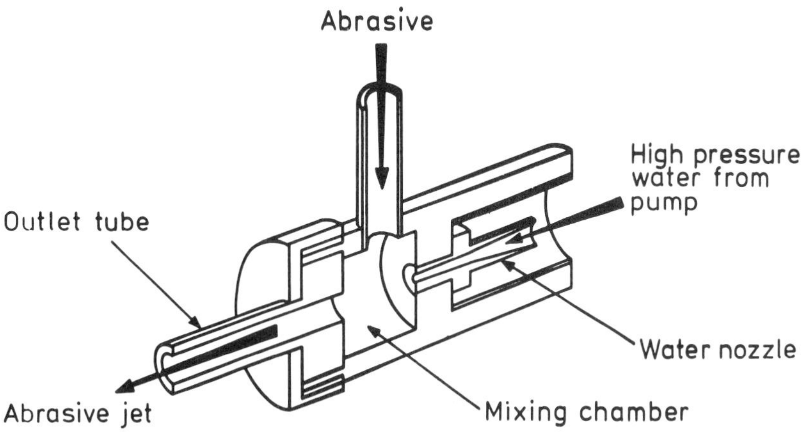

Figure 6.1.13. Schematic of abrasive-jet cutting head

Figure 6.1.14. Reinforced concrete cut by abrasive water jet.

Figure 6.1.15. Mild steel plate, I-beam and cast iron cut by abrasive water jet.

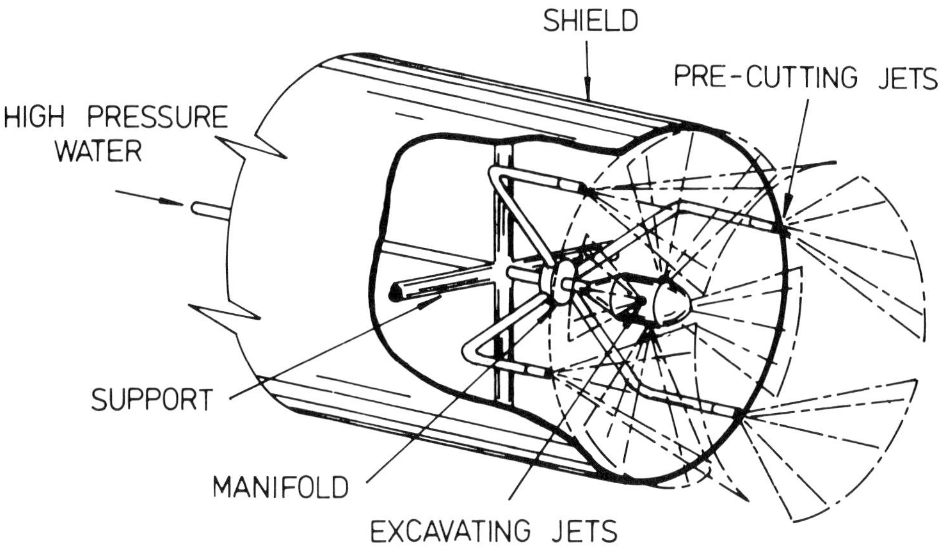

Figure 6.1.16. Fan-jet shield.

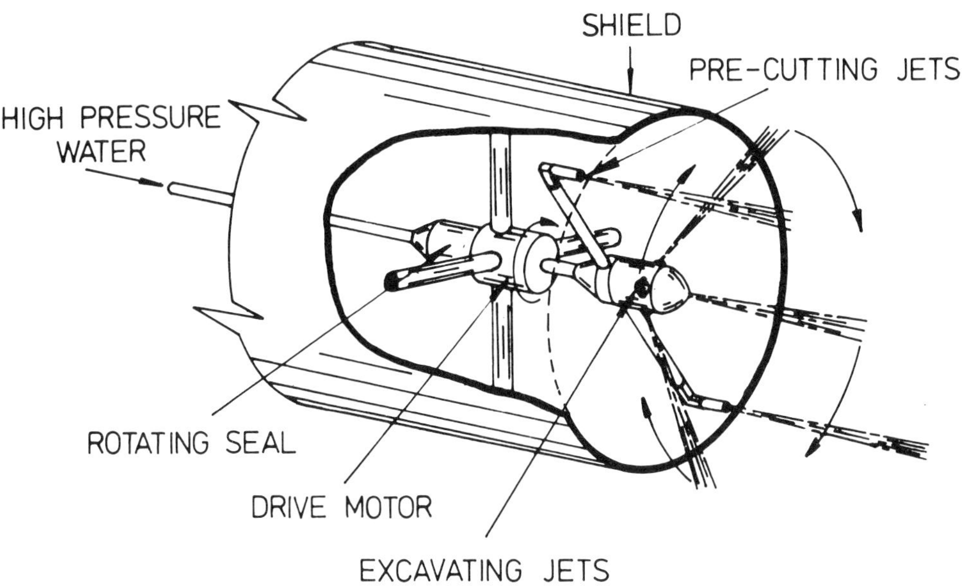

Figure 6.1.17. Rotating-jet shield.

Author's Introduction

Mr R M Fairhurst: Water-jetting techniques: existing and potential uses for underground services

Water jets are used for a wide range of cutting and cleaning activities, both in their own right and in association with mechanical techniques. This paper discusses work, much of which is credited to other authors, on the application of water jets to areas which can be related to trenchless construction. It must be treated as a speculative paper, the main aim being to make delegates aware of some of the possible applications.

Water jets are capable of cutting many stones and rocks, and offer an excellent transport medium for the waste materials. Figure 6.1.1 shows a pure water jet drilling head which has two conical entry nozzles, usually set at different angles. This is rotated to produce holes in sandstone and granite. Typical operating pressures would be 10 000psi (690 bar) to 60 000psi, with rotational speeds of 200rpm.

Performance can be improved and harder rocks drilled if the water is combined with some mechanical means, as in Figure 6.1.3. Here the jets are an integral part of the cutting tip, providing lubrication and cooling of the tool as well as hydraulically fracturing ahead of the tool to reduce drilling forces. Applications for this type of equipment include drilling of blast holes and for roof bolting.

A recent application of drilling in soft ground has been for cable installation and replacement. Water jets may be used to assist impact moles, showing up to fourfold improvements in penetration rates. Reduced vibration and ground disruption should also result.

A steerable boring tool for installing new cables over long distances is shown in Figure 6.1.4. This uses a single jet within a separate anvil and jets in the desired direction of travel, thus aiding steerability. Because the impactor is separate and slightly larger than the main body, an in-hole advancing system is necessary as a means of counteracting the force of the piston on the return stroke.

A pure water jet cable-following tool has also been developed (Figure 6.1.5). This uses an array of jets to excavate the soil around the buried failed cable, and is traversed along by the machine using a double-acting gripper system.

Field trials in unconsolidated sands have required the use of drilling mud as the jetting fluid to prevent collapse of the hole behind the tool. Tests in compacted soils and clays have required the fixed jets to be replaced by rotating jets, at pressures of around 4000psi (275 bar). The rotating jets provide efficient area coverage for the 5-inch bore without the need for too many jets.

Rotating water jet drilling heads have also been used at depth underground for methane drainage in coal mines, and for the linking of vertical process wells in underground coal gasification (Figure 6.1.6). The procedure is described in detail in the paper.

When we start talking about larger diameter holes, then we are really going on to tunnelling machines. The Hydrojet shield, shown in Figure 6.1.8, is a very good example of how relatively low-pressure slurry jets have been used for excavating soft soils and sands. This shield is used for pipe jacking down to 1.5m diameter. It uses three oscillating nozzles which sweep across the work face of the machine, thus excavating the soil.

However, most tunnelling to date has usually been in combination with mechanical cutters, for example, disc cutters with two or more jets spaced on either side of the disc. The jets put free surfaces into the rock and this results in a reduction of shearing force required by the disc. Full-face heading machines have been produced on this principle. The machine shown in Figure 6.1.9 used pressures of 60 000psi and showed a reduction in the feed force of over 50 per cent as a result of using water jets.

There are also four assisted boom-type road headers undergoing field trials in the UK at the moment (Figure 6.1.10). These use an array of jets, usually 15, in the forward part of the machine where most of the force is encountered. The machines have shown reduced feed force and also reduced vibration when cutting harder limestones which would normally have been beyond economical excavation. They are at present undergoing field trials in coal mines, where they have shown most of their advantages in limestone, and also they offer good dust suppression. Obviously similar advantages could be realized in mechanized pipe jacking applications.

The sewer cleaning industry is probably the largest user of high-pressure water-jetting pumps in the UK at the moment. It was thus logical that water-jetting techniques would be applied to trenchless sewer rehabilitation. Present methods usually require the initial removal of any obstacles from the sewer and this includes lateral pipe connections which must be cut with a precise and controlled water jet. Special carriages have been developed which can traverse the inside of the pipe and thus cut such lateral connections. The water jets can also be used later to re-open these connections in the new pipe using similar equipment. Figure 6.1.11 shows an example of a machine for re-opening a lateral connection.

Relatively recently abrasives have been added to water jets to improve their performance and this allows many materials to be cut at pressures which can be reliably generated. The method can also be used safely in environments which are potentially hazardous. The abrasive is usually entrained into the water using a jet-pump type of principle. Figure 6.1.13 is a schematic cut-away of a jet-cutting head. The high-pressure water passes through an accelerating nozzle and across a mixing chamber to an outlet tube. The action of the jet allows the abrasive to be added to the mixing chamber where it is then accelerated along the outlet tube by the water. The abrasive is usually fed to the cutting head in slurry form at low pressure and to do this we use a suitable abrasive feed system comprising a hopper and a screw feeder (Figure 6.1.12). The abrasive is metered in with water and fed through a low-pressure slurry pump at 2 to 6 bar pressure. Such jets have accurately cut many materials, for example, reinforced concrete and steel (Figures 6.1.14 and 6.1.15).

The cold-cutting ability of the jet has shown itself to be safe in hazardous environments and this is obviously ideal for the flammable and dust-laden atmospheres frequently encountered underground.

Looking to the future now, it would seem that there are many possibilities for water jets to be applied to trenchless construction. As far as pure water-jetting machines are concerned, pipe jacking and small-bore tunnelling could possibly benefit from the use of high-pressure water jets. If some of the advantages that have been seen in other applications could be applied, jacking forces could be considerably reduced if water jets were used. A hypothetical fan jet shield is shown in Figure 6.1.16. In soft soils and sands all that may be necessary is an array of fixed fan jets, but for compacted soils or even rock and stone, it is expected that rotating or oscillating straight jets would be necessary. The pressures are likely to be around 10 000psi for this sort of operation, and if the jets are allowed to ream a slightly larger bore ahead of the shield, this could potentially improve steerability. The jets should also increase break-up of the waste materials and therefore aid transportation to the surface.

The ability of abrasive water jets far exceeds that of plain water jets, but it is unlikely that plain water jet machines could be adapted for use with abrasives, and we would therefore think that this would require the design of completely new machinery. One application, however, could be for tunnelling in hard and abrasive rocks which are currently beyond economical excavation. The abrasive jets could be used for profiling, for instance, and this would result in a reduction in the cutting forces at the work face and would also reduce over-break. Some of the benefits would be the elimination of backfilling while still giving full support.

The good quality cuts that can be obtained with fairly low-pressure abrasive jets in brittle materials can also be applied to trenchless rehabilitation of sewers. As well as an increase in the cut quality, the ability to use such low pressures should result in reduced damage to the sewer and surrounding soil. It is therefore obvious that water-jetting techniques offer a very exciting area for development in trenchless construction.

Discussion on Paper 6.1

C E G Bland (Director of Engineering Services, Clay Pipe Development Association): Water jetting, as Mr Fairhurst rightly says, has been going on for many years. It is interesting that he is now coming forward with boring machines that can be used for NO-DIG purposes.

He gives an accuracy of ±800mm per 100m for the Cornering Water Jet Drill. I think that is excessive compared with the specification of ±25mm per 100m laid down by the original Trenchless Pipelaying Working Party set up by this Institution.

I should be interested to know how long it takes to cut out a lateral by this method. Would it depend on the pipe material — clay, concrete, PVC or polythene?

How does the use of abrasives affect the diameter of the jet being used? Surely the abrasive will cause wear and open up the jet, so reducing the pressure. How long does a jet last?

The use of fan jets seems a very good idea. Presumably they could have a steerable head, and, by using the techniques described in the previous three papers, they could be traced accurately through the ground. Is that the way water jetting is likely to go?

J Mercer (Senior Project Manager, FlowMole Corporation): FlowMole (formerly FlowJet) developed some of the jet-cutting tools described in Mr Fairhurst's paper. The paper covers our work up to about 1982. I would like to bring the account up to date.

FlowSystems, the sister company of FlowMole, manufacturers ultra high-pressure water cutting equipment with pressures up to 4000 bar. The two companies have a great deal of experience with jet cutting, and have developed quite a few tools using both pure water and abrasives.

The cable follower is now in commercial operation replacing underground utility cable. The tool excavates around the old cable, and the new cable is pulled in as the old cable is pulled out. The two guided tools we are developing, called the GuideDril™ and the Guided Mole, do not require a cable in the ground.

The GuideDril™, which is just being introduced commercially, uses a bentonite solution in what we call a soft-boring concept. The pressures are low enough for the jets not to harm other utilities — they will not cut insulation, wire, metal or plastics. The tool itself is smooth, and has no cutting edges to damage other services. It operates in a wide range of soil types, from pure sand to hard clay, and from homogeneous soils to fairly rocky soils. It produces a 40mm diameter hole and reams holes up to 150mm. The tool can be started from a small slit in the ground or directly from the surface. It is fully locatable and guidable. We have complete steering on the tool, with a turning radius of about 10m.

The drill can drill a range of greater than 60m, and currently the average daily production is about 120m per day. The working accuracy is within a radius of 15cm along the path, and at the exit it can be controlled down to 8cm or better. It has been used for installing water lines, gas lines, electric cables, telephone cables and conduits.

The Guided Mole uses the same bentonite soft-boring concept. It produces a 75mm bore and has a fully automated guidance and control system, giving increased accuracy and productivity. It has a drilling range of about 120m, and is expected to be in commercial demonstrations later this summer.

As an illustration of what the GuideDril™ can do, in one of our demonstrations we launched the tool from the surface and controlled it down to a depth of 75cm. At that point we steered the tool to make it travel in a level line. We advanced it 30m, then steered it up, and in 3m it came back up to the surface. Later, we excavated to measure the accuracy of the hole, and found that everywhere along the course it was within 5cm of the desired path.

Mr R M Fairhurst: Mr Bland commented on the accuracy of the Cornering Water Jet Drill. This machine was actually undergoing field trials in about 1983 at a simulated coal seam 100ft below ground. The instrumentation was still being developed, and there were problems such as water getting into the equipment. That is why the accuracy was only within ±5in on 50ft.

For cutting laterals I have outlined some of the advantages of being able to use a pressure of 2000psi with abrasives instead of 8000psi, which is probably the recommended pressure for cutting lateral connections with pure water. Cutting time is much about the same for any material. Clay pipes require about six minutes maximum, and PVC pipes rather less. But compared with the time needed to winch the tool into position and get set up, the actual cutting time is a very small proportion of the overall procedure.

The effect of abrasives on cutting heads is a tricky subject. Wear-resistant materials are used, but correct design is the key factor. With suitable design and tungsten carbide nozzles — easily replaceable parts — the cutting head will last for 200 hours or so before major modification is necessary. The design of the cutting head — the diameter of the outlet tube — will determine the particle size of abrasive which it can use most efficiently. As far as nozzle wear is concerned, it can be seen from Figure 6.1.13 in the paper that the abrasive does not pass through the nozzle itself, which may be under 10 000psi, but only through the outlet tube. So the head will wear relatively slowly, and it will produce a jet which is fairly consistent in diameter for at least a couple of days' cutting, which is very adequate.

Fan jet accuracy depends very much on the pressure, and there is a relationship between the pressure and the traverse rate. The equipment must be advanced at a fairly constant rate so that the jet is using the first few inches as a cutting edge.

R Tharby (Chairman, National Joint Utilities Group, Session Chairman): I would like to ask a question which I am sure is in many people's minds. While a water jet system is being used inside a guided tube, problems of user safety may not arise. But once it is being used for independent moles, producing relatively small diameter holes, and appearing, one hopes, in the right place, there must be problems of user safety because the energy levels are so high. Can you tell us something about that?

Mr Fairhurst: Any water-jetting contractor who uses jets probably in the region of 3000 psi for sewer unblocking knows about the basic dangers. The jet is not generally operated until it is underground, because most of these nozzles have rearward-firing retrojets which would do considerable damage. The equipment is, of course, properly designed to hold the high pressures — possibly up to 15 000psi. This is basic engineering practice.

An Association of Water Jetting Contractors has been formed, one of whose major roles is to provide Codes of Practice for the safe operation of this type of equipment. These cover such matters as the proper position of barricades, handling of equipment, use of the right controls on the equipment, and so on. Even greater care is needed in handling abrasive water jets, which are much more hazardous. But if the equipment is maintained and used correctly, there is no reason why it should pose any more problems than a diesel or petrol engine, where very high pressures are also involved.

A Coulthart (Public Health Engineer, Property Services Agency): Two speakers at the recent Imperial College Symposium on sewer renovation stressed the dangers associated with the use of both high- and low-pressure jetting for cutting and cleaning in sewers. The problem highlighted was of the jets, through misalignment, being directed at cracks or damaged joints and causing flushing out of the soil behind the pipes, thereby creating a void.

Given the extensive current use of jetting, can Mr Fairhurst, or anyone else, give me any reassurances that these fears are unfounded?

Mr Fairhurst: I have already said how good water is as a transporting medium, so I can hardly deny that it will flush out soil. There is a threshold pressure above which damage will start, and the damage will obviously be much greater at 8000psi than at 1000 or 1500psi. The trouble is that the lower pressures will not cut the pipe. That is the main reason why we have gone to using low-pressure abrasive jet cutting heads for this type of use, because these will cut at pressures of 1000 to 1500psi. This, combined with the greater accuracy of the abrasive jet, reduces the risk of damage quite considerably.

P B Rumsey (Water Research Centre, and IPHE Organizing Committee): The Water Research Centre has developed a low-pressure abrasive cutting machine which is currently undergoing trials in the field by a contractor. As part of that work we have carried out damage evaluation trials to compare the damage caused by high-pressure water jets, and low-pressure jets using abrasives. It is not yet possible to publish the results, but I can give some information to anyone who is interested.

Certainly, damage is recognized as a potential problem, and we do not know what happens when the jets are moved around in sewer pipes or elsewhere. Soil is a highly erodible material, and a great deal of care, skill and accuracy must be exercised in using water-jetting equipment.

6.2 Development of a guided horizontal-boring tool

Steven R Kramer

Project Manager
Gas Research Institute Chicago, Illinois, USA

To minimize the high cost of pavement removal and restoration, utilities prefer to bore under roadways or other paved surfaces. Existing tools are adequate for boring short distances up to 60 feet (20 metres), but are not sufficiently advanced to provide directional control for longer distances. In 1984, the Gas Research Institute (GRI) initiated two programmes on a parallel path to evaluate various technologies applicable to guided horizontal boring. The ultimate objective of these programmes is to design and demonstrate a boring tool capable of drilling small-diameter holes up to 120 feet (40 metres) with a directional accuracy of one foot (0.3 metres) in all planes. First, the team of Maurer Engineering Inc. and Illinois Institute of Technology Research Institute, with funding from GRI, is retrofitting a commercially available horizontal-boring tool with tracking and steering systems. Second, Ingersoll-Rand Research, Inc., with GRI funding, is developing a new guided horizontal-boring tool with steering incorporated. During 1984, the key elements of the guided boring tool were designed and the operating characteristics demonstrated. This paper summarizes the accomplishments achieved to date.

INTRODUCTION

The installation and replacement of underground utility lines are costly operations. Conventional practice is to excavate soil by trenching. In order to reduce the high costs of pavement removal, excavation, and site restoration, utilities and contractors often use underground piercing and boring tools to dig horizontal holes below the surface for the placement of pipes, cables, wires and conduits. Currently available tools include ground-piercing impactors, pipe pushers and soil and compacting augers. These tools are adequate for boring short distances up to 60 feet (20 metres), but are not sufficiently advanced to provide directional control over longer distances. Once the tool has deviated from a straight course, there is no mechanism to correct it. Local soil conditions or rocks may also cause a boring tool to follow a non-linear path that cannot be detected from above the ground. As a result, the tool may miss its target, bury itself, or in the worst case, damage adjacent utility lines.

A survey of US gas utilities was conducted to determine the applications, performance goals, and design specifications for a guided horizontal-boring tool. The personnel from the responding utilities stated that deflection of the piercing tool from the desired path is a major problem. Only 30 per cent of the installations for gas distribution pipes were installed with horizontal-boring tools. Additional results of the survey were: 1) 33 per cent of the percussion-bored holes were not successfully completed in one attempt; 2) 89 per cent of the respondents consider the ability to continuously monitor the path of the boring tool highly useful; 3) Almost 90 per cent of the respondents have significant use for a boring tool that could be accurately guided to 120 feet; 4) 60 per cent of the utilities expected or desired to increase the use of boring tools in the future.

In 1984, Gas Research Institute (GRI) initiated a research programme to develop a horizontal-boring tool that can be accurately tracked and guided. Sensing capabilities to stop the boring tool before it damages other underground utilities are also being explored.

TECHNICAL APPROACH

Two alternative technologies are being investigated to develop a practical, reasonably-priced, guided horizontal-boring tool that could be used to significantly improve productivity in the installation of underground gas distribution piping. First, the team of Maurer Engineering Inc. (MEI) and Illinois Institute of Technology Research Institute (IITRI), with funding from GRI, is retrofitting a *commercially-available* horizontal-boring tool with tracking and steering systems. Second, Ingersoll-Rand Research Inc. (IRRI), with GRI funding, is developing a *new* guided horizontal-boring tool which incorporates steering.

The objective of the first phase of research was to design, fabricate, laboratory test and demonstrate the key operating characteristics and features of an improved horizontal-boring system capable of drilling small-diameter holes, one to eight inches in diameter (25mm to 200mm), for distances up to 120 feet (40m) with a directional accuracy of one foot (0.3m) in all planes. The successful application of the tool requires the development and integration of three subsystems. These subsystems are: 1) a mechanical steering system to direct the tool; 2) a tracking or orientation system to monitor the tool's position; 3) a proximity/obstacle detection system that can sense the presence of obstacles in the path of the tool. The third component will not be the emphasis of the research effort in the early phases. The funding of two alternative technologies with the same ultimate objective was undertaken to increase the overall probability of technical and commercial success. Upon the completion of the first phase in early 1985, GRI will select one concept or a combination of appropriate technologies for further funding.

The first phase of research was divided into the following tasks:

1. Define the state-of-the-art in technologies appropriate to guided horizontal boring.
2. Establish feasible design specifications and prepare design drawings.
3. Design and fabricate the key components for the guided horizontal-boring tool.
4. Test the laboratory model in controlled field situations.

RESULTS

A literature search was completed on percussion borers, augers, and drilling methods. Table 6.2.1 summarizes the performance of currently-available boring systems. Percussion moles have the fastest steering response of the six boring tools examined. The reasons for this performance are that percussion moles are short in length, typically four to seven feet (one to two metres), self-propelled, and do not require multiple drill strings. Therefore, percussion moles were selected for development because of the strong steering potential. While positive displacement motors offer faster penetration rates, these systems are expensive and not used by the gas industry.

Using the information obtained from the literature review, MEI and IRRI were able to prepare preliminary design drawings for their proposed systems. MEI decided to retrofit an existing tool (the Allied Hole-Hog piercing tool) with a guidance and steering system, and IRRI elected to develop a new guided piercing tool. This approach addresses two technically feasible alternatives for accomplishing the objective. The retrofit approach has the advantage of fast implementation since only the guidance and steering mechanisms need to be developed. However, the commercially-available systems were not designed to be steered.

Mechanical steering systems

Both contractors generated lists of candidate steering systems (Table 6.2.2). From these lists, the most promising concepts were designed and fabricated.

MEI retrofitted passive assemblies on a pneumatic piercing tool. The highest rated system, in terms of performance, cost, reliability and adaptability, was a slanted nose piece and lockable or movable tail fins (Figure 6.2.1). The front of the tool was mounted with a slanted nose cone (anvil).

Various curvatures and areas of anvils were tested to find the optimum design for turning. The results indicated that the turning capability of the anvil was repeatable within a few inches. In addition, the turning radius varied linearly with anvil angle (Figure 6.2.2) and the turning radius decreased in direct proportion to an increase in area of the anvil. Tail fins were screwed into the back of the tool to assist in changing the direction of the tool or to offset the effect of the slanted anvil. These components required no internal modifications to the tool. When the slanted anvil and tail fin assembly was field tested in a clay soil, it was demonstrated that the boring path of the tool could be redirected in a predictable, controllable manner at turning rates of two degrees per foot. The components added to the tool did not affect its penetration rate.

Meanwhile, the IRRI research team developed a new and original design for a piercing tool. This approach had the advantage of designing a tool with steering as one of its primary functions. The IRRI tool is shorter than existing equipment, and therefore it should be easier to steer. The drilling mechanism for the IRRI guided boring tool is based upon the Ingersoll-Rand (IR) valveless down-hole drill. IR down-hole drills are currently used for blast-hole drilling, drilling in rock, and oil well drilling. Some of the significant features of the tool (based upon computer modelling of the valve cycle and piston/anvil forces) are: 1) the tool should be easier to steer due to lower stresses and higher frequency than commercial piercing tools; 2) additional horsepower will allow the tool to move forward at a faster rate (5.3ft/min, 1.6m/min) than commercial tools (1 to 4ft/min, 0.3 to 1.2m/min). Steering will be accomplished by an internal eccentric cam which repositions the rear of the tool relative to four orthogonally-positioned pads which remain in contact with the side of the borehole. The cam is rotated within a steering ring by a hydraulic motor. A roll sensor was devised to ensure the operator understands the orientation of the tool at all times. The theoretical turning radius of the steering device is one foot per five feet (0.3m per 1.5m) of travel (Figure 6.2.3). All of the key components have been fabricated. Field trials will be conducted by IRRI in early 1985 to confirm the predicted performance.

Both systems have the potential of adequately achieving the desired objective. The turning radius for both tools is satisfactory.

Tracking and detection systems

Several systems were investigated for tracking the boring tool. These were: conventional pipe locators, electromagnetic induction trajectory system, acoustical accelerometers, and emplacement of coil(s) in the boring tool with surface and/or target receiving coils.

The technical feasibility and practicality of a magnetic field attitude-sensing and rotation system were demonstrated. Magnetic fields generated by energized coils on board the boring tool are detected by sense coils located in the launching and/or retrieval pits. The coils are intermittently excited by a low frequency (25Hz) generator. The output of the coils is proportional to the sine of the angle in the vertical direction, and the sine of the angle in the horizontal direction between the axis of the tool and the axis of the boresite. Any turning or deviation by the tool from the desired boresite path (reference point) is immediately detected by the sense coils. The signals are processed to provide a graphic display to the tool operator depicting the attitude and rotation of the tool. Figure 6.2.4 is an artist's conception of the tracking system in operation. Results of the laboratory experiments demonstrated that the sensing system could detect and interpret a signal at distances up to 100 feet (30m).

An alternative approach for tracking the boring tool is to use conventional electromagnetic conduction pipe and cable locators. These instruments are currently utilized for locating and tracing buried metal objects. This approach involves attaching a signal generator to the mole by a wire cable or to an air hose (Figure 6.2.5). The radiated signal is detected by a hand-held receiver which locates the mole. The advantages of this technique are: 1) ease of use, 2) low cost, and 3) readily available. However, these instruments may have difficulty in precisely locating the end of the tool and in determining its depth.

Neither of the above approaches has been integrated with the mechanical steering system.

Also, the field worthiness of the tracking systems has not yet been proven. Integration of the tracking system into the piercing tool is planned for early 1985.

CURRENT DESIGN SPECIFICATIONS

Table 6.2.3 summarizes and compares the specifications of the two guided horizontal-boring tools under development.

The Table illustrates that there are notable advantages for each system. The MEI design can be easily retrofitted onto existing piercing tools. The key components (steering and tracking) have been demonstrated in the field or laboratory. On the other hand, the features of the IRRI tool were carefully designed to improve the ability to steer the tool accurately. Therefore, it is possible that the IRRI tool will be easier to turn and have a higher degree of steering accuracy.

The estimated production costs to retrofit an existing piercing tool with the steering and tracking components under development by MEI are $7,000 (US). Obviously, these costs will depend on quantity, market conditions, and the mark-up by the manufacturer. At the present time, IRRI has not calculated the cost to manufacture the new piercing tool.

FUTURE PLANS AND PROSPECTUS

Based upon the significant results of both programmes in the first phase of research, GRI plans to continue to sponsor the development of a guided horizontal-boring tool. In the next phase, the prototype boring tools will be field-tested in co-operation with US gas utilities. The steering and tracking systems will be integrated into the prototype model. Further, the tool will be tested under a broader range of soil conditions. Parameters to be assessed include ruggedness, reliability of the tracking and steering systems, and user acceptance. Assuming the second phase is successful, the next step will be to build several (10) pre-production models of the guided horizontal-boring tool. These tools will be field-tested in all major soil types and weather conditions. An obstacle detection system (proximity device) will be developed and added to the boring tool in later generations.

Key US manufacturers of boring equipment have been actively involved in this research to ensure the successful commercialization of the guided boring tool. In later phases, the manufacturers will co-ordinate testing with the research organizations.

A preliminary survey of US gas utilities indicates a strong market potential for a guided horizontal-boring tool. With the completion of this programme, utilities will be provided with a practical, economic guided horizontal-boring tool that can burrow accurately to distances of 120 feet or greater. A guided horizontal-boring tool should significantly increase the utilization of boring equipment. The tool has the potential to reduce the cost of installation and restoration for underground utility lines by at least 25 per cent (as much as 50 per cent in urban areas) when compared to trenching, backfilling and reinstatement costs.

ACKNOWLEDGEMENTS

The author acknowledges the significant accomplishments of the contractors — Maurer Engineering Inc. (MEI)/Illinois Institute of Technology Research Institute (IITRI) and Ingersoll-Rand Research Inc (IRRI) — during the past year. Special recognition is given to the following individuals: W J McDonald and G T Pittard of MEI, J E Bridges and J O Enk of IITRI, and A J Ehrat of IRRI.

REFERENCES

McDonald, W J *et al., Development of a Guided Horizontal-Boring Tool — Phase I,* Annual Report GRI-84/0236, Gas Research Institute, March 1985.

Ehrat, A J, *Development of a Guided Horizontal-Boring Tool — Phase I,* Annual Report for GRI Contract No 5083-271-0857, Gas Research Institute. To be published April 1985.

TABLE 6.2.1 Performance summary of horizontal-boring tools

Type of Boring Tool	Maximum Boring Distance, feet (m)	Penetration Rates, ft/hr (m/hr)	Steering Radius[a] feet (m)
Percussion moles	150-200 (45-60)	60-300 (20-90)	28-60[b] (8-18)
Positive-displacement motors	3000 (915)	120-1800 (35-550)	300-400 (90-120)
Rotary drilling	300-500[c] (90-150)	60-300 (20-90)	1000[d] (300)
Spoil augering	300-500 (90-150)	60-240 (20-90)	1000[d] (300)
Compact augering	200 (60)	180-300 (55-90)	1000[d] (300)
Pipe pushers	200 (60)	120-240 (35-75)	—

a The steering radius is the radius required to turn the tool 90° when starting from a straight orientation.
b The steering radius was determined from field tests and analysis.
c The boring distance is based on the use of a typical, properly-sized earth auger machine.
d Steering radius is limited by operational considerations. Excessive wear of the drill-string tends to occur in high-curvature holes.

TABLE 6.2.2 Candidate mechanical steering concepts for guided horizontal-boring tool

Type of system:	Tail fin(s) Pad(s) Slanted anvil/bent sub Wedge(s) Articulated joint Rotating body Combination of any of the above systems
Location of system:	Front Rear Front and rear
Motion of system:	Pivoted Rotated Wedged Retractable Articulated Disengageable
Power supply:	Hydraulic Pneumatic Electric

TABLE 6.2.3 Guided horizontal-boring tools under development

	Retrofit Tool (MEI)	New Tool (IRRI)
Tool Type	Pneumatic Piercing Tool	Pneumatic Piercing Tool
Diameter of Tool	3.75 inches (95mm)	3.4 inches (85mm)
Steering System	Slanted anvil and movable or fixed tail fins	Rotating eccentric cam and four orthogonally-positioned pads
Method of Activating Steering System	Pneumatic or hydraulic	Pneumatic
Tracking System	Magnetic field attitude sensing system	Electromagnetic signal generator with hand-held receiver
Tool Length	59 inches (1.5m)	40 inches (1.0m)
Frequency	400 cycles/minute (cpm)	1345cpm
Horsepower	1.8 horsepower (hp)	2.0hp
Air Consumption	90 cubic feet/minute (cfm)	51cfm
Blow Energy	150 ft-lb/cycle	50 ft-lb/cycle

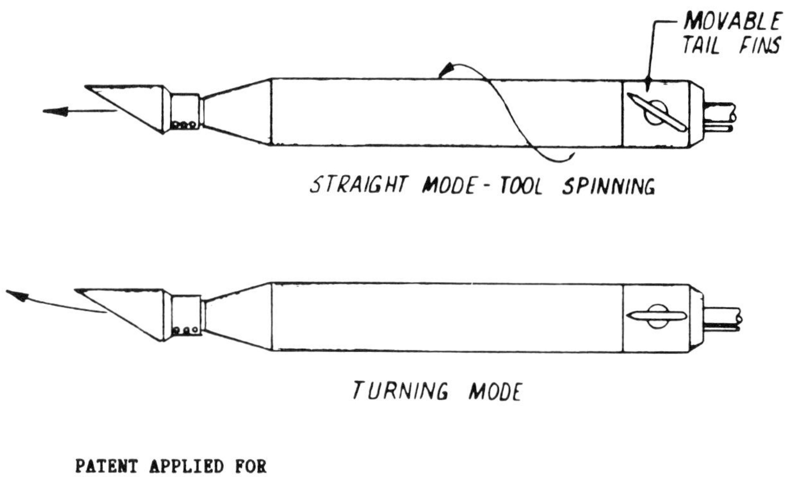

Figure 6.2.1. Steering assembly retrofitted on an existing piercing tool.

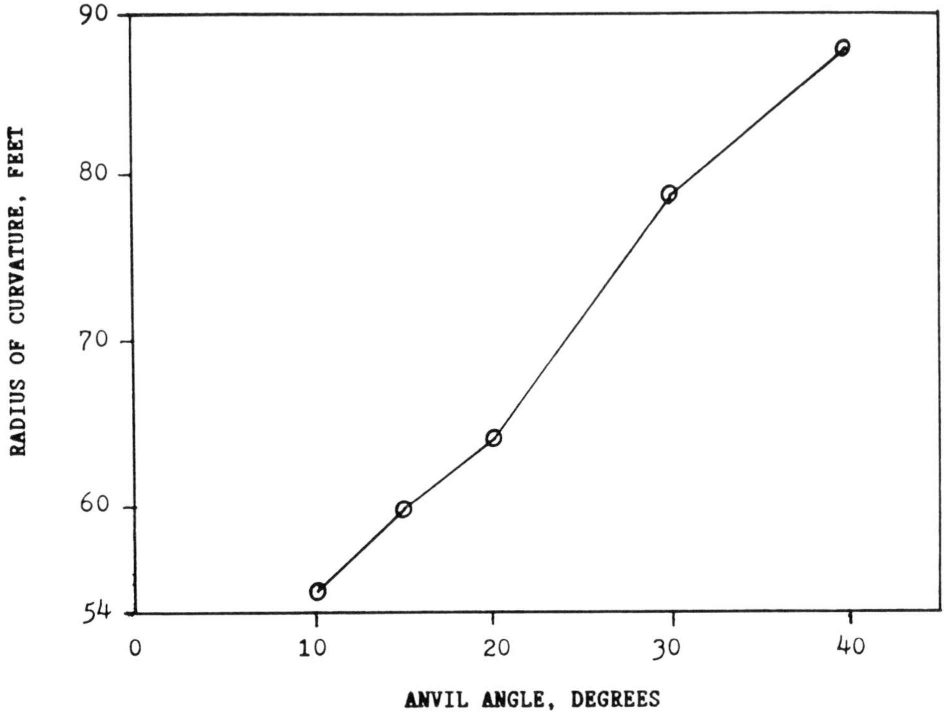

Figure 6.2.2. Radius of curvature v. anvil angle for a pneumatic piercing tool retrofitted with a 2.5 inch diameter anvil.

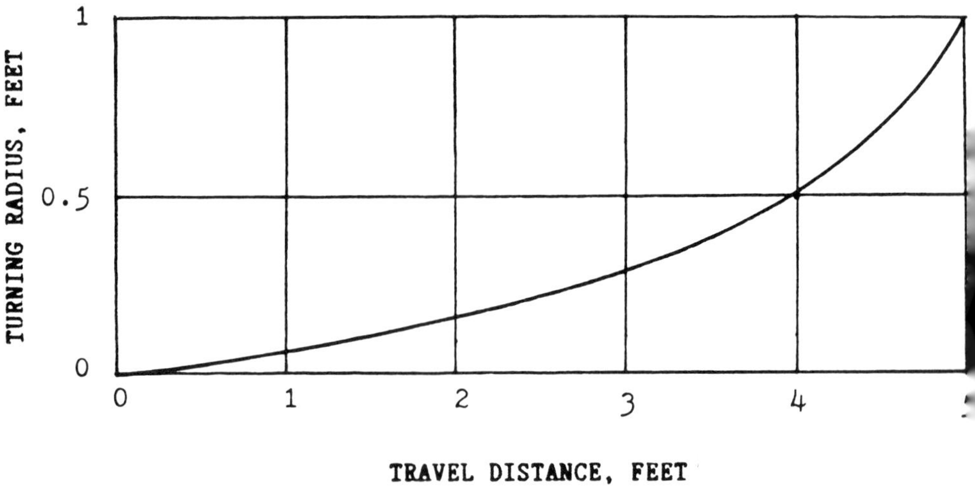

Figure 6.2.3. Theoretical turning capability of prototype horizontal boring tool.

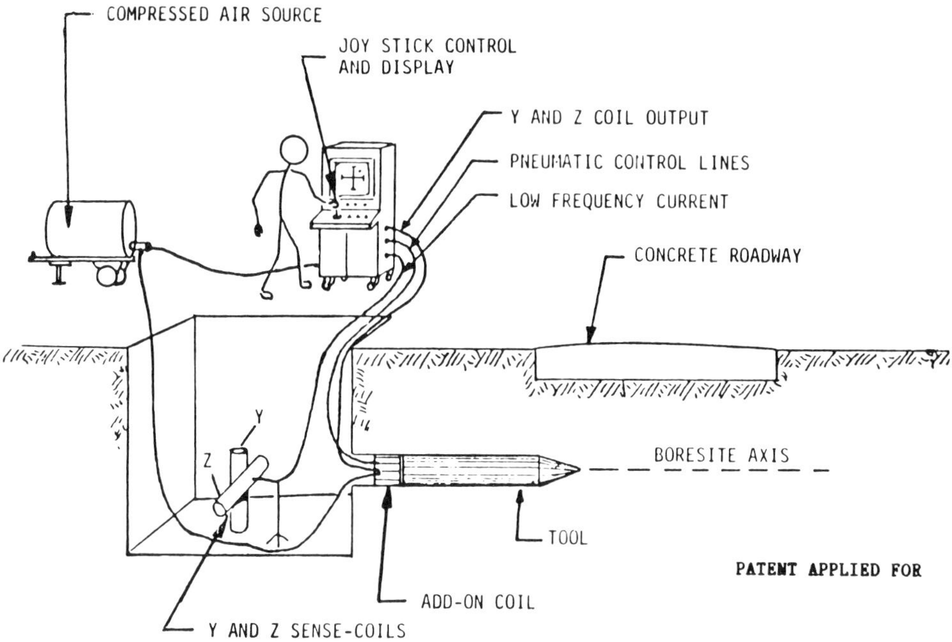

Figure 6.2.4. Conceptual design of the magnetic field attitude-sensing system.

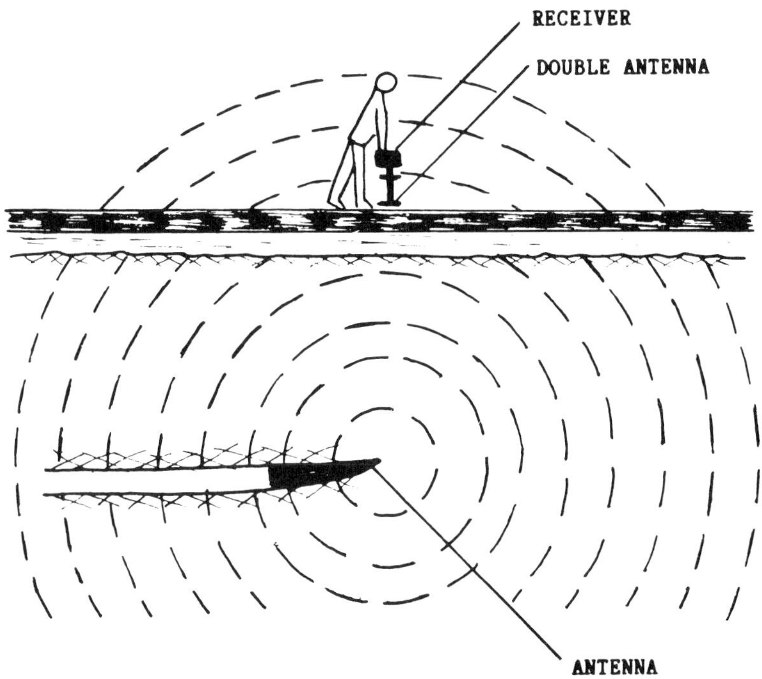

Figure 6.2.5. Conceptual design for electromagnetic tracking/detection system

6.3 *In situ* concrete microtunnels

Dr H P Uffmann *Technical Director*
Krings International GmbH

In the last few years, a great deal has been written about systems of underground working in sites that are inaccessible to personnel, especially in regard to developments in Japan.[1,2,3,4] The time has now come to take a look at the pipe-injection system of Krings International of West Germany. When this system was being prepared, one of the main intentions was to offer contractors an economically viable alternative to open-trench digging. The result was not simply a piece of drilling equipment, but a complete cost-saving system with drills, jacks, and in fact everything needed to take care of the whole pipelaying process from start to finish.

CURRENT CONTRACTUAL COSTS

If we consider the cost structure for pipelaying in impassable ground, we shall easily discover that, in view of today's contract prices of *ca.* 2000-3000 DM (some £575-£850) per linear metre (of 500mm dia. pipe) the process in question is available only in exceptional cases. The exceptional cases include tunnelling under thoroughfares, railway lines, houses and rivers. For the most part, in the light of current pricing and costing, this environmentally sound and energy-saving method is unlikely to be employed during the next few years, however, unless there is some urgently needed rethinking on the part of those concerned.

A price basis is required from the manufacturer and contractor which would enable the promoter to specify the method in question when issuing contracts. As matters stand, even the most ardent environmentalist has no option but to choose conventional open-trench pipelaying at 1000-1500 DM (some £300-£450) per linear metre for 500mm pipe at a depth of 5m rather than to pay prices of the order mentioned above.

VOLUME OF INVESTMENT

In order to explain the value of what is being proposed, a few words must now be said on the future volume of investment.

In January and February 1984, a thesis[7] was prepared in the Institute for Construction Machinery and Management (IBB) at the Aachen Polytechnic, having for its subject the anticipated West German requirement up to the year 2000 for laying pipes of diameter ‹1000mm. The writer of the thesis visited the relevant promoters in the 30 most populous cities in the Federal Republic of West Germany. Enquiries were made in regard to the requisite conduit extensions and replacements and in regard to pipe diameters and lengths, to line laying at depth, to the proportion of new estates and other developments, and to materials and soil conditions. Also noted were the estimated works for which, according jto the persons questioned, finances would have to be earmarked for the periods 1984-1990 and 1991-2000. The total values were as follows:

1984-1990: 2711km ‹1000mm
1991-2000: 2435km ‹1000mm

If these values (based on findings for some 26 per cent of the entire population, ie the inhabitants of the 30 largest cities) are extrapolated for the whole West German Republic, the total length of piping to be laid of diameter ‹1000mm comes to an estimated 20,000km. Then if the costs are averaged at *ca* 1000 DM (some £290) per linear metre, the investment required to the year 2000 works out at 20 milliard DM (some £5,800,000,000) or at 1.25 milliard DM (some £359,000,000) per annum.

THE SYSTEM

In order to achieve a realistic and competitive price, all the essential cost-generating factors in pipe injection have to be studied with a view to seeing how their effects can be reduced, with special reference to the initial and final shafts, to the actual injection process and to the raw materials utilized.

Initial and final shafts

In this connection there are two points that can contribute to cost reduction: an increase in the distance between the initial and final shafts to *ca* 100-150m, and the employment of tunnelling systems which are quick to install and are often reusable (in contrast to the current employment of horizontal timber sheeting and sheet pile walls).

Tunnelling systems should exhibit the following features:

- Frame construction without stiffeners;
- Watertightness;
- Integrated wall and face;
- Easy and safe handling;
- Compatibility with large area linings.

With the exception of the requirement for watertightness, the above-mentioned points are already met by the proven double-slide rail-lining system (Figure 6.3.1). For the contractor who is already equipped with Krings large-area linings all that is usually needed is the purchase of the newly developed angle slide rails (Figure 6.3.2). With the installation of plates of different lengths it is easy to line shafts of any length and width.

Pipe Injection

Cost reduction in the injection process itself may be attained by: lower prime costs for equipment, fewer operators, improved injection performance.

The latter point appears to be especially important. The usual daily output of between 5 and 15m of piping laid has to be increased to 20-30m per day by using more efficient equipment and, among other things, by cutting down drastically on idling time for the borers, currently 50 per cent of the total drilling time. In order to solve this problem, we have either to develop pipes, which, as they are joined together can be automatically coupled with all utility lines, or else to develop half-round pipe sections in which the utility lines can be introduced without coupling (Figures 6.3.3 and 6.3.4). Because of their expense and also because of their later use, these sections cannot be left in the ground permanently but are employed purely as pressure pipes to protect the excavation. They have to extracted when the final safety measures are being installed.

The KR-VS-864 drilling equipment consists of a shield (external diameter = 864mm), into which are introduced the whole-surface drilling head and motor drive and, in addition, a trailer in which are carried the dredging pump and the direction controller — an inductive laser table. At the articulation, four jacking cylinders are positioned at the 45-degree points, working automatically in pairs or capable of being controlled individually from the console (Figure 6.3.5).

The drilling head (Figure 6.3.6) is tipped with kennametal pins for breaking up harder soil formations and occasional obstructions. The wet rock cuttings flow through 20 circular holes (39mm dia) behind the drilling head. Here they are mixed with raw water and carried away by hydraulic action. An activator behind the drilling head continuously mixes the slurry. At the same time, cleaning pins unblock the holes in the drilling head when necessary.

Behind the drilling head, a water level indicator and a pressure cell enable the quantity of water used to be determined accurately for optimum results.

The drilling head rides back on a cushion of oil if the main station thrusts forward too suddenly. At a preset position of the head, the main station is automatically disengaged to avoid any possibility of damage to the drilling equipment.

Steering is carried out on the basis of information relayed by the inductive laser table on the roof of the trailer. The laser beam by which the drilling axis is determined strikes a target covered with photoresistors, and there is a cyclic readout of the latter by the computer. A system of coordinates corrected for movement out of line, plus the point of laser impact, are also shown on a monitor for the benefit of the operator — who can also gauge the magnitude of movement out of line, the inclination of shield and trailer and the water level and/or pressure at the working face. The steering cylinders are fitted with non-contacting, electrical odometer systems for detecting steering errors. The positional values every 10cm for a connected pressure pad are automatically recorded on a strip chart.

Considerable spare capacity has been built into the equipment. Thus the jacking force of 5000kN is certainly very high for a proposed segment of 100-150mm. In the same way, the ratings of the tractive force, the strength of the controls and, in particular, the turning moment, are also very high. The latter is especially important since, thanks to this high turning moment and to the hard-tipped drilling head, obstructions can be thrust aside and hard soil formations can be broken up. Where solid rock has to be negotiated, a drilling head tipped with disc rollers is available. Of course, the reactive force from the high turning moment has to be accepted by the drilling head and drill stem, and the latter has therefore been reinforced against the stresses involved.

The tensile strength of the pipe section jointings is an additional advantage, as it enables the drilling equipment to be withdrawn and recovered in the initial shaft if obstacles are encountered which are completely resistant to drilling, or if something goes wrong with the machinery (neither possibility can be entirely ruled out). At the same time, the filling-in of the cavity can be supervised with the help of the pressure-load cell behind the drilling head.

To sum up, here are some of the advantages possessed by the system just described over other equipment now on the market:

- It is an integrated system with half-round sections and other equipment covering individual stretches of 100m;
- The turning moment is very high (18,000Nm at 18 rpm);
- Different drilling heads are available;
- The holes in the drilling head can be cleared with cleaning pins;
- By regulating the raw-water pump with the help of a water level or pressure indicator, any desired pressure may be exerted at the working face;
- The dredging pump is inside the drilling equipment;
- There is an inductive laser table linked to a screen at the console. The laser target point displayed has already been corrected for movement out of line. Also displayed are the movement out of line itself, the tilt of the shield and of the trailer, and the water level or pressure as the case may be;
- The drilling is continuously monitored;
- When controlled filling of the cavity is being carried out, it is possible to withdraw the machine into the initial shaft;
- Progress of over 20m/day is possible in normal circumstances owing to the easy reduction of drill idling time with the use of half-round sections;
- Different internal diameters are possible with the same machine.

Raw materials — lined pipes

Sequence of operations

Before the new process — the manufacture of a so-called *in situ* poured concrete composite pipe or 'lined pipe' — is elucidated, a few words must be said about the injection pipes ‹1000mm currently on the market, as these too can of course be used with the injection system described here. Two negative points should be mentioned in contrast to their undeniable good quality[6]. These are, on the one hand, the very high price (500-800 DM per linear metre for 500mm) and, on the other hand, the technical problems presented. With the employment of ready-made injection pipes, the drilling machinery lies idle for prolonged periods while new joints are being added to the drill stem, and supply lines and hydraulic power lines are being coupled on. The non-drilling time amounts to *ca* 50 per cent of the total drilling time.

The very high price is due partly to the fact that the maximum injection force (up to 3000kN) has to be transmitted. In addition, because the possibility of steering mistakes cannot be excluded, we have to reckon with an off-centre application of force. It is this, in connection with the very expensive jointing, which leads to the high price of the pipes. The piping is designed for the relatively short but stressful time of the injection process, and is consequently completely oversized for its primary use covering a period of 50-100 years.

With these points in mind a steel injection pipe consisting of upper and lower half-round sections has been constructed as already mentioned. This pipe can easily be jointed on in a relatively short space of time without uncoupling the supply lines to the drilling equipment (Figure 6.3.8) and requires only the temporary support of the side wall bracing.

At the request of Krings International tests were carried out on this type of pipe at the IBB of Aachen Polytechnic from 1980 to 1984.

The following sequence of operations (Figure 6.3.9) was put to the proof in six full-scale tests in the test road of the IBB (Institute for Construction Machinery and Management) with the first results coming through in time for the Hanover Fair of 1984.

When the stage of boring from the initial shaft to the final shaft is completed, the drilling equipment is recovered in the final shaft. A liner of the requisite internal diameter is then introduced into the steel pipe section. In the final shaft the liner is centred with face formwork applied to the steel pipe. Then the steel pipe insert is withdrawn into the injection shaft. At the same time, the space left between the adjacent soil and the liner is filled with concrete. The continuous filling is monitored by a pressure capsule mounted in the ridges of the face of the formwork. Work can be carried out at higher or lower pressures according to the height of the overburden and to the type of soil, so that ring closure and a tidy filling of the annular space can be ensured every time. The liner accepts the support of the fresh concrete so that, in comparison with what was found using a slip form[5] some years ago, quick and economical working is guaranteed.

Figure 6.3.10 shows the steel preliminary pipe insert with the face formwork. The concrete intake is visible below, and above is the pressure capsule that monitors the annular closure. Figure 6.3.11 illustrates three sections through *in situ* concreted compound pipes.

As a general rule, any desired pipes may be employed as liners, such as for example pipes made of stoneware, steel, glass-fibre reinforced plastic or HDPE. Figure 6.3.12 shows the insertion of a stoneware liner (500mm).

The cost of an *in situ* concreted pipe with HDPE liner amounts to *ca* 300DM (some £86) per linear metre. This price has been worked out for a reach of 100m.

Special application

In addition to the normal applications, the process has further special advantages to offer, such as in the use of steel pipes when impeccable external protection against corrosion is required. This would be necessary in oil or gas pipelines.

In mining subsidence areas or in regions where for some other reason allowance has to be made for settlement of the soil, the process described can, with certain small changes, yield optimum results. Here, instead of concrete for coating the liner, a non-setting material can be used. For instance, the cement can be removed from the mixer to be replaced by filler. In this case HDPE piping should be employed as the liner since, owing to its flexibility, this can withstand larger soil movements without damage.

A further application would be drainage of a deep-lying layer. Here a screen pipe could be covered with a drainage material.

Advantages of in situ *concrete microtunnels*

What then are the advantages of this alternative technique?

- favourable price;
- seamless piping;
- choice of liner;
- subsequent filling if too much soil removed during injection;
- different internal diameters possible with the same piece of drilling equipment;
- other applications (steel pipes with corrosion protection, pipelaying in areas of subsidence, groundwater drainage);
- higher performance thanks to the recoverable half-round sections. The high-tensile strength of the jointing of the pipes made with these sections guarantees not only the ability to withstand torque during the exigencies of the injection process, but also the recovery of the drilling machine;
- optimum conditions for subsequent underground laying of branch lines to buildings.

Branch lines to buildings

The last-named and hitherto intractable problem, namely the underground laying of branch lines to buildings, is certainly the most important consideration after the need for more reasonable pricing of the process itself, and it is crying out for an early solution. Research into the problem is currently in progress at a number of different centres and, among others at the IBB of Aachen Polytechnic in collaboration with Nordrhein-Westfalen and Krings International. A viable solution is expected by 1985.

A big advantage of the technique described where the laying of house connections is concerned is the development of the seamless pipe. The relatively small-bore pipe (150-200mm) required for house connections need be absolutely trued for position only on the level. A deviation of up to 50cm is permissible in the horizontal, if so required by the shallow gradient of the collecting drain. Therefore the drilling equipment for branch lines must be manoeuvrable only in the vertical plane. Where the customary injection pipes are employed (length 1.5 to 3.0m) in the construction of collecting drains, it is imperative to avoid connection with a vertical seam of the collecting drain, as a watertight joint is impossible at this point. Consequently, in the laying of the collecting drain with conventional injection pipes, the branch line drilling equipment has to be manoeuvrable in the horizontal as well as in the vertical plane. This means that the equipment will be more expensive.

SUMMARY

The above considerations deal firstly with the extent of pipelaying work in the diameter range of under 1000mm to be carried out in West Germany between now and the year 2000 and, secondly, with the possibility of considerable cost reduction in the raw materials. Thanks to the described improvement in the efficiency of the injection process following the introduction of half-round sections and the attendant cutting down in the time drilling equipment lies idle, thanks too to the use of systematic excavation of the initial and final shafts, total costs are reduced to the point where underground laying is hardly more expensive than open-trench work. If a viable solution of the problems of the underground laying of branch lines to buildings is also obtained in the near future, the environmentally responsible promoter, who is also interested in what will be most economical in the long run, will see to it that a great proportion of pipelaying work is done underground.

REFERENCES

[1]Stein, *Utility and other lines in sections inaccessible to personnel*, TIS 3 + 4/82.

[2]Weber, Uffmann, *Manoeuvrable horizontal-drilling equipment for inaccessible pipelines,* TIS 2 + 3/83.

[3]Schlick, *Controlled injection of piping in inaccessible ground,* BMT 12/83.

[4]Uffmann, The HBG 701 horizontal-drilling equipment, initial site testing, *Hoch und Tiefbau* 12/83.

[5]Uffmann, *The stabilization of side walls in making horizontal cavities,* TIS 3 + 4/82.

[6]Stein, Bielecki, Underground conduit sections in terrain inaccessible to personnel, *Tunnel* 3/84.

[7]Wubbe, *Determination of pipelaying requirements for diameters under 1000mm until the year 2000 in the Republic of West Germany*, Thesis at the IBB of Aachen Polytechnic (unpublished).

Figure 6.3.1. Initial shaft with Krings linings.

Figure 6.3.2. Angle slide rails.

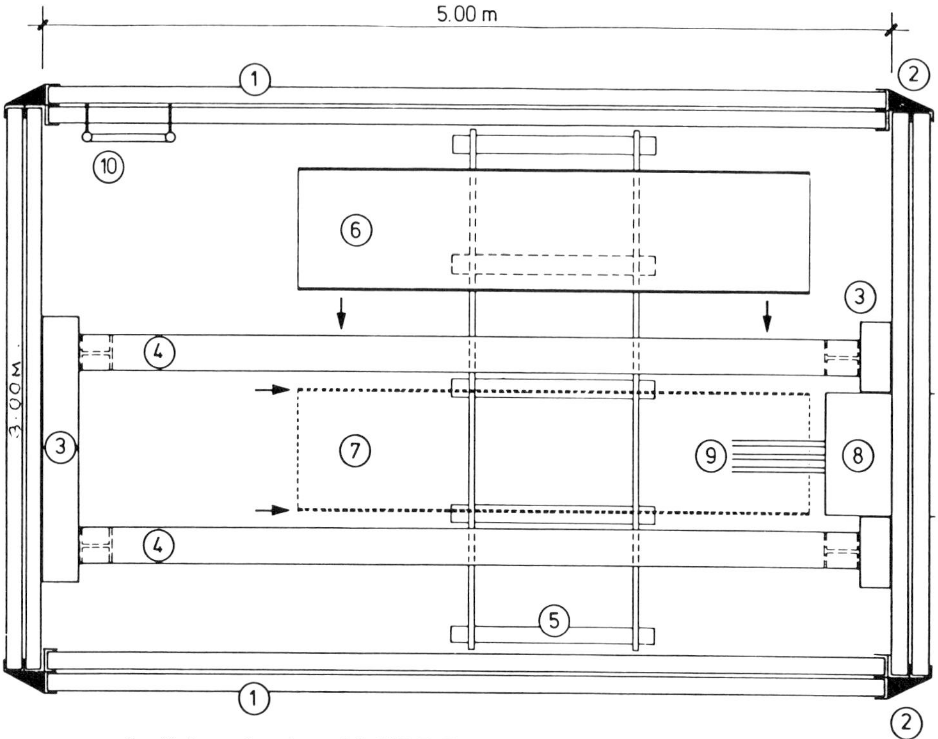

1. Driven sheeting with KR DG
2. Corner posts
3. Abutment
4. Driving equipment
5. Track
6. Lower half-round section of pressure pipe in servicing position.
7. Lower half-round section of pressure pipe in insertion position.
8. Bored pipe section
9. Supply lines
10. Manhole ladder
11. Upper half-round section of pressure pipe.

Figure 6.3.3. The system in plan.

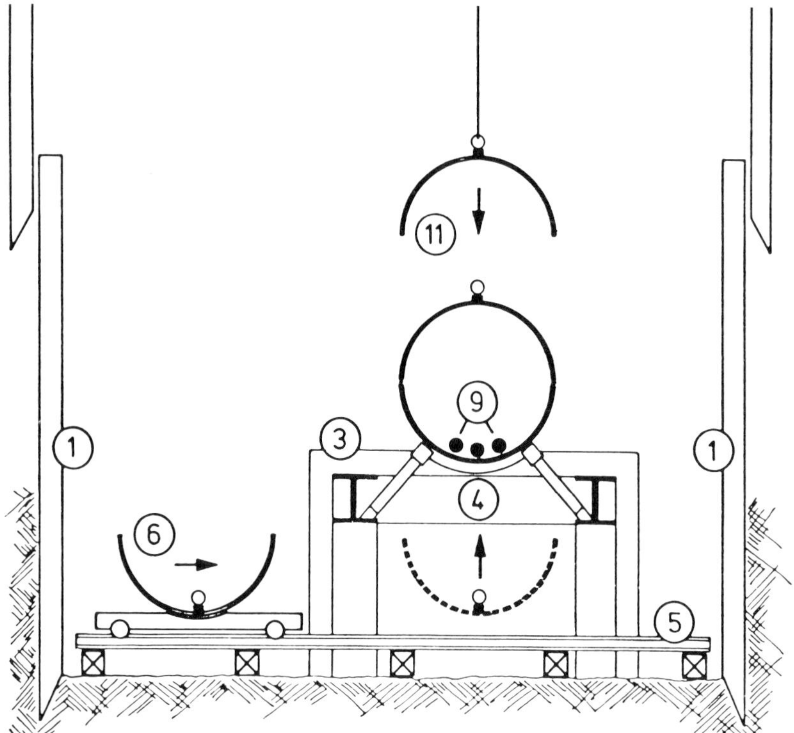

Figure 6.3.4. The system in cross-section.

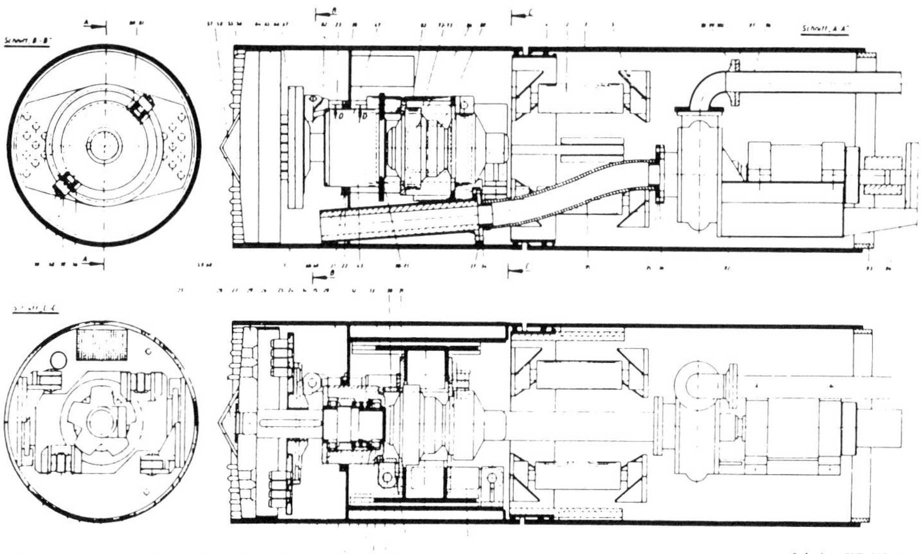

Figure 6.3.5 Sections through the KR VS 864

Figure 6.3.6. The KR VS 864 drilling head.

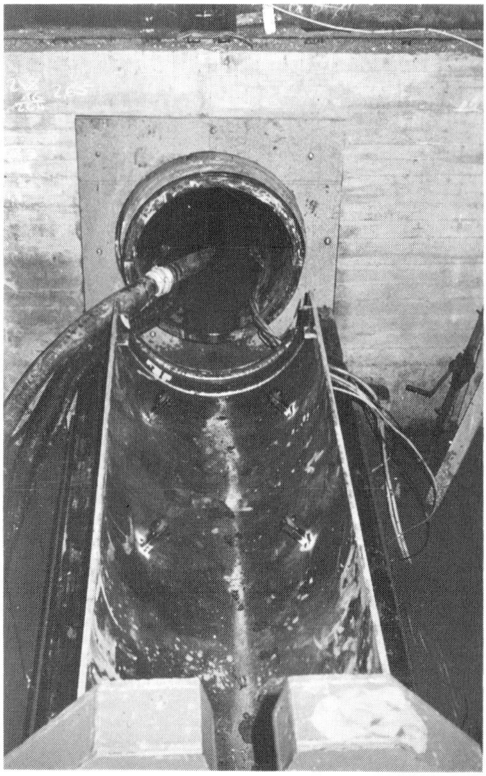

Figure 6.3.7. The console.

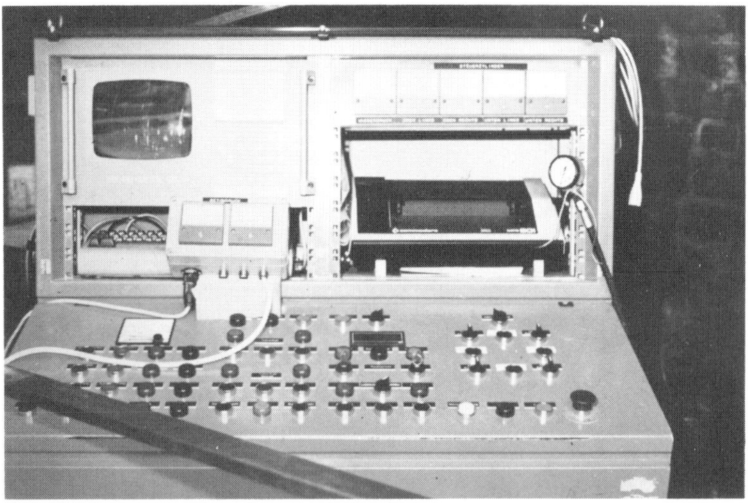

Figure 6.3.8. The lower half-round section loaded with supply lines.

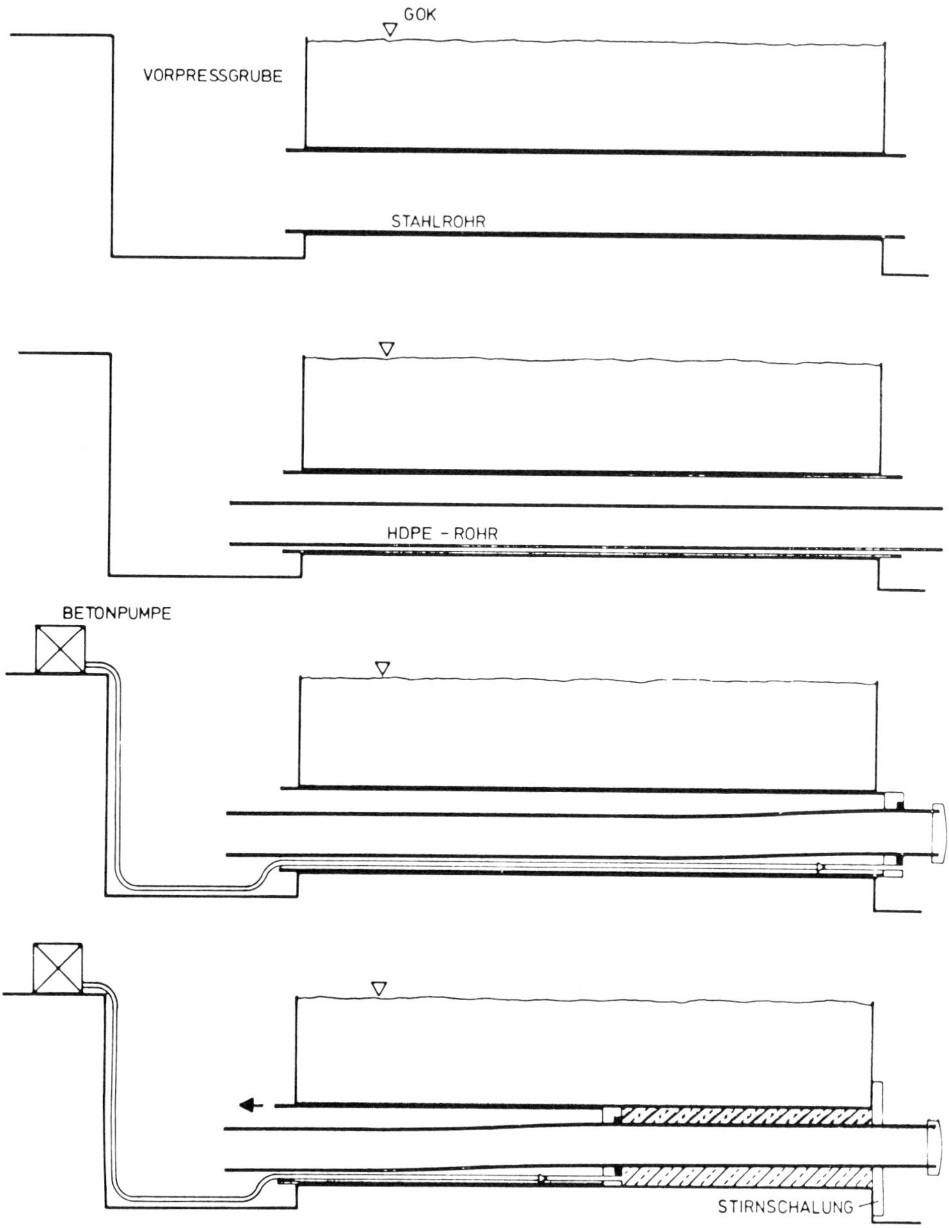

GOK	=	ground level
Vorpressgrube	=	initial shaft
Stahl Rohr	=	steel pipe
HDPE-Rohr	=	HDPE pipe
Betonpumpe	=	concrete pump
Stirnschalung	=	face formwork

Figure 6.3.9. Sequence of pipe lining operations.

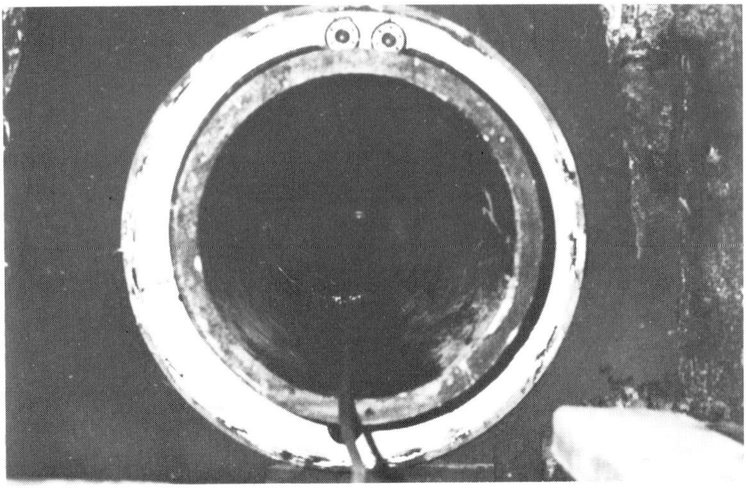

Figure 6.3.10. Face formwork with stoneware pipe inliner.

Figure 6.3.11. Three sections through in situ *concreted compound pipes.*

Figure 6.3.12. Insertion of stoneware liner.

Authors' Introductions

Mr S R Kramer: Development of a guided horizontal boring tool

During this presentation I will describe a project being sponsored by the Gas Research Institute to develop a guided percussion mole. We initiated this programme because of problems with the existing technology, including augers, pipe pushers and piercing tools. All these tools cannot be guided, they are unable to bore distances typically longer than 20 metres, they like to deflect off rocks, they miss their targets, they can bury themselves or, in the worst case, damage adjacent utility lines. What does all this mean? — unsuccessful bores. These are expensive when they happen and they are not necessary.

With this need identified, in 1984 the Gas Research Institute initiated a programme to develop and demonstrate a percussion mole that could be accurately tracked and guided. We set some strict performance goals for this system. It should drill small diameter holes, ideally from 25mm-200mm, to distances of at least 40m, with a directional accuracy of ±0.3m in the x/y plane. Also it should detect and avoid obstacles in its path, and of course it needs to be operated by typical utility crews. Finally it must be cost-effective and practical. That means cheap. I have heard the words 'least disruption' and 'lowest cost' throughout the presentations. This tool can satisfy that need.

We set up the programme with two contractors, Maurer Engineering Inc. (MEI) in Houston, Texas and Ingersoll-Rand Research Inc. (IRRI) in Princeton, New Jersey. They were to evaluate two different types of technology. MEI retrofitted a commercially available tool, the Allied Hole-Hog, with the necessary component for tracking and steering of the mole. IRRI selected a different approach, to design the tool from the ground up, because these tools were not made to be steered, so possibly they could not be simply adapted.

There are three sub-systems necessary to meet the performance goals identified. First, a mechanical steering system to turn the tool; second, an electronic tracking or detection system so you know where the tool is; and third, the most difficult one of the three, a proximity or obstacle detection system so that you know what is in front of the tool. This theory is being addressed in a separate research programme that I will not touch upon today.

As explained in the paper, percussion moles were selected because they have the fastest steering response of the horizontal boring tools in distances up to 30m. Other systems may have faster penetration rates but they are much more expensive.

First I will describe the mechanical steering systems. We looked at various concepts: where the system should be located, the type of system necessary, the motion of the system and how it should be powered. From this we selected the most promising technologies and fabricated them and tested them under real field conditions. MEI elected to retrofit the commercially available tool with a slanted anvil (Figure 6.2.1). As Figure 6.2.2 illustrates, the steering response varied linearly with the anvil angle. From this we were able to decide on the optimum design for turning. In this case it had a turning radius of less than 10m. If you add the anvil at the front of the tool it is no longer going to go straight. Therefore we added fins at the back to offset the anvil and also to assist in turning the tool. This is more than a concept, it is a reality right now.

On the other hand IRRI was going to design a tool from the ground up, with steering in mind. They shortened the length of the tool because a shorter mole should be easier to turn. Also the tool has lower stresses and a higher frequency, which should help in steering. The steering is accomplished by an internal eccentric cam which repositions four orthogonally-positioned pads at the back of the tool which remain in contact with the borehole at all times.

Now that we know how to turn the tool, the next step is to understand the attitude in rotation of the tool — which way is up. Illinois Institute of Technology Research Institute (IITRI), working with MEI, developed a magnetic field attitude-sensing system. In this system magnetic fields are generated by energized coils and are detected by sense coils located in the launching or retrieval pit. This same system can also be used for rotation by adding an additional coil at the back of the tool.

On the other hand IRRI selected a different approach. They were going to use conventional electromagnetic pipe locators, similar to those dsescribed by Mr Garnett in paper 5.2, for tracking the position of the tool, but even though you understand the position you must still understand the tool face orientation or rotation. IRRI devised a roll sensor for understanding which way is up.

To summarize briefly the two developments, MEI's commercially available tool, retrofitted with the necessary systems for steering and guidance, is simple to operate and is further advanced than the IRRI tool. However, the tool developed by IRRI, because it was designed with steering in mind, should be a more accurate tool.

To conclude, this first phase of research has shown that percussion moles are capable of producing small-diameter guided boreholes. The necessary components for mechanical and tracking systems have been developed and successfully tested in the field. The results indicate that the boring path of the tool can be redirected in a predictable, controllable manner. Finally, all of these systems can be manufactured inexpensively. What does this mean? In the field, improved productivity, increased boring accuracy, expanded applications for these tools and a reduction in the pavement removal costs, excavation and restoration.

Dr H P Uffmann: *In situ* concrete microtunnels

For some years now, systems for underground pipe jacking in non-accessible diameters have been under study in the West German Republic. Several kilometres of pipe jacking have already been successfully completed using this equipment. It has been found that the technical problems of guided pipe jacking in this diameter range can be solved for various soil conditions. The cost of the process is still some 50 to 100 per cent more than that of laying in open trenches. The problems regarding the house connections are still unsolved.

The Krings pipe-jacking machine is operated by remote control and has hydraulic soil transport. As in other systems excavation is carried out with a full-face drilling head (Figures 6.3.5 and 6.3.6). The drilling head has a torque of 1.8 tonne-metre at 18rpm. The high torque in conjunction with the kennametal pins on the drilling head enable harder formations to be traversed and big stones to be broken up. Quick and therefore economical drilling is ensured by the high rate of revolution and the powerful hydraulics. In the region behind the drilling head there is an activator disc that provides for continuous processing of the slurry of soil and water. A combined water level and pressure indicator, used together with an adjustable raw-water pump, makes it possible to have different water levels and pressures in the mixing chamber. In addition, when the activator disc is advanced, the cleaning pins clear any blocked inlet holes.

Behind a waterproof wall, we have the engine room. Here are the hydraulic drives for the drilling and the activator disc as well as the soil/water pump and the steering cylinders. With the help of the four steering cylinders a controlled force of 36 tonnes in each direction can be achieved.

In the trailer there is an active laser table that relays to the control desks the relative position of the laser target point as well as the inclination of the machine. This enables the necessary corrections to be made by means of system-controlled steering or individual steering. The driving equipment installed in the starting pit generates a jacking force of 5000 kilonewton. For economical construction of starting and target pits Krings International offer a shaft-lining kit using the tried and tested Krings trench-lining system (Figures 6.3.1 to 6.3.3). This, together with the long pipe-jacking distances of up to 100m, reduces the total cost per metre.

Before going into details on the *in situ* concrete composite pipe, a few reasons for the departure from the conventional pipe jacking should be given. The price is very high because of the high pressures involved. When a pipe is being newly jointed, all the leads to the drilling equipment have to be disconnected, threaded through and reconnected. Therefore the down time is quite long, making inefficient use of the drilling equipment.

The pipe-jacking equipment must be adapted to the external diameter of the pipe. Therefore a steel jacking pipe with upper and lower half-round sections has been designed. This pipe can be jointed on without trouble in a relatively short time, with no uncoupling of the supply lines to the drilling equipment, and is only required for temporary support of the soil. Upon completion of drilling from the starting pit to the target pit, the drilling equipment is recovered in the target pit.

A liner with the required diameter is introduced into the steel pipe section (Figure 6.3.9). In the target pit, the liner is centred with the face formwork fixed on the steel pipe. The steel jacking pipe is now retracted into the starting pit. At the same time the space produced between the *in situ* soil and the liner is filled up with concrete. The continuous filling is monitored by a pressure box in the top of the face formwork. The liner supports the fresh concrete. In principle pipes of any desired material can be used as a liner. For example, we might mention pipes of stoneware, steel, HDPE.

The *in situ* concrete composite pipe offers the following advantages. A low price; seamless piping; choice of liner; different diameters can be produced with the same drilling equipment; higher pipe-jacking performance is possible using the half-round sections. Thanks to the tensile strength of the connections of this pipe in addition to the torque counterbalance, drilling equipment can be recovered if necessary at any time.

The above considerations show the feasibility of considerable cost reduction in pipe materials. As a result of the use of half-round sections and the introduction of shaft-lining kits for the starting and target pits, the total cost of underground working is brought down to a figure not much in excess of that of open-cut pipelaying. If an economical solution for the problem of house connections can be found, the process I have described will be used for a large part of the inner city pipe jacking.

Discussion on Papers 6.2 and 6.3

J H Spiller (Distribution Engineer, Bristol, South Western Electricity Board): Our experience at Bristol is that damage to other plant has certainly increased with the greater use of moling. I am glad to hear from Mr Kramer that directional control has at last been achieved. Has he seen a reduction in damage to other utilities' plant as a result of using the steerable mole?

One of the commercial costs that we have had to face recently has occurred where damage to our plant has been caused by moling and a small factory has lost its electricity supply for several hours. The cost to that company of the loss of production can be substantial and should be offset against the cost of moling. I suggest that we do not always take into account all the supporting or consequential costs of our techniques.

The South Western Electricity Board do use moles, and they are economic for road crossings and similar work. However, moling along footpaths, parallel to other utilities' plant, is just not practicable.

J Thomson (Consultant): I am very encouraged to hear of the developments Mr Kramer has described. Could he give an indication of what he would expect to pay for a normal tool compared with an 'improved' version? How far can he go underground — two or three hundred metres?

Would Mr Uffmann give some information on the costs of the system he described, compared with the costs of traditional methods in Germany?

A L Coe (Distribution Engineer, Thames Water): How does Mr Kramer intend to negotiate underground obstructions with his steerable mole? Does the mole sense obstructions in advance of meeting them and steer a suitable course, or merely stop on impact, to be withdrawn and redirected?

A Rooke (Director, European Sales, Iseki Poly-Tech Inc): Will Mr Uffmann explain how the *in situ* lining method avoids the formation of voids in the surrounding grouted concrete? How does he ensure that the liner pipe remains in the centre of the surrounding concrete?

Mr Kramer: There has been a long history of failures in evolving the directional mole. As to whether we can 'see' obstacles without hitting them, we are developing a system for doing this, using radar, and we hope this will be able to integrated into the mole.

Percussion moles are very widely used in US industry. A large gas utility may have hundreds. However, they are only used for short installations of, say, ten metres.

I would foresee a run of up to 70 metres without great difficulty with the new mole. Our goal on price is between $10 000 and $15 000.

Dr H P Uffmann: The cost of the complete system described in my paper would be about DM 900 000 (about £225 000). Auger boring equipment is estimated at around DM 400 000 up to 1.2 or 1.3 million DM (about £100 000-£300 000).

The price per metre, under normal soil conditions, with a 500mm pipe laid from 4 to 5m deep is around DM 1600 (say £400), compared with open trench costs of DM 1200 (£300). It is more expensive at the moment, but we expect to bring the costs down in a year or two.

Mr Schmidt (Krings International): In answer to Mr Rooke, the concrete is spread from the invert of the system and builds up a ring all round the inliner and joins together at the top. In the crown of the pipe there is a pressure gauge which reacts to the pressure of the concrete. When that pressure is registered, it means automatically that the ring has been completed. If there are any voids in the soil due to the excavation or jacking that has taken place, they will be filled with concrete.

The centring of the inliner is done automatically during the process of drawing back the steel pipe. On the face of this pipe there is a plate with a centre hole through which the inliner has to pass, so that the line is automatically centred before the concrete is poured.

Professor P F Stott (Session Chairman): What is the range of pipe sizes for which this method is suitable?

Dr Uffmann: Any diameter up to 600mm, but from 300 to 600mm is the best range economically.

Professor Stott: Would Mr Kramer say how he sees his programme developing?

Mr Kramer: The key to progress is to get the manufacturers involved early in the programme. We help the commercial timetable greatly by doing this. At some time, we are going to have to cut back on the development of other tools. At present, the retro-fitting approach seems more feasible.

A F Elvidge (Manager, Methods Group, WRc Engineering): Would Dr Uffmann or Mr Schmidt explain the technique used for connecting house drains to the newly formed pipe?

Mr Schmidt: At present, correct house connections can only be carried out if the inliner of the main pipe is made out of HDPE. By using a small auger boring equipment we can bore from the house to the sewer. When the boring reaches the sewer, the boring head and conveyor screw are taken out of the lining and a diamond rockdrill crown is inserted, with the help of which the OBV pipe can be holed precisely. After that an HDPE pipe of 100 or 150mm diameter is fed into the house connection boring and welded waterproof with the inliner of the sewer. After welding the house connection and the sewer, the gap between the soil and the house connection is filled with concrete.

Mr Kramer: Over 30 per cent of percussion boreholes in the USA are not completed on the first bore, which makes it very expensive. Have people in other countries had the same problem of having to do two operations before getting a successful bore?

D A Jameson (Senior Engineer (Development), Southern Gas Board): I am very surprised to hear that. In Southern Gas we are operating about 150 minimoles of 45mm diameter, and our success rate is 91 per cent first time bores. These are usually road crossings of about 6 to 8 metres in length.

Mr Spiller, from the South Western Electricity Board, spoke of increased damage done by moling. We have no firm statistics, but the indications are that *less* damage is done by moling. The most important point, however, is that if we do hit a cable when moling the operator is safe. If a cable is hit during open-cut excavations, then the man is at great risk, especially if using hand tools.

R Smith (Mains Engineer, Bristol Waterworks Company): We have heard views from the gas and electricity industries on the use of moling, and I wish to make a plea on behalf of the water industry. If the gas people damage a water main, the worst that can happen is that we get water leaking. But if we strike a gas main there is a risk of explosion and damage on a very large scale. My plea is that the various utilities should get together, in conjunction with WRC and NJUG, and prepare and issue a Code of Practice. Such a code would not only enable Water Authorities to follow a common agreed procedure, but, should they cause damage to an adjacent utility, they could demonstrate to society that they were carrying out work in a responsible manner to an agreed code.

A F Elvidge (Manager, Methods Group, WRc Engineering): I can assure Mr Smith that WRC has embarked on a major study of the effect of percussion moling on adjacent services, and we would be pleased to advise him. Our aim is to put several moles through different types of pipe and measure the strains that are developed in other pipes at right angles. WRC intends to approach British Gas to discuss the possibility of preparing a joint Code of Practice.

Mr Kramer: On the question of damage to other utilities, the experience of the USA gas industry is that approximately 70 per cent of gas-pipe damage incidents are caused by backhoes and trenchers. So moles are not the main cause of damage in the USA, either.

6.4 Guidance systems for microtunnelling

Peter M Zollman BSc, PhD, DIC, *Managing Director*
ZED Instruments Ltd

Conventional tunnel surveying methods are either impracticable or undesirable in microtunnelling. There is a very strong case for instrumental guidance unless the tunnel is very short or accuracy is not critical. The operator needs continuous and accurate information about the position and attitude of his machine to reduce friction, to save intermediate jacking stations, to avoid costly, time-wasting corrections and stoppages.

First we shall consider the general requirements for the remote or automatic guidance of microtunnelling machines.

Next we examine three distinct microtunnelling operations and the appropriate guidance principles to suit these:

A. Straight pipe jacking under 150-250m.
B. Straight pipe jacking over 150-250m.
C. Curved pipe jacking.

After a brief survey of existing and potential alternative guidance methods a few conclusions are drawn.

INTRODUCTION

The social and economic significance of microtunnelling is widely recognized today. There are, of course, many problems to solve — not all technical — before this construction method may occupy the position it deserves among its rivals.

As defined, microtunnelling implies that numerous operations in the tunnel, from the face right back to the shaft, are remotely controlled. One of these operations, an essential prerequisite of microtunnelling, is remote guidance, the subject of this paper.

Every guidance system must perform at least two basic functions:

(a) a continuous or sufficiently frequent survey to establish the position and attitude of the tunnelling machine in space with respect to a convenient frame of reference, usually the designed tunnel axis;
(b) immediate processing of the results of the survey and presentation of the processed data in a format which is most appropriate for manual or automatic steering.

In addition to these two essential functions a guidance system may provide a wide range of useful ancillary services, for example taper selection, warnings, alarms, or a permanent record of the operation in the form of a survey report automatically printed at regular intervals.

It is reassuring to know that such guidance systems have been in service since 1977 with a proven record in tunnelling and mining. ZED Instruments Ltd has developed and manufactured a comprehensive range of computerized tunnel guidance systems installed in six continents. Systems have been supplied, for example, for driving curved and inclined metro tunnels, road and railway tunnels, water and service tunnels, vertical and inclined shafts, sewer tunnels and special projects such as EUROLEP, the giant circular electron-positron storage tunnel at CERN, Geneva.

Much of the experience was gained in manned tunnels. Even the dedicated microtunnel guidance systems were first tested in larger, accessible tunnels. But pipe jacking and in particular microtunnelling applications have also been prominent right from the start.

In 1977 the Tekken Construction Co, Tokyo, Japan was the first to install a guidance system of this kind on a Kawasaki slurry mole, with immediate success. (Tokyo Water Supply Bureau, Kita-ku utility tunnel, 985m long). This was followed by a second system in 1978.

In the late 1970s the German Federal Ministry for Research and Technology initiated a major project for the development of a remotely-controlled sub-metre-diameter tunnel-boring machine suitable for a wide range of strata. To meet the requirements of the machine manufacturers, Wirth GmbH of Erkelenz, an especially compact guidance system was provided. The reliability of the new concept was confirmed through a series of test drives conducted by the Institute of Construction Machinery at the University of Aachen in closely controlled ground conditions. A total length of 130m was jacked using 711mm steel pipes[1]. Since the successful conclusion of the project the Wirth TB 700 Tunnel Boring Machine and other small-diameter machines have driven several thousand metres of inaccessible service tunnels by remote guidance in Germany and Switzerland.

SURVEY REQUIREMENTS

To determine the position and the attitude angles of a tunnelling machine six independent measurements have to be made. As shown in Figure 6.4.1, X and Y are the horizontal and vertical position coordinates, while roll, lead and look-up are the principal rotations. Not shown in the diagram is the sixth independent variable, Z, which defines the position along the tunnel axis.

In the case of straight tunnels it may not be necessary to measure Z. It is important, however, to measure the three angles of rotation. Unless X and Y are measured on the axis — and this is not practicable in most cases — roll correction is a necessary part of the survey. Not essential but extremely useful is the knowledge of the lead and look-up angles to minimize steering errors and to establish the optimal crabbing attitude of the machine .

STRAIGHT PIPE JACKING UNDER 150-250m

In conventional tunnel boring the tunnel wall is stationary; in pipe jacking it moves with the face. Hence, in pipe jacking the only stationary base where a datum laser can be mounted is at the jacking shaft as long as the designed tunnel axis is straight and the tunnel is not too long for the laser beam to be safely received — typically 150-250 metres. Extreme care must be taken of course to ensure that the laser base is adequately stable and not affected by ground movements caused by jacking forces. The use of self-levelling or self-aligning lasers is strongly recommended.

As seen in Figure 6.4.2, the laser, located at the shaft, emits a datum beam parallel to the designed tunnel axis. A combined Target/Inclinometer mounted on the tunnelling machine receives the beam and measures continuously the exact position — X and Y — and the lead angle of the machine with respect to the laser beam, while look-up and roll are measured with respect to gravity. The Engineer's Unit/Monitor is located usually at the shaft. It is certified for coalmining and has a proven record of reliability in tough environments. It contains a special purpose computer which receives digital data from the Target/ Inclinometer and has facilities for the input of numerical values. From these it calculates the following quantities:

Present position — X and Y —
Lead and look-up
Roll
Predicted position — X and Y —

and displays these in millimetres on its built-in digital monitor. In its supervisory role the Engineer's Unit checks the validity of all inputs, the presence of the laser beam, the continuity of the cables etc. In case of a fault the appropriate warning symbol is shown on the Monitor.

The system measures continuously the position and the attitude angles of the tunnelling machine, *en route,* relative to the designed tunnel axis. Actual and predicted deviations are instantaneously displayed in millimetres on a Monitor in front of the driver who steers the course with the minimum of correction. The digital information is also available, if required, in the site office and can be automatically recorded by a printer. Automation can go further still. With a suitable electrohydraulic interface the system will steer the machine automatically or semi-automatically without relying on operator judgement.

STRAIGHT PIPE JACKING OVER 150-250m

Pipe jacks over typically 150-250 metres between shafts present some additional problems. Firstly, the distance between laser and Target may be too long for the safe reception of the beam. Conditions are worse when excessive dust, mist or temperature gradients are present or in the case of pressurized tunnels, where the beam has to pass through one or more windows before reaching the Target. Secondly, intermediate jacking stations may be used introducing further variables to be measured, transmitted to the computer and incorporated in the calculations.

Figure 6.4.3 shows the schematic arrangement of a system configured from standard units for a pipe jack under the River Rhine at Cologne, Germany by Hochtief A G, Essen. For the first few hundred metres the laser remains fixed at the jacking shaft. The machine is guided in the usual way and its position is precisely recorded as a function of its distance from the shaft. After a drive of, say, 150 metres the Laser Base is mounted in the rearmost pipe. Some 40 metres later the Rear Station is inserted and the laser is transferred to the already prepared Laser Base establishing a moving Laser Station. From this point onward the principle of the guidance is as follows:

The respective distances between Target, Laser Station and Rear Station are known and, if intermediate jacking stations are inserted, continuously updated. The positions of the Laser Station and the Rear Station are known from 'historical' data ie from the previously recorded positions of the macchine. The orientation of the Laser Station, however, is not based on the recorded data because even a small angular error would give rise to a significant beam offset on the Target screen. For this reason the optical transmitter-receiver link between the Rear Station and the Laser Station measures the actual orientation of the laser beam and transmits this information to the Engineer's Unit (with optional automatic roll correction, if required). Having established the position and orientation of the laser beam the present and predicted positions of the machine as well as its attitude angles can be computed and displayed to the operator on the Monitor in the usual format.

CURVED PIPE JACKING

In the case of curved-axis pipe jacking there is no direct line of sight between the machine and the jacking shaft. In general it must be assumed further that the curvature of the designed tunnel axis does not remain constant along the drive. Consequently guidance with a laser fixed at the shaft or somewhere in the travelling lining is not practicable.

The solution of this problem has already been seen in a simplified form with the introduction of the Rear Station and the principle of using 'historical' data captured at an earlier stage of the drive (see STRAIGHT PIPE JACKING OVER 150-250m) and verified in over 1km curved pipe jacking in Hamburg. (ARGE Kuhmeuhle Nebensammler, pressurized shield by Herrenknecht GmbH.)

As seen in Figure 6.4.4, the position and the attitude angles of the machine are established by a polygon survey from a known fixed origin at the jacking shaft and extended to the tunnelling machine. The polygon is formed by special survey stations spaced so as to ensure a safe direct line of sight between neighbours. The length of each chord is recorded when the respective station is inserted. The effect of intermediate jacking stations is measured and accounted for in the computation. Each station consists of two optical transmitter-receiver pairs, one facing forward, the other backward, and a set of inclinometers to provide automatic correction for roll and look-up errors. The stations operate under the supervision of a computer, situated above ground or at the shaft. This computer is also responsible for storing the coordinates of the designed tunnel axis, processing the measured values and calculating from these the present and predicted positions of the machine for display on the operator's Monitor, remote display, printout etc.

It is evident that the position of the machine axis — ie the last chord — can be determined if all angles and chords of the polygon are known with sufficient accuracy but in the case of long tunnels and tight curves the number of stations may be prohibitive. It can be shown, and the Hamburg results support this, that the number of stations can be significantly reduced if use is made of the earlier-mentioned 'historical' data.

ALTERNATIVE METHODS — Existing and potential

The systems described above are suitable for most microtunnelling machines and applications. Some machines are offered with the manufacturers' proprietary guidance instrument. Contractors have also developed special-purpose devices.

Visual laser targets are widely used because of their low initial cost and simplicity. They are difficult to read accurately. Roll correction, attitude angles and predicted position are not obtainable.

The visual readout problem can be overcome by a small TV camera mounted near the target and transmitting the image to the shaft. This system cannot provide attitude angles or prediction either.

Hydrostatic gauges (differential liquid level detectors) are useful in monitoring level and grade but cannot give information about the lateral position and lead angle of the machine. In special circumstances metal detectors above ground can be of assistance.

Inertial guidance appears to offer an attractive alternative to optical systems because it does not depend on a direct line of sight. Apart from their high initial and maintenance cost commercially available systems have very limited stability by tunnelling standards.

CONCLUSIONS

Instrumental tunnel guidance is an essential part of microtunnelling. It achieves major savings of time and operating costs, enhances accuracy and improves safety. It has made the precise remote control of tunnelling machines practical and real. It leads the way to further automation for unmanned operation in hazardous or inaccessible conditions. Instrumental guidance helps the tunnelling industry to remain competitive.

REFERENCE

[1]Weber, W and Uffmann, HP, *Tiefbau, Ingenieurbau, Strassenbau* Nr 2 + 3, 1983.

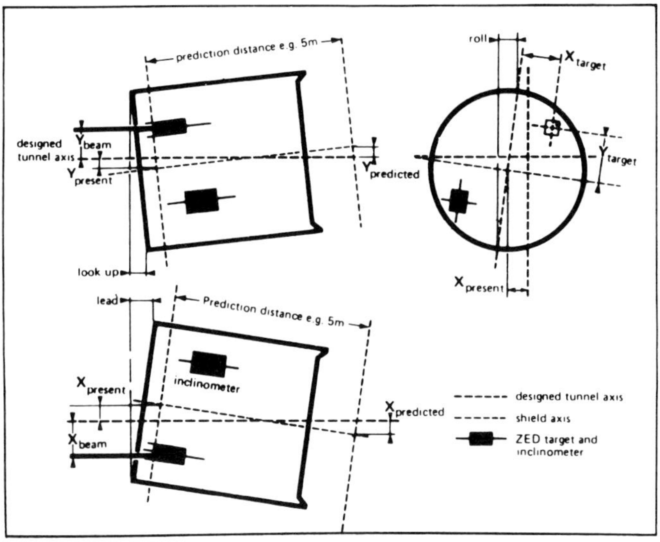

Figure 6.4.1(a). Definitions.

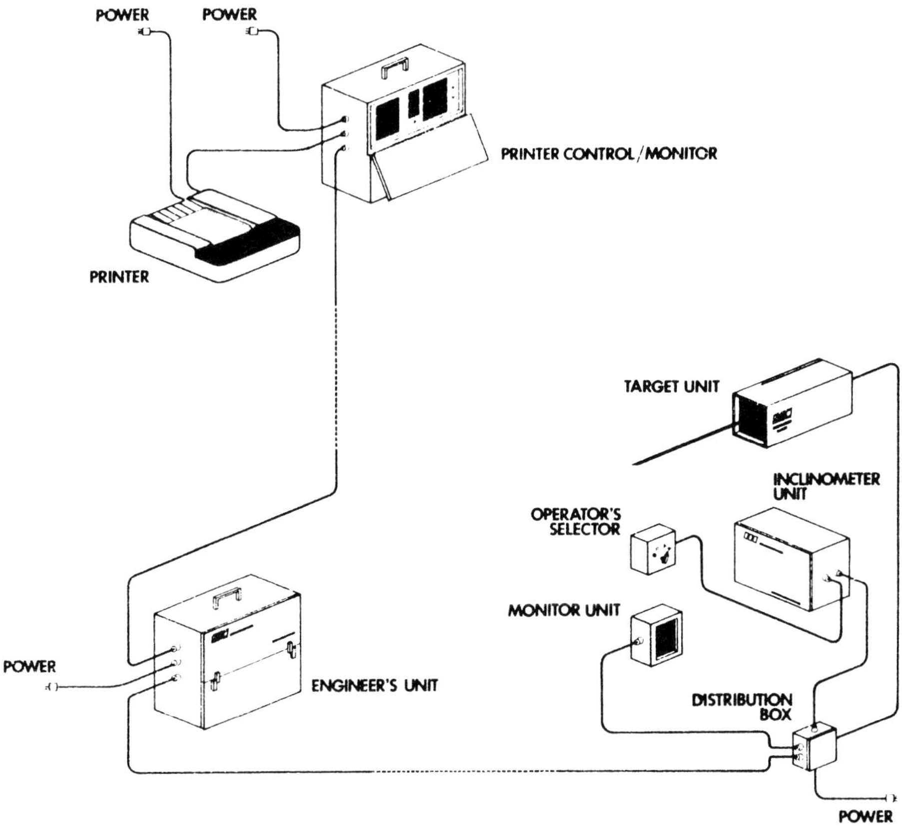

Figure 6.4.1(b). TG-26: unit schematic.

Figure 6.4.1(c). Monitor unit.

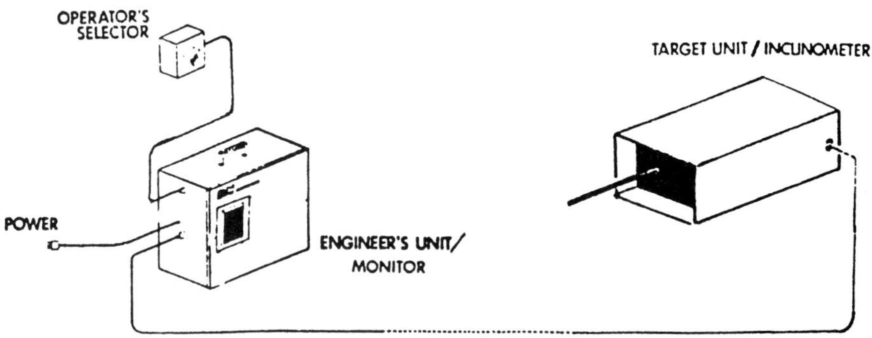

Figure 6.4.2(a). TG-26/E system: unit schematic.

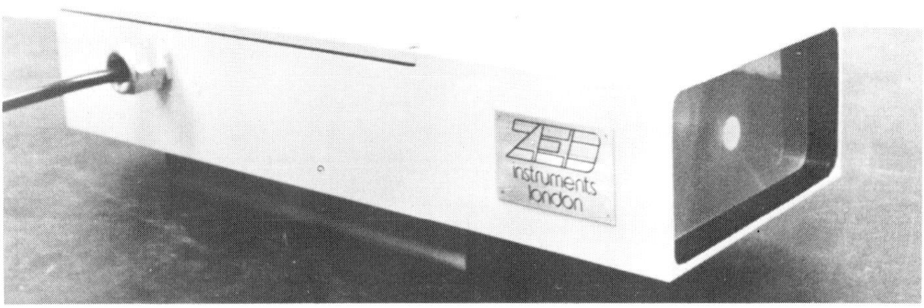

Figure 6.4.2(b). TG-26/E system: target/inclinometer.

Figure 6.4.2(c). TG-26/E system: target/inclinometer on Wirth TB 700

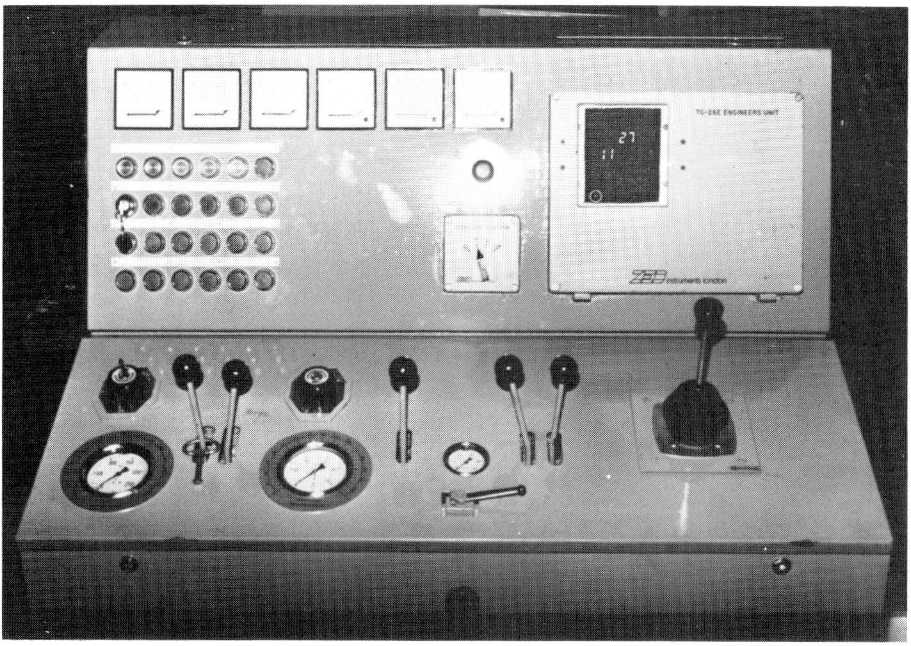

Figure 6.4.2(d). TG-26/E system: Wirth TB 700 control unit with engineer's unit/monitor and operator's selector.

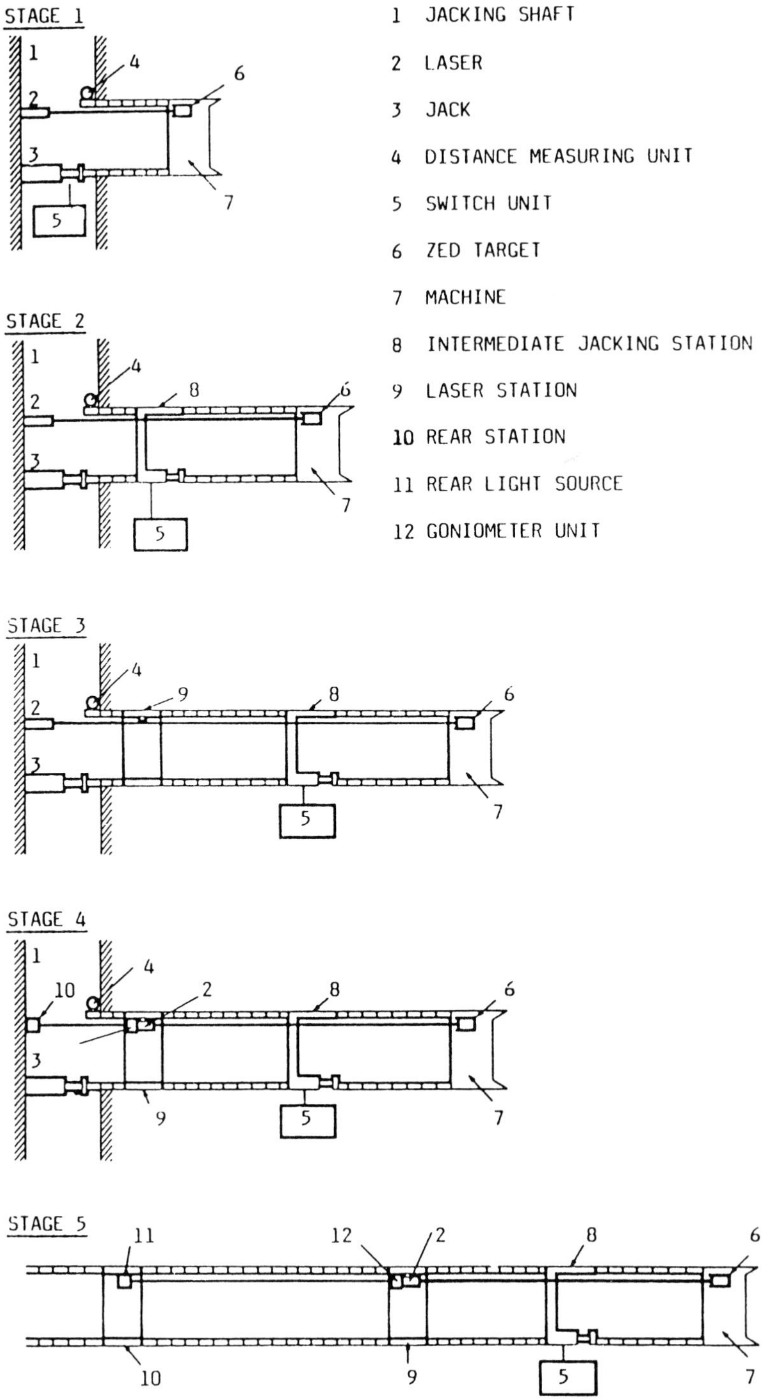

Figure 6.4.3(a). Straight pipe jacking: instrumentation stages.

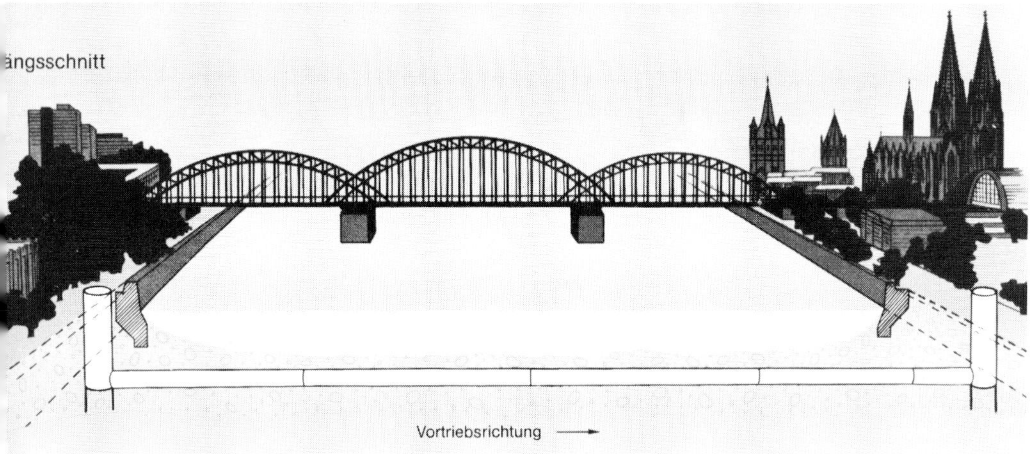

Figure 6.4.3(b). Straight pipe jacking: Rheindueker Cologne.

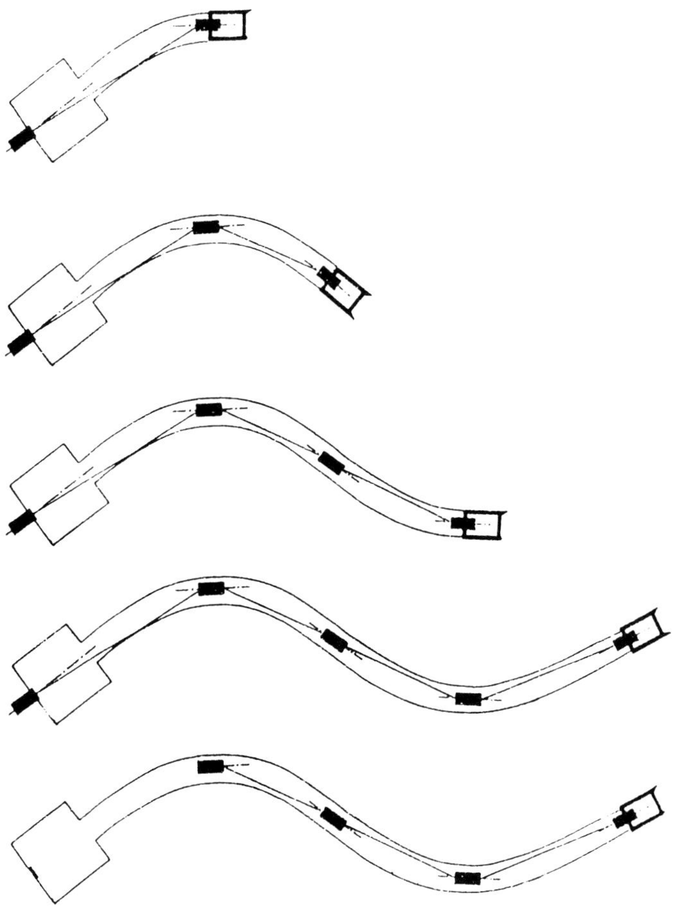

Figure 6.4.4. Curved pipe jacking: polygon survey.

6.5 Microtunnelling for communication services

Hiroshi Yamamoto, *Senior Staff Engineer*
Nippon Telegraph and Telephone Public Corporation

Yoshio Hashimoto, *Staff Engineer*
Nippon Telegraph and Telephone Public Corporation

Takaaki Fukumoto, *Engineer*
Nippon Telegraph and Telephone Public Corporation

The characteristic features of our method of thrusting small-diameter (300 to 600mm) steel pipes are as follows.

The most suitable mechanism for boring and for removing soil can be chosen according to the diameter of the steel pipe; the position of the boring head can be controlled with high accuracy; and steel pipes can be thrust for distances of 100 metres. These methods depend on the application of soil mechanics (for reduction of the jacking power required, hydraulic transmission of soil, and stress analysis of the surrounding soil) and the most advanced techniques of shield tunnelling and direction control, based on electrical engineering. We have carried out field experiments to test the capabilities of the machines. As a result, we find that we can thrust a steel pipe for about 100 metres.

INTRODUCTION

In the modern information-orientated society, electric or electronic communication is constantly progressing and changing and becoming more and more a part of social activity. It must not be interrupted at any time. The best use must be made of space in the modern urban environment, and new construction methods must be developed to avoid interference with essential activities.

For these reasons, Nippon Telegraph and Telephone Public Corporation (NTT) is developing various methods of trenchless pipelaying. The pipe-jacking methods introduced in this paper depend on past experimental techniques, modern soil mechanics and advanced mechatronics, and meet the following specifications.

1. A directional control system permits pipe jacking for distances of more than 100m with an accuracy of 1 in 1000.
2. Boring methods can be selected according to pipe diameter and soil conditions (range of Standard Penetration Test N-value 0 to 40).
3. Steel pipe joints are of the socket type, allowing rapid construction (10 to 15m/day).
4. The D301 system incorporates accurate directional control and can lay pipes in curves of radius of curvature R ⩾150m.

Outlines of the three systems are shown in Table 6.5.1. Details of the D401 system will be given in this paper to show the basic ideas of the system, the specification of the equipment and the results of trials.

TREND OF TRENCHLESS CONSTRUCTION TECHNIQUE

Recently the field of data communication has experienced a revolution — the Information Network System (INS). NTT has to promote digitalization and the use of optic fibre cables for long-distance transmission, and to renovate deteriorating facilities in urban areas.

However, changing environmental conditions mean that greater care must be exercised in road reinstatement, the control of traffic and protection of the public (Figures 6.5.1. and 6.5.2.). This necessitates efficient methods of reinstatement, the use of low-noise equipment and keeping the public informed. It is essential to develop trenchless construction methods instead of cut-and-cover methods in the interest of environmental protection.

In urban areas the roads provide space not only for traffic and for essential services such as gas, water, sewage, electricity and telecommunications, but also for those who live and work in the city and wish for a pleasant and safe environment. The cost of reinstatement after open-cut excavation is constantly increasing. It is therefore desirable to extend the use of trenchless construction and to reduce its cost by modern techniques.

BASIC IDEAS OF THE D401 SYSTEM

Decrease of thrust force

Most of the design conditions for a pipe-jacking system are determined not by static load (earth pressure, etc) but by the *in situ* thrust force. A decrease in the thrust force required is effective in reducing the cost of construction and in simplifying equipment, especially in bores of more than 100m.

The D401 is a small-diameter (450mm) pipe-jacking system which uses lateral vibration and a lubricant to reduce thrust and frictional resistance. The combination of water injection with vibration liquefies dense sand, while the injection of bentonite lubricant results in a considerable reduction in friction and cohesive resistance.

Figure 6.5.4. shows a dynamical model of the vibration head, and Figure 6.5.5. indicates how the optimum operating conditions are determined. The experimental results plotted in Figure 6.5.6. lead to the relationship shown in Figure 6.5.7. between thrust force and vibrational acceleration. The ranges of values are comparable to those in pile-driving. It is important to maintain the vibration within the optimum to avoid disturbance of the ground surface.

Directional control system

Previous pipe-jacking methods, which rely upon the accurate alignment of the jacking device in the thrust pit, and on the experience of the operator, are unable to drive distances of more than 100m, because of various unstable and unpredictable soil conditions. It is necessary to provide an accurate directional control system to adjust the driving direction of the head. This system consists of a device to measure geometric position, a mechanism to control direction, and a guidance principle.

Device to measure geometric position

The factors involved in tracing the path of the driving head are the head itself (position, angle, etc), the method of measurement (optical, magnetic, sonic, etc) and the measurement procedure (interval between readings, starting point, etc). The method used should be chosen to suit the particular conditions (accuracy required, type of application, etc).

The D401 system uses flexibly-jointed pipes, which may cause pipe meandering by progressive buckling (Figure 6.5.8.). Close control is necessary to prevent this meandering. For straight driving a laser beam is used, since it can give an accuracy of 1 in 100 000. This allows one-man operation at a desk, using a CRT indication of geometrical position and angle computed from the input data. (Figure 6.5.9).

Mechanism to control direction

This mechanism is indispensable for long-distance pipe jacking. In small-diameter pipe jacking the driving head must accommodate a direction control mechanism of high accuracy as well as equipment to excavate and convey soil and to measure the geometrical position.

The D401 employs a mechanism which controls direction by the application of a turning moment of up to 3.8 ton-metre, with a maximum controllable angle of 3 degrees.

Guidance principle

Control of the driving direction requires a knowledge of the geometric position of the head. However, feedback control takes some time to adjust the position because of the uncertainty of ground conditions. This causes divergent meandering of the machine. The D401 therefore employs a feed-forward control which limits the divergence of position and angle within allowable values.

Mechanism to convey excavated soil and to limit ground disturbance

In the D401 system the excavated soil is slurried at the driving head by the rotation of the head and the injection of water. The slurried soil is removed by a piston mechanism and conveyed under pressure through a pipe to a storage point near the starting pit, in the same way that concrete is pumped. In this way the narrow starting pit is kept clear and no local pollution is caused.

It is possible to control the process so that there is no interference with the ground surface. Figure 6.5.12. shows the theoretical calculation of the influence zone, and Figure 6.5.13. shows the results of some field experiments.

The excavating and soil conveying system has openings which will allow soil to pass, but not gravel above a specified size. Larger stones are pushed to one side by the vibration and rotation of the driving head (Figure 6.5.14.).

THE D401 PIPE-JACKING APPARATUS

The main features of the D401 are shown in Figure 6.5.15. It consists of the driving head, the main jacking unit and the units on the surface. Steel pipes are pushed from the starting pit, while pipes and cables are connected every two metres.

Driving head

The driving head (Figure 6.5.16) is equipped with mechanisms for vibration, rotation and control of direction, and with a soil-conveying system, laser receiver and concentrated control apparatus. The specifications are detailed in Table 6.5.2.

Main jacking unit

The main jacking unit (Figure 6.5.17) is equipped with a driving jack, a pushing plate, a jack to adjust the axial centre of the pipe, a device to transmit reaction, etc. Details are given in Table 6.5.3.

Unit on the ground

This comprises an operation desk, hydraulic power unit, control panel and bentonite injection apparatus (Figure 6.5.18).

FIELD EXPERIMENTS WITH THE D401 SYSTEM

The D401 has been driven through sand and clay containing gravel for distances of about 100 metres. The results of one such test are shown in Figures 6.5.19 to 6.5.22.

CONCLUSION

NTT, with the Hitachi Construction Machinery Company, has applied for world patents on the D401 system. In addition to the pipe-jacking process described, NTT has developed a shield-tunnelling method M-11, which uses Hume pipe of 1200mm diameter and can drive 150 metres, with the ability to work to a radius of curvature R $\geqslant$200m. NTT is also developing a shield-tunnelling method M-2 to drive 500m of 1200mm internal diameter tunnel with automatic *in situ* concrete lining, also with a possible radius of curvature R $\geqslant$200m. Progress is being made on the thrusting method D101, which lays conduits of diameter about 100mm up to a distance of 50 metres.

NTT is also developing methods for the trenchless repair of deteriorated pipes.

TABLE 6.5.1 NTT pipe-jacking systems

Item	System:	D601	D401	D301
Ground condition	Soil	clay, sand, gravel (up to 100mm)	clay, sand, gravel (up to 100mm)	clay
	N-value	30-40	30-40	10-15
	Max. water pressure	0.5kg/cm^2	0.5kg/cm^2	0.3kg/cm^2
Driving distance		100m	100m	100m
Curves		straight	straight	curve R$\geqslant$150m
Driving method		horizontal auger	rotation with vibration	compacting
Outer diameter of pipe, D		600mm	450mm	300mm

TABLE 6.5.2 Driving head specification

Apparatus	Item	Specification
Driving head	Diameter	480mm
	Length	5214mm
Excavation	Frequency	0-50Hz
	Rotation rate	0-40rpm
	Maximum torque	165kg-m
Directional control	Control jacks	10t x 15mm x 662kg/cm^2 x 4
	Controllable angle	max. $\pm$ 3°
Soil conveying	Conveying capacity	1.5m^3/h
	Conveying volume	2.6 l/stroke
	Conveying cycle	8 times/min

TABLE 6.5.3 Main jacking unit specification

Apparatus	Item	Specification
Main jacking unit	Driving jacks	50t x 500mm stroke x 2
	Driving speed	less than 15cm/min

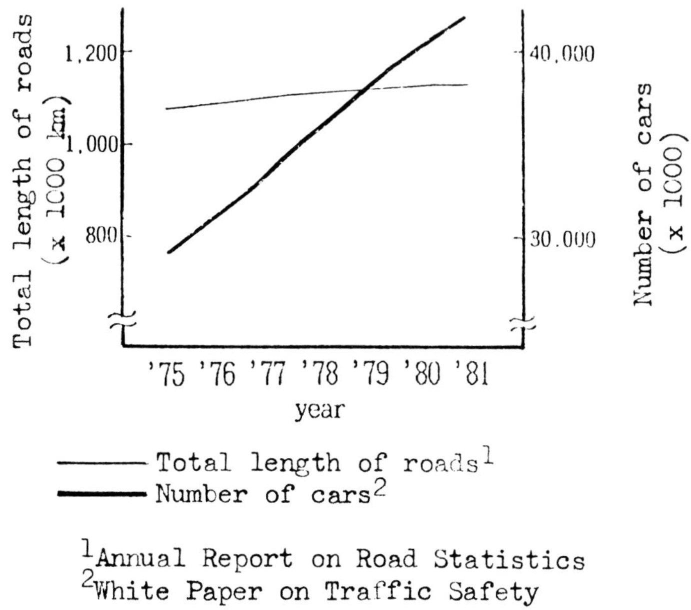

Figure 6.5.1. Total length of roads and number of cars, 1975-81.

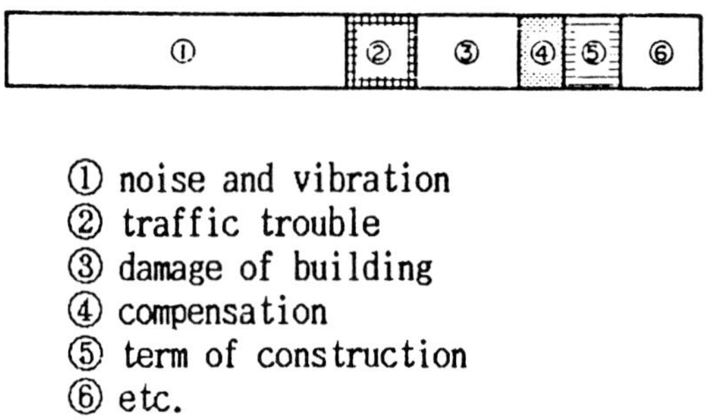

Figure 6.5.2. Causes of reported damage, 1981.

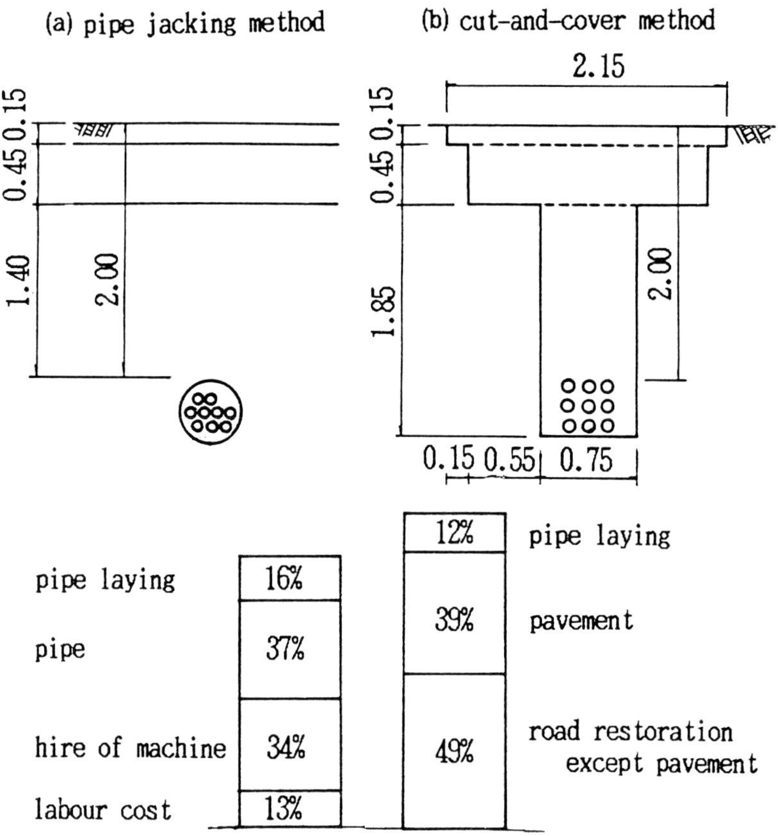

Figure 6.5.3. Cost comparison between pipe jacking and cut-and-cover.

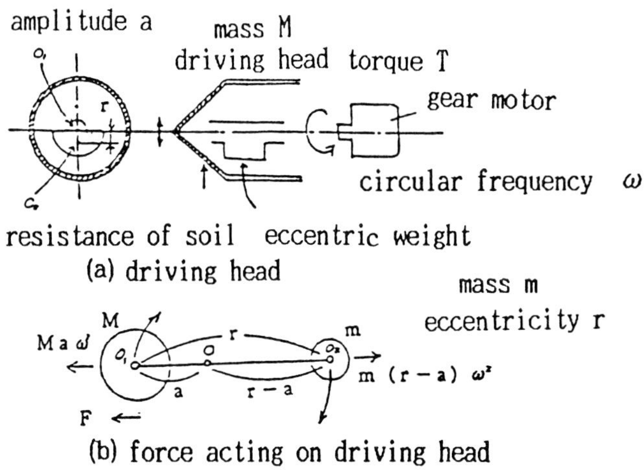

Figure 6.5.4. Model of driving head.

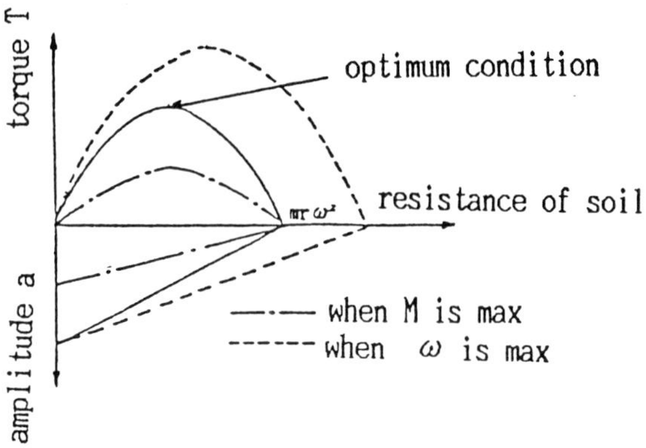

Figure 6.5.5. Change of torque and amplitude caused by resistance of soil.

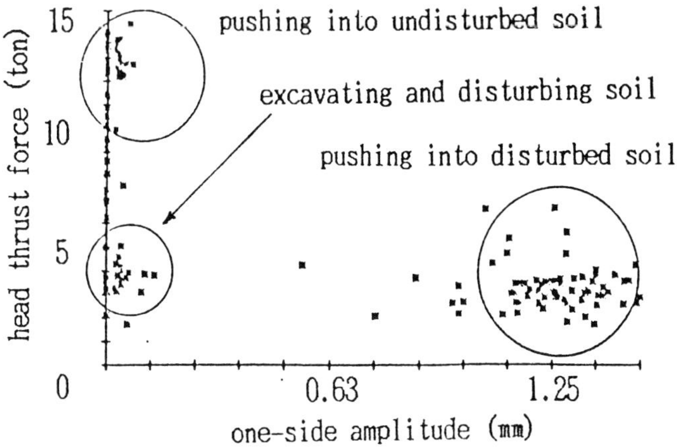

Figure 6.5.6. Relation between amplitude and thrust force.

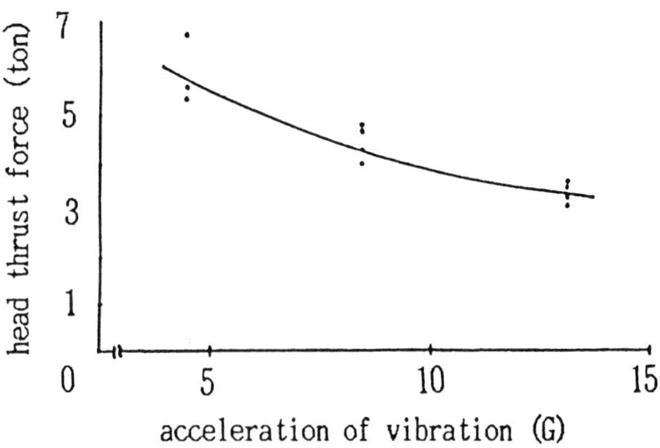

Figure 6.5.7. Relation between acceleration of vibration and thrust force.

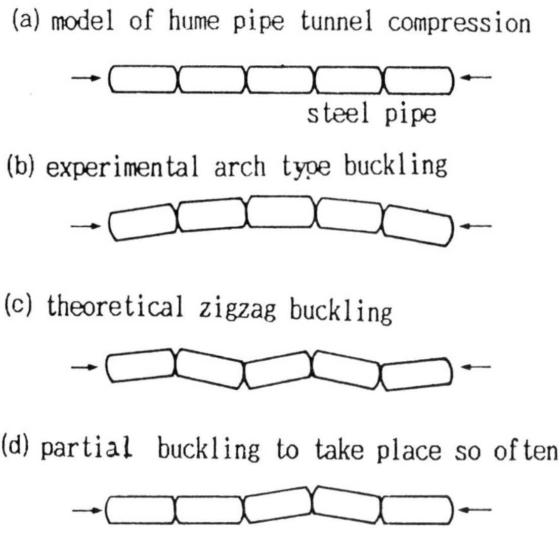

Figure 6.5.8. Pattern of meandering.

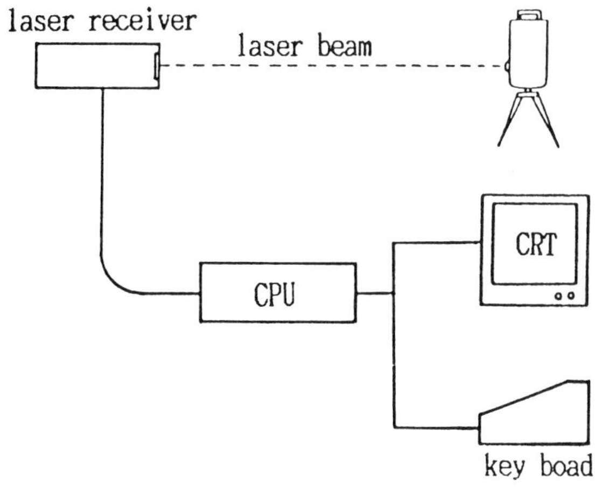

Figure 6.5.9. System to measure geometric position.

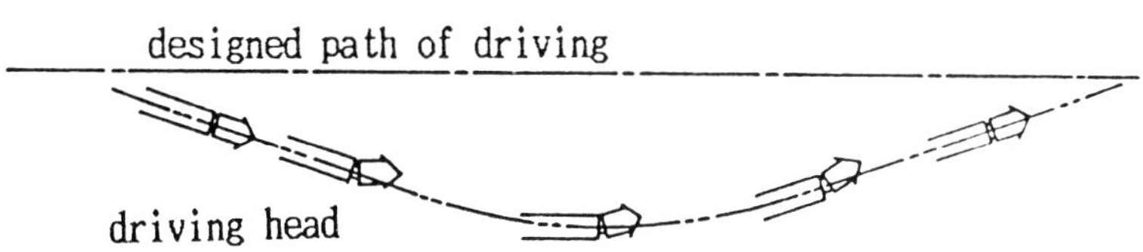

Figure 6.5.10 Outline of directional control.

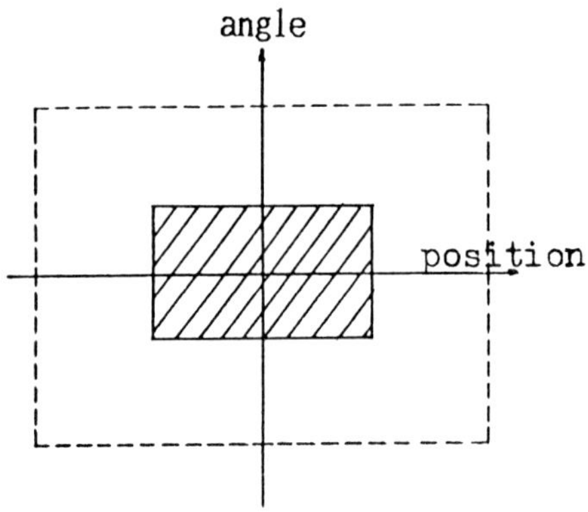

Figure 6.5.11. Limits to divergence of geometrical position and angle.

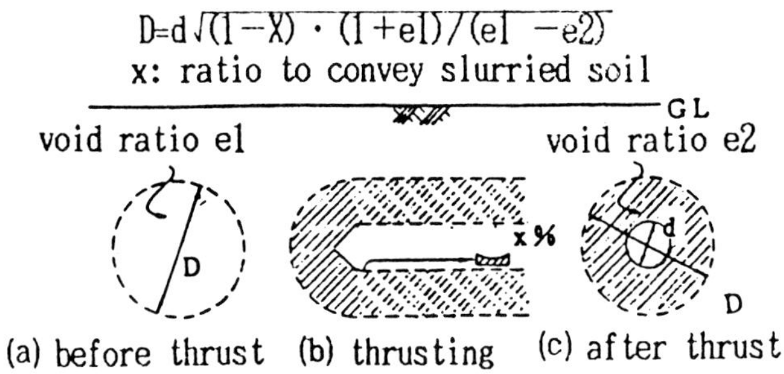

Figure 6.5.12. Influence area of excavation.

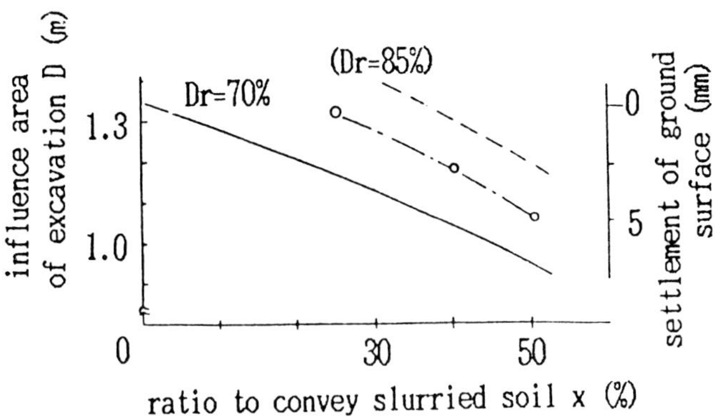

Figure 6.5.13. Influence area of excavation and settlement of ground surface v. ratio to convey slurried soil.

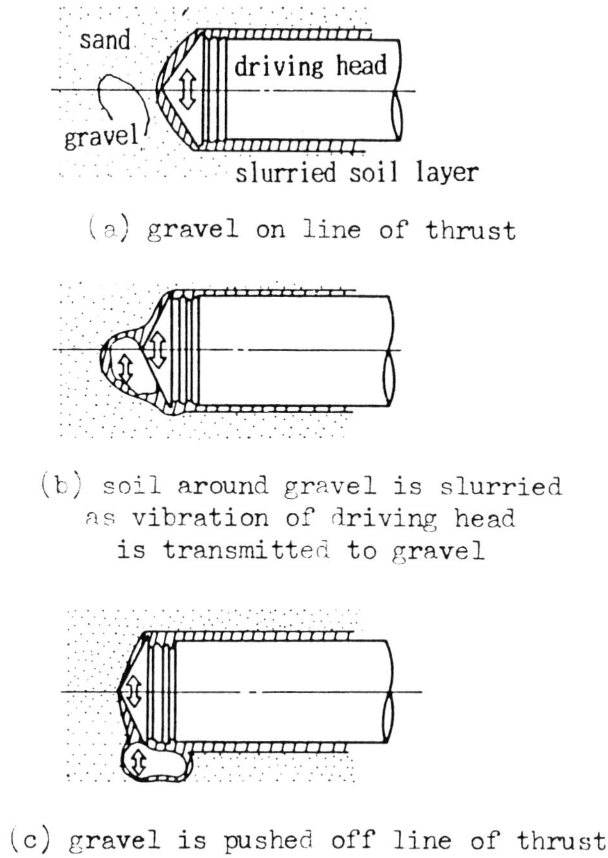

Figure 6.5.14. Mechanism for deflecting large gravel and stones.

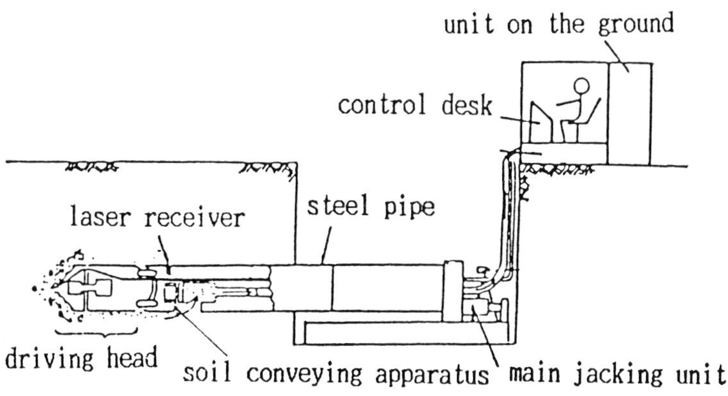

Figure 6.5.15. Outline of D401 pipelaying system.

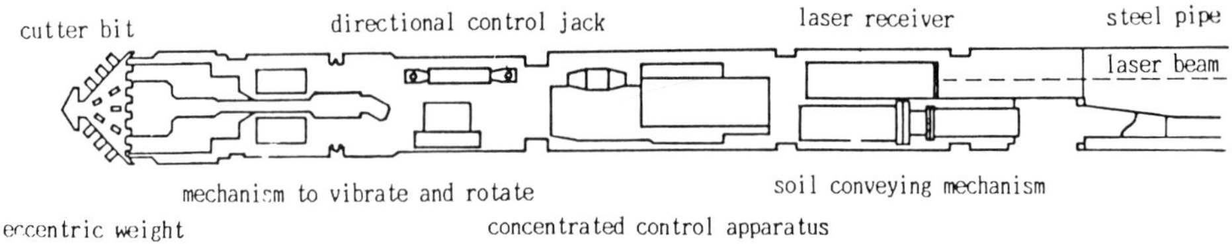

Figure 6.5.16. Outline of driving head.

Figure 6.5.17. Main jacking unit.

Figure 6.5.18. Unit on the ground.

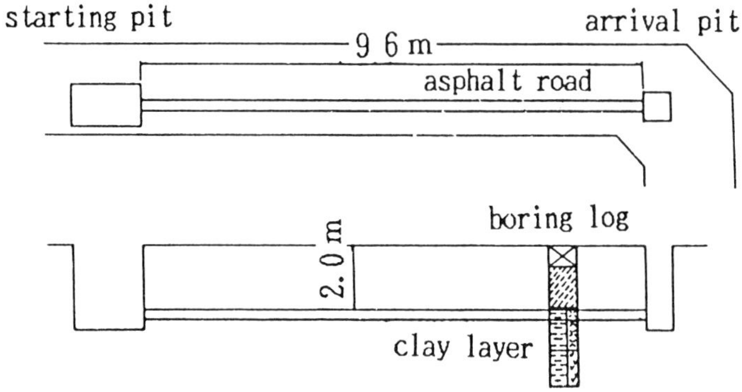

Figure 6.5.19. Outline of driving test.

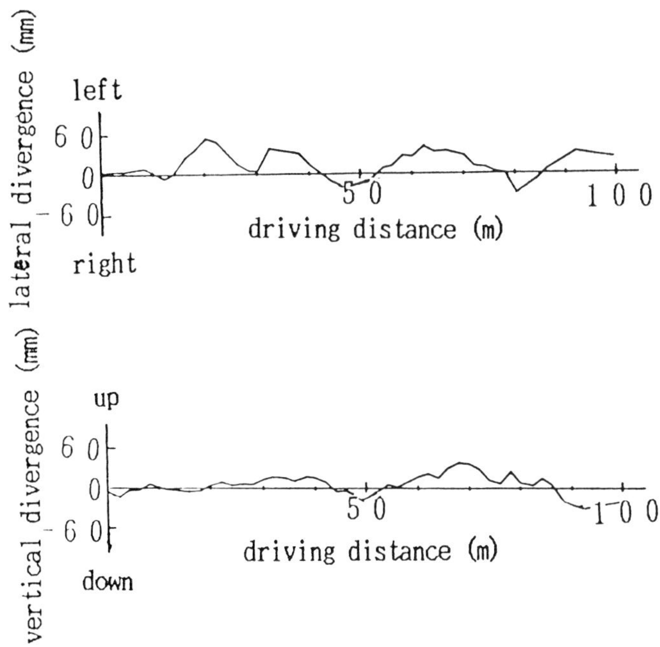

Figure 6.5.20. Locus of driving.

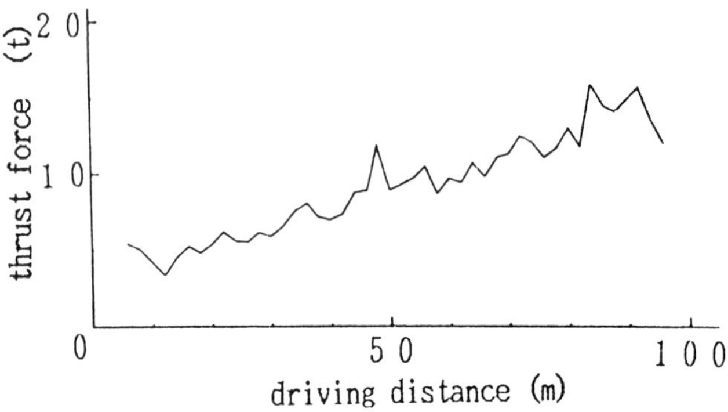

Figure 6.5.21. Thrust force v. driving distance.

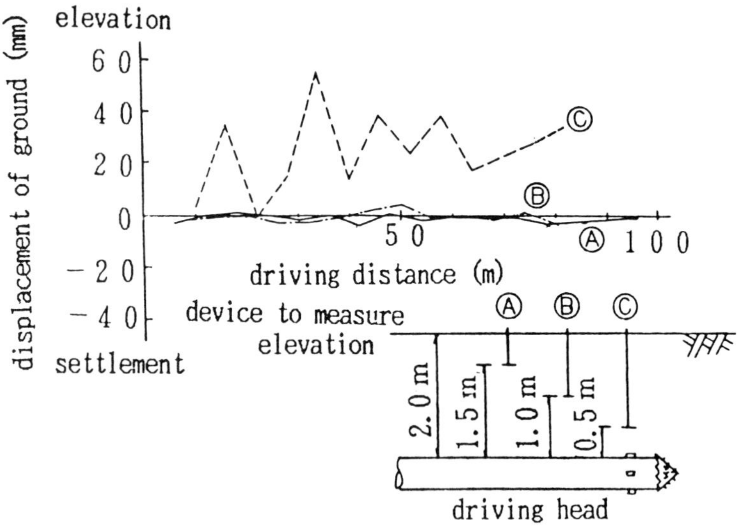

Figure 6.5.22. Displacement of ground v. driving distance.

Authors' Introductions

Dr P M Zollman: Guidance systems for microtunnelling

Coming so late in the proceedings, I have had a chance to sense where the emphasis will lie in connection with the subject which I am going to discuss. We have heard about the need to avoid kinks and wiggles, the necessity to reduce friction, the reduction of strain on joints, and of deformation of the ground during and after pipe jacking. I think all these aspects lead to the conclusion that guidance is a very important and essential part of microtunnelling. Costs are of course a vital question, and the problem is to combine performance with an acceptable price. I would stress right from the beginning that my subject is very specifically one of guidance of pipe-jacking machines and moles not below 500mm or so in diameter, and as such it has its limitations within the scope of microtunnelling. At the same time I believe it does provide an attractive solution to several of the problems with which we are confronted.

Figure 6.4.1 (b) illustrates very broadly a practical piece of equipment and shows its basic simplicity because all the vital components are shown there. On the bottom is a combined target and inclinometer unit which continuously produces a precise survey of the position in space of the cutter head, say, of a steerable pipe-jacking machine. It gives the position and all the attitude angles, that is to say it will measure five degrees of freedom of the total six degrees that a system of this kind may have. Above it is a monitor combined with a computer. This can be situated either in the tunnel in a large diameter system or at the pit.

The monitor shows in clear digital values the deviation of the axis of the machine from the designed tunnel axis. If the error is up then the number appears above the centre; if the error is to the right it appears on the right. At various commands, in addition to the present position of the axis, the monitor will show the predicted position of the machine. This is vitally important because corrective measures can be taken before the machine deviates from the axis and accurate steering means reduced steering. The less steering you have to do the more precisely you can guide and the better you will meet all the requirements. It is also true that the more accurately one can guide, the longer the length of drive can be, all other factors being considered. Mr Moss mentioned the importance of risk avoidance or risk limitation. It is quite obvious that with an accurate guidance system the risk of a job going wrong or of hitting existing services in the ground is reduced. The possibility of knowing exactly where the machine is, plus where the machine is heading, is of vital importance.

Figure 6.4.1 (a) shows the fundamental principle regardless of what instrumentation package is provided. The machine is shown with all the misalignment errors that the human mind can imagine. A laser beam and gravity provide the basic references to the system. The laser beam defines where the desired tunnel axis is, and is detected by a target. This target measures both the horizontal and vertical deviations of the laser beam from its centre, and also measures the angle that the laser beam makes with the normal to the target surface. An inclinometer, which is sometimes built into the target, sometimes independent, provides information on the roll of the machine about its own main axis, and what we term look-up, the vertical angle of inclination of the machine axis itself. With the continuously updated measurement of these five variables and occasionally, if necessary, the measurement of the distance that the machine has travelled from the pit, one can continuously establish the position of the axis in space. This is done by all measured data being processed by the computer, plus the consideration of all the constants of the geometry of the stystem, for example the position where the laser beam was mounted with respect to the tunnel axis.

The first guidance system of this kind was in fact developed for a 1.2m-diameter minitunnel, but then, mainly under market pressures, the application went towards bigger diameter machines. To date about 50-odd systems have been introduced worldwide, mainly for tunnel-boring applications with various kinds of machines, but also for pipe jacking down to 700mm diameter, shaft boring and a few special applications.

[Mr Zollman showed several slides of both large and small machines fitted with the guidance system. The examples included a small full-face tunnel-boring machine, a large metro construction system in Lyon working along a curved axis, a vertical blind shaft boring machine guided to within 2cm accuracy over a total depth of 600m and a pressurized pipe jacking machine in Hamburg, which, after 350m of curved driving, arrived with a deviation of less than 2cm.]

The system I have been describing has its limitations. The laser beam has to be within direct line of sight, and cannot be very far away from the target, particularly if dust or other obstacles spoil the reflection of the laser beam. In order to improve on that a system has been designed which is at present producing a tunnel under the river Rhine at Cologne of a total length of 470m between shafts (Figure 6.4.3 (b)). This is a pipe-jacking application and Figure 6.4.3 (a) shows the five stages of the procedure as far as guidance is concerned. The five steps enable three stations, the machine station, the laser station and the rear station, to travel down the system and provide accurate guidance regardless of the actual length of drive one has to do. The total distance between the laser and the target in this case is not greater than 40m.

The principle of the curved pipe-jacking system, with which we obtained such excellent results under Hamburg, is described in the paper, and illustrated in Figure 6.4.4. At the moment this system has been used on 1000m of driving.

In conclusion, I can say that for applications where this kind of system can be accommodated, it produces accuracy, reliability and continuous drivage.

Mr Y Hashimoto: Microtunnelling for communications services

The Nippon Telegraph and Telephone Corporation (NTT), for which I work as a staff engineer, was converted from a public corporation into a public enterprise on 1 April this year in order to cope with the modern demands of an information-orientated society.

In order to protect the urban environment, NTT have been developing various shield tunnelling methods including small-diameter systems. New pipelaying methods described in this paper have been developed to cope with the spread of optical fibre cable and to construct small diameter pipes.

Outlines of these systems are shown in Table 6.5.1 and are as follows:

The D601 system is based on the conventional auger jacking method which excavates by means of a cutter and discharges by means of an auger. Compared with conventional methods there are several advantages. One is the method for preventing water gushing, and another is the method for controlling direction.

The D401 system uses lateral vibration, rotation of the head and injection of a lubricant. This system is so unique that we have described it in our paper in detail.

The D301 system is a small-diameter pipe-jacking method suitable for long and curved drives. Control of direction from the surface is achieved by both a directional adjusting mechanism and a position measuring system. Its action is based on soil compacting, so its operation is simple and very speedy.

These systems have been tested in various places in Japan.

I would like to call attention to three technical features of the D401 system. The first is the method of decreasing the thrust and frictional forces by lateral vibration, rotation and the injection of a lubricant, which is discussed in detail in the paper.

The second feature is the method for controlling direction, which involves both a directional control mechanism and a laser position-measuring system. The most difficult part of the design is to produce an algorithm to represent measurement and control of direction and position. The characteristics of the algorithm depend on the type of boring head and soil conditions as well as on the methods of measurement of position and control of direction.

In developing our systems, we have had to consider the following questions.

1. How often should we control the direction of the machine to keep it from meandering?
2. How much force should be used to adjust direction?
3. What is the relationship between the target accuracy required and the allowable limits of adjustment?
4. What accuracy is required in measuring the geometric position?
5. Which is the more important factor in control, position or angle?
6. What are the effects of machine length and diameter on the tendency to meander and on the ease of adjustment?

The answers to these questions will greatly influence the cost of a system.

[Mr Hashimoto used slides to illustrate the following points.]

We have used probability methods to analyse data recorded in test drives. By plotting the occurrence frequency of various differential changes of direction, we produce a probability curve in which the mean absolute value corresponds to ease of control and the standard deviation measures the tendency to meander. It appears that we can use a normal probability curve to represent the phenomena which occur during practical thrusting.

When we plot geometric angle against position, for various intervals of control, we can see how the meandering increases as the time interval between corrections increases. Further analysis using various allowable limits of position and angle enable us to estimate the probability of reaching the driving target under given conditions. We are, therefore, sure that driving control technique can be generalized by the accumulation of measured data and interactive operation, though in the past it has depended on the operator's skill and judgement.

The third main feature of the D401 is the technique of transporting the excavated soil, which is described in detail in the paper.

We at NTT are willing to exchange technical ideas and news on trenchless pipelaying, and hope our paper will be of value to others. We will be happy to give more information on our work.

Discussion on Papers 6.4 and 6.5

A F Elvidge (Manager, Methods Group, WRc Engineering): There is a general move towards smaller and smaller diameter tunnelling, and we now have tools that will go down to diameters of 250mm and perhaps even smaller. Does Dr Zollman think his company will move in this direction, and, if so, will their methods achieve the same degree of accuracy as with the larger diameters? How will working with the smaller sizes affect the cost of guiding such tunnels, particularly where there may be a need on curved tunnels for more intermediate stations?

Dr Zollman referred briefly in his paper to automatic systems which could 'close the loop' between guidance and progress. Has he any direct experience of such systems? Is he considering working on any other guidance systems, and, in particular, has he made any progress with inertial guidance systems?

P B Rumsey (Water Research Centre, and IPHE Organizing Committee): What is Dr Zollman's opinion on the feasibility of incorporating a guidance system such as he has described — or any other type — in a percussion mole? What would be the cost implications of introducing such a system?

Could Dr Zollman discuss the viability of sensing techniques which might be incorporated in a moling system to allow advance warning of obstructions and services during operation?

A year or so ago there was talk of a steerable mole having been produced in Japan. Could Mr Hashimoto confirm whether this is correct, and whether the mole is available in the Japanese market?

Mr Hashimoto: Some steerable mole methods are popular in Japan, and these applications have been about 50km per year recently.

Dr P M Zollman: It is very likely that within the next six months we shall have a tried and tested system which will go down to an internal diameter of 250mm. One is actually working at the moment, but it would be premature to say much about it. Up to now, we have been building one family of systems in which all units are compatible, but, as a result, we have accumulated an increasing cost burden. Having now to rethink our programme to take account of the very small sizes, we have done some spring cleaning, and I seriously hope that we will be able to get our selling prices down to 40-50 per cent of the present prices. The typical systems I have described cost from £18 000 to £62 000. This is a wide bracket, but one cannot be more specific without going into a lot more detail.

Mr Elvidge asked about automatic systems. In the last few years, we have expanded our activities into designing hydraulics which interface with customers' existing operating rams and power packs. We have one large tunnel-boring machine which we have automated to close the human loop, and this works quite well. There is no reason why automation should not be used.

This leads me to a very fundamental question — the biggest question to come out of this Conference, in my opinion. Is there really room for an independent equipment guidance system supplier in this field, with smaller and smaller machines coming along? Will it remain possible to supply standard equipment which will suit any size of tunnelling machine? The small machines require such an integration of the various design features that I believe far better results would be achieved by specialist manufacturers. However, I do not quite know how the market will go.

In reply to Mr Rumsey, I believe that in principle guidance systems could be used with percussion moles, but such a system would have to be designed specifically for a particular design of percussion mole.

I am sure that inertial guidance systems will come into use, though I think they are not yet ready for practical application. Size, cost and stability still present problems, but they should be available in five years' time, and possibly sooner.

Professor Stott: Mr Hashimito's system uses a laser for guidance, whereas Dr Zollman uses both a laser and an inclinometer to give line and attitude. Are both of these instruments still used on the smaller machines?

Dr Zollman: The smaller systems do not use lasers. They still work on an optical principle. However, the final objective of knowing the position and the attitude of the machine you are steering still remains. It is not sufficient to know where the machine is — it is also important to know which way it is pointing.

It seems to me that Mr Hashimoto's method must rely on dead-reckoning principles — he knows what increments have been made, but this is not the same as knowing exactly where he is. I am surprised that, since he has the chance of receiving a laser beam, he does not use it as an angular reference as well as a positional reference.

Mr Hashimoto: The D401 system uses a laser for horizontal and vertical measurement not only of position but also of angle, and I have explained the characteristics of directional control depending on these four parameters. Furthermore this system uses an inclinometer to determine its pitching.

We have tried curve driving with a laser as in Figure 6.4.4. of Dr Zollman's paper, but it has not been successful in microtunnelling, where a man cannot enter, because the development of the laser relaying instrument is too difficult. So now we use the laser for straight driving only. However, we can successfully drive curves using the D301 system as mentioned at the beginning of our paper and in Table 6.5.1.

A Delegate: Would Mr Hashimoto explain further about soil removal in his method?

Mr Hashimoto: Soil at the driving head is slurried by the injection of a bentonite lubricant, assisted by the vibration of the head. The slurried soil is drawn into a chamber about 4m behind the head and pumped by hydraulic cylinders through a pipe to the surface near the starting pit, as illustrated in Figure 6.5.15 in the paper.

6.6 Concluding address: the way ahead

E W Flaxman

Binnie & Partners
(Chairman of NO-DIG 85 Organizing Committee)

The increasing demand and the potential for trenchless methods are currently leading to rapid developments in guided microtunnelling and associated fields. Attention is drawn to technical aspects in which further advances may be particularly significant and to the need for social costs to be assessed and apportioned when comparing trenchless construction with trenching. The Institution's future NO-DIG plans are summarized and details given of a special Medal which will be awarded in 1986.

The task of concluding an International Conference as important and wide-ranging as NO-DIG 85 is not a simple one. In the time at my disposal — and I am well aware that many delegates will wish to leave punctually by 4.30 — I propose to pick out what seem to me the most important issues likely to govern development in the field of trenchless construction in the future and to propose ways in which the keen interest generated here may be developed worldwide.

I say worldwide advisedly. This first International Conference on Trenchless Construction has shown how engineers in many different countries have been making notable advances over the past decade. They have made good use of advances in technology but at the same time have often been spurred on by public opinion, exasperated by the disruptive effect of open-cut trenching on everyday life.

Delegates to this Conference do not need to be reminded of these disruptive effects. But we now have a much clearer idea of the extent of the problem. Perhaps as much as 500 000km of mains and services buried worldwide each year — the great majority by open-cut trenching.

The digging of trenches for installing conduits or other services below ground is a civil engineering method of great antiquity. Despite modern improvements in mechanical plant, trenching inevitably suffers three fundamental disadvantages:

1. It is slow and inefficient because of the large volumes of soil that have to be removed and then replaced;
2. It usually results in disturbance of the surrounding ground;
3. It sterilizes the ground surface for any use while the trench is open.

This last factor usually causes most widespread public concern, but all three have significant adverse effects in modern urban society.

Our modern urban society is heavily dependent on buried services — water, gas, electricity, telecommunications and sewerage. Yet if these utilities are provided or renewed by open-cut trenching there is every chance of disrupting the other vital element of modern urban life — transportation — and also of maximizing local inconvenience to residents and businesses.

So we, the Public Health Engineers, the Gas, Electricity and Telecommunications Engineers have all created a vast underground infrastructure. Graham Cox's slide showing the density of services in one London street (Figure 6.6.1), though doubtless an extreme example, does illustrate the way in which addition after addition has been made to the systems. Rarely is a main or service actually removed: more often it is just added to.

Those of you who studied the maps of parts of Kensington and Melbourne showing evidence of recent trenching may perhaps have wondered whether they are typical. They certainly are typical of many cities but they give no indication whatever of the total extent of existing underground services. If one assumes that there is only one main in each street for each of the five principal services — water, gas, electricity, telecommunications and sewerage — then the map of a small part of Kensington begins to look like an extraordinarily complex spider's web (Figure 6.6.2).

The demands of modern society have obliged us to create this underground web of services. And until recently the only trenchless methods available to us for installing mains and services underground were non-steerable and therefore restricted to relatively short distances. But now, within the last decade, we have seen the development of steerable microtunnelling machines with greatly improved capabilities both as regards distance and accuracy. We have heard during the Conference of exciting new developments taking place in, for example, Germany and Japan. But these developments cannot yet achieve their full potential for four main reasons:

1. The heavy development costs of steerable microtunnelling machines tend to make the construction costs of trenchless methods relatively high;
2. Insufficient account is often taken of social costs in comparisons between trenchless and trenching methods of construction, and even when they are assessed decisions are needed on who pays;
3. We do not know accurately the positions of existing underground services;
4. The problems of making service connections to new mains.

I propose to discuss the future for microtunnelling briefly under each of these four headings, dealing first with costs and technical aspects of development.

There can be no doubt that unit costs will fall as use increases and development costs can be written off. But further technical progress will continue. Great though the improvements have been already, there will doubtless be striking further advances during the next few years in accuracy, in length capability and in smallness of size.

There are two areas of technical advance which seem to have major potential for the future. The first is the creation of a pipeline or duct *in situ* in a trenchless hole The second is the elimination of the need to remove soil in order to create the hole.

The paper by Dr Uffmann this afternoon has demonstrated one idea for forming a pipe *in situ* using concrete. Other methods of *in situ* lining either with quick-setting materials or with concertina tubes or with helical strip seem likely to develop rapidly in the future in order to eliminate the costs and disruptions of pushing or pulling factory-made pipes into position.

In relation to displacement techniques of trenchless construction, the Conference has discussed impact moling and pipe bursting in cases where the size of the utility to be installed is fairly modest. Despite the obvious risks of damage to road surfaces and other services from these expansive methods they seem bound to receive wider attention in the future because of their great advantage in not requiring the excavation, removal and disposal of soil.

The current generation of excavating microtunnelling machines can cope with most ground conditions other than hard rock but the removal and disposal of soil is a complication which would be much better avoided. In some cases the speed of advance of existing machines is controlled not by excavation but by the limitations of separating soil from bentonite.

As already mentioned, more widespread use of the existing generation of microtunnelling machines will undoubtedly reduce unit costs in the future but at present the construction cost of trenchless methods is often much more than the construction cost of trenching.

However, simple comparisons between the construction costs of trenchless and trenching methods can be misleading if social costs are ignored, particularly in the urban situation. Social costs need in some way to be taken into account before we arrive at decisions on which method of construction to use.

The question is very topical in the UK because of Professor Michael Horne's forthcoming report to the Minister, on possible changes to legislation governing the work of Utilities in Public Streets. It is to be hoped that any new legislation will take account not only of the availability of trenchless methods but also the need in many cases to assess social costs before deciding which construction method to use.

One simple step is to forbid trenching altogether in certain classes of road, particularly in sensitive urban areas. This has been done in Australia, Germany and elsewhere, and in the UK no services are permitted under one class of road — the motorways. But such restrictions could not be imposed universally because in some cases they might lead to unnecessarily expensive construction in order to save quite small social costs.

What is needed is information. Utilities proposing to carry out street works in certain key areas could be required as a first step to prepare estimates of construction cost, or even obtain tenders, for both trenching and trenchless methods. If trenching proved to be more expensive the choice would usually be straightforward (Figure 6.6.3). The problem arises where the construction costs of trenchless methods are substantially higher than for trenching but it is not known whether differences in social costs are likely to be large. The only solution to that problem is to estimate also the social costs likely to be incurred by each method of working — trenching and trenchless (Figure 6.6.4).

In some cases the choice would still be straightforward — the cost of trenching would remain less even when social costs were taken into account. The problem case would be that in which the trenchless method was more expensive in construction cost but less expensive in total cost to the community. The key question if the trenchless method were adopted is who should pay for the increased construction costs?

The only way I can suggest for dealing with this very difficult question is to adopt a compromise. If a Highway Authority insists, because of high potential social costs, that a Utility's works must be carried out by more expensive trenchless methods, then perhaps the additional cost should be shared between the Highway Authority and the Utility. I realize only too well that this suggestion may not immediatley commend itself to Highway Authorities who sometimes regard the Utilites' works as an unmitigated nuisance, causing damage to their streets and obstructing the traffic which is the reason for the street's existence.

But is that so? Is traffic the only reason for the existence of streets? Maybe it was many years ago, but today I believe we should regard streets in urban areas as vital arteries not just for traffic but also for the mains and services operated by the Public Utilities.

The recent revelation that the closure of one lane on a three-lane motorway for maintenance purposes may impose social costs on the community exceeding £50 000/week shows only too clearly the importance of this whole question to the general public and to Government. A national approach to the problem is essential in order to tackle the complex issues raised by the multi-purpose use of our streets in urban areas. The appropriate Government Departments must be encouraged to address the problem on a national basis.

Turning now to the third matter of vital importance to the future of trenchless construction — the location of existing underground services — this morning's Session provided a useful insight into likely future developments in this important field.

It would ill become a Civil Engineer, such as me, to comment in detail on the highly specialized techniques now under test. But I am confident that somewhere, somehow, someone will develop a dependable method of accurately mapping existing underground services. Bearing in mind the remarkably powerful developments during the last decade in such diverse locational techniques as side-scan sonar, geophysical methods and satellite imagery I have no doubt that this major outstanding problem will be solved. The considerable efforts now being made to crack this problem certainly deserve our strongest support.

With regard to the fourth important problem, that of making service connections to new mains, it was unfortunate that we did not have room for a paper on this subject in the programme. Great progress has been made in the United Kingdom with the remote connection of sewer laterals and with minimum-excavation methods for making other service connections.

Because we are fortunate enough to have in UK a sewerage system more comprehensive and complete than any other country in the world and because parts of the system are relatively old we have concentrated our efforts during the last five years largely on renovation and on-line replacement techniques. Methods are now available for making connections remotely without excavation to sewers and drains, and minimum excavation techniques for other types of service connections are developing rapidly. Perhaps at a future NO-DIG event we should devote one entire session to papers on this important subject, so closely linked with the development and use of trenchless techniques for mains.

And so what of future events? The response to this first NO-DIG International Conference and Exhibition has been so dramatic, with delegates coming from more than 20 different countries and the Conference booked full to capacity, that there is clearly keen interest in this subject worldwide. Many suggestions have been made to me that we really must continue to communicate within this fast developing industry both by meetings such as this and by the regular publication of information on microtunnelling and other NO-DIG subjects.

We have accordingly come to the conclusion that the Institution should sponsor another major international event in either two or three years' time. At present we have an open mind as to where such an event, which will obviously be called NO-DIG 87 or NO-DIG 88, should be held. The Institution would prefer not to have to organize single-handed NO-DIG 87 or NO-DIG 88 if it *is* held outside UK and we would be pleased to hear from engineering institutions elsewhere who might be interested in sharing the organization of the event. A decision on the venue will have to be taken within the next few months and we should be glad to receive such approaches before the end of June this year.

In order to provide for the regular exchange and publication of NO-DIG information an international journal is essential to cover the closely related fields of first-time microtunnelling, on-line replacement and non-intrusive mapping. We are therefore proposing to form an International NO-DIG Society which will publish technical articles and information on a wide range of specialist NO-DIG techniques, equipment and achievements. We shall shortly be issuing to all those who have attended this Conference or the Exhibition a short note on this new International Society and you will be invited to join and to signify which are your particular interests in the NO-DIG field.

Finally, ladies and gentlemen, I am pleased to announce details of the award that was mentioned by our President, Mr Peter Banks, when he introduced the Conference on Tuesday. The Institution has decided to award a special Medal for the most notable advance in trenchless construction or related fields of guidance, of lining or of the location of existing underground services. This award will be presented by our President towards the end of 1986. Any advance in performance, in invention, or in technology related to trenchless construction during the 12-month period following this Conference will be eligible for consideration.

1986 NO-DIG AWARD

The Institution of Public Health Engineers is pleased to announce that a special Medal will be awarded in 1986 for the most significant achievement in the NO-DIG field during the 12 months May 1985 to April 1986.

This international prize will be awarded to the individual, firm or authority making the most notable advance in this field during the 12 months following the NO-DIG 85 Conference and details should be lodged with the Institution before the end of June 1986. Achievement may be in the field of distance attained between excavations: in smallness of diameter: in on-line replacement: or in any invention or technical development which will enhance the usefulness and/or economy of NO-DIG methods in the future, including non-excavation techniques for locating, identifying and mapping existing underground services.

A small international panel of adjudicators will make a recommendation for the award by September 1986 and the winner will be announced by the Institution in October 1986. The Medal will be presented by the President of the Institution at a ceremony to be held towards the end of 1986 and the winner will be given an opportunity to present a paper at the next International NO-DIG Conference to be organized by the Institution of Public Health Engineers.

(Reproduced by courtesy of Thames Water)

Figure 6.6.1.

Figure 6.6.2.

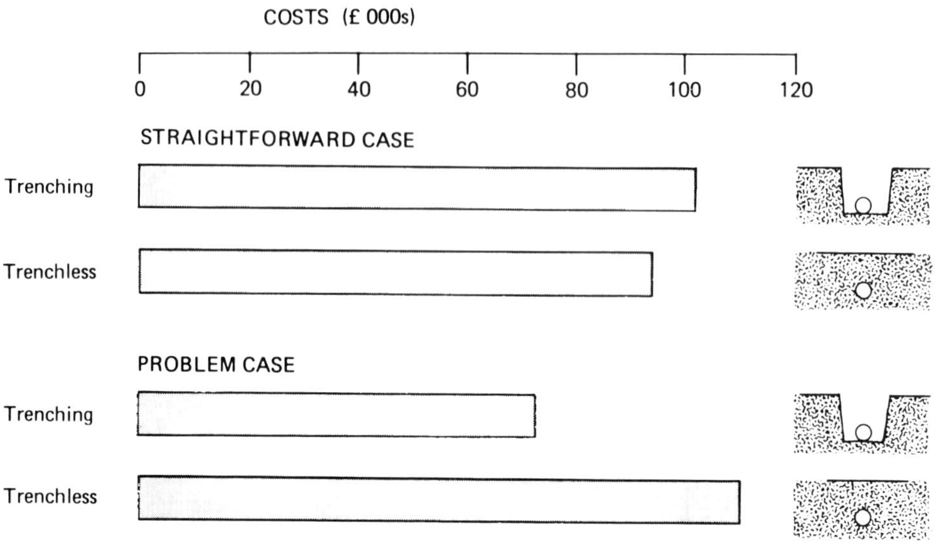

CONSTRUCTION COSTS

Figure 6.6.3.

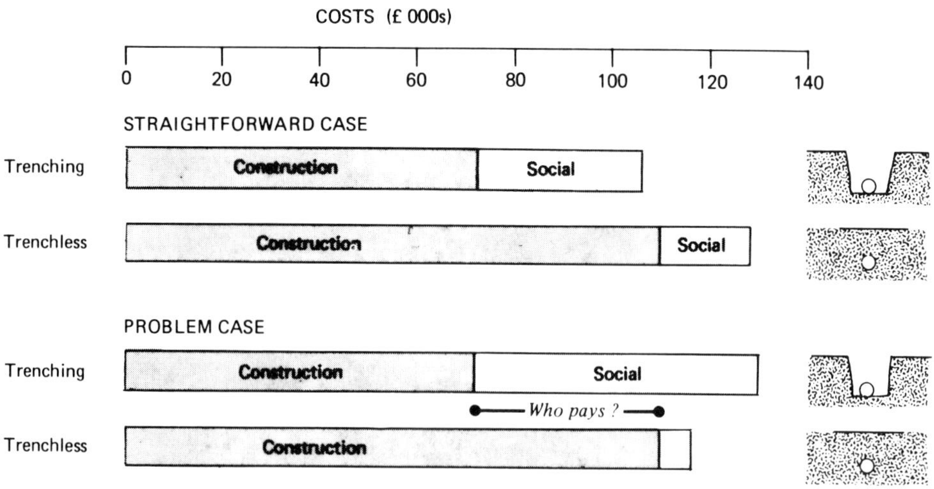

COST TO THE COMMUNITY

Figure 6.6.4.

BIOGRAPHICAL DETAILS OF SESSION CHAIRMEN AND AUTHORS

Session Chairmen

Peter A Banks is President of the Institution of Public Health Engineers for 1984-85. He is a partner in the firm of John Taylor & Sons, Consulting Engineers, specializing in public health engineering. He has been responsible for many large water supply, sewerage and sewage treatment schemes in the UK, Middle East and South-East Asia.

Dr Leslie S Blake has been with the Construction Industry Research and Information Association (CIRIA) for nearly twenty years, formerly as Research Director, and now as Director. CIRIA has sponsored research on pipe jacking and other trenchless methods.

Priv-Doz Dr Ing D Stein is an Associate Professor in the University of Bochum, West Germany. He has for many years worked on the scientific aspects of trenchless methods, including microtunnelling and the development of pipes specially designed for pipe jacking.

Professor Peter F Stott is Nash Professor of Civil Engineering at King's College, University of London. He was formerly Chief Engineer of the London County Council, Controller of Planning and Transportation for the Greater London Council, Director General of the National Water Council and Secretary General of the International Water Supply Association. He is an Honorary Fellow of the Institution of Public Health Engineers.

Ron Tharby is a Telecommunications Consultant, in which capacity he advises British Telecom. He is a founder member, and current Chairman, of the National Joint Utilities Group (NJUG), which co-ordinates the activities of all utilities in the UK, and is therefore a key body in the orderly management of underground services.

Prof ir H P S van Lohuizen is a Professor in the Technical University of Delft, Netherlands, and Vice-President of the International Tunnelling Association. He has devoted much attention to the problems associated with the effective use of underground space in urban areas.

Authors

Toshio Akesaka is Director and Chief Engineer of Iseki Kaihatsu Koki Inc, Japan.

R N Craig is a Principal Engineer with Sir William Halcrow & Partners, whom he joined in 1961. He has specialized in tunnelling for various purposes and has written reports for the Transport and Road Research Laboratory and for CIRIA on tunnel linings and pipe jacking.

Neil Cullen (presenter of paper 5.3 on behalf of the authors) is a Project Leader with the Water Research Centre Engineering at Swindon. After experience with consultants and Water Authorities, he joined WRC in 1978 to work on sewer dereliction and rehabilitation. He is Technical Secretary to the Sewers and Water Mains Committee of the Water Authorities Association.

David Daniels is Assistant Division Manager at the British Gas Corporation's Engineering Research Station (ERS), having had previous radar research experience with Elliot Radar and Plessey. His present interests also include optical sensors and acoustic leak detection.

Philip Etherton is the founder and Managing Director of Merstan Impact Moling Ltd. He has been responsible for developing moling techniques including on-line sewer replacement using pneumatic hammers.

Mark Fairhurst is a Higher Research Engineer at the British Hydromechanics Research Association, the Fluid Engineering Centre.

Takaaki Fukumoto is an Engineer with the Nippon Telegraph and Telephone Public Corporation, Tokyo.

Andy Garnett is Managing Director of Electrolocation Ltd which he joined after working overseas for fifteen years. He has been very active in developing a range of equipment for locating underground plant.

Edward Glennie is a Project Leader at the Water Research Centre, where he has worked on the structural design of sewer linings, sewer collapse statistics and, more recently, economic aspects of sewer rehabilitation. He was formerly at the British Gas Engineering Research Station.

John Hair is the Manager, Contracts and Engineering, River Crossing Division, Reading & Bates Construction Company.

Michael Hardy is County Surveyor, Hertfordshire County Council, and a former President of the County Surveyors Society and of the Institution of Highways and Transportation. He is a member of the Civil Engineering Subcommittee of the Science and Engineering Research Council, and is Chairman of the TRRL Working Party on Highway Maintenance Research.

Yoshio Hashimoto is a Staff Engineer with the Nippon Telegraph and Telephone Public Corporation, working in the Underground Facility Division.

Malcolm Howe's early career was in the shipbuilding industry, followed by research in engineering science at Durham University. Since joining British Gas ERS he has been responsible for research on distribution engineering, including environmental effects on distribution systems.

Peter Hunter has been a Principal Engineer with the British Gas Corporation since 1980, having previously been a Senior Engineer working on leak control and the development of gas conditioning processes.

Doug Irvine is Technical Services Manager, Central Engineering Department, Tarmac Construction. He is concerned with design and construction methods for both temporary and permanent works, and has served on committees of CIRIA and the Concrete Society.

Steven Kramer is Project Manager, Distribution Operations, at the Gas Research Institute, Chicago. He has been carrying out research on new or improved equipment for gas utilities, including a water jet cutter and guided horizontal boring tools.

Dr-Ing E Kuntze recently retired from his post as Head of the Sewerage Department in the City of Hamburg after twenty years' service.

W D McMurdie, now retired, was formerly Area Construction Engineer with the Melbourne and Metropolitan Board of Works, and was principally responsible for the development of the auger boring system.

A F Moss is a Director of a leading specialist tunnelling contractor, Delta Construction Company, and is immediate Past-President of the Pipe Jacking Association. He is an authority on compressed air tunnelling, and has represented the industry on committees of CIRIA and the British Standards Institution.

Alastair Munro is Acting Engineer for the Metropolitan Works Construction Division of the Melbourne and Metropolitan Board of Works, where, for the past twenty years, he has been engaged on pipeline construction using the auger boring method.

Tom O'Rourke is Associate Professor in the School of Civil and Environmental Engineering, Cornell University, USA, and has specialized in the problems of excavation, tunnelling and pipelaying, including the effects of earthquakes on pipelines.

Kristian Olimb is Managing Director of the firm of Kristian Olimb, which he founded in Norway in 1958. The firm specializes in equipment for pipe jacking and drilling in rock and soil, and operates mainly in Norway, Sweden and Finland.

Nick Pearce, of the Water Research Centre Engineering, Swindon, has been engaged in the WRC study of digital mapping for underground services.

Tony Poole has been a contractor for most of his working life, and, as Divisional Manager with the contractors Avon Lippiatt Hobbs Ltd, has had considerable experience of the pipe-bursting system for water main and sewer replacement.

Ken Reed joined the Water Research Centre in 1983 after working with consulting engineers for fourteen years. He is Section Leader, Method Group, and has mainly been involved in the monitoring of developing sewer renovation techniques and research on on-line replacement.

Donald Rees OBE has pioneered many developments in connection with underground services, including small-diameter tunnelling, television inspection, infiltration remedies and structural renovation. He served on the National Water Council's Standing Committee on Sewers and Water Mains, and is now an independent consultant.

John Reynolds is a Senior Engineer in the Hampshire Division of the Southern Water Authority, concerned mainly with capital programming. He worked previously with consultants on sewage disposal systems.

Alan Rooke is Director, European Sales, for Iseki Poly-Tech Ltd of Japan, manufacturers of a range of remote-controlled tunnelling machines. He has had wide experience in civil engineering generally, and pipelaying in particular.

Ron Rosbrook is Development Engineer with the Southern Water Authority, with responsibility for the development and promotion of new methods of design and construction.

D H Saunders is a Principal Engineer at the British Hydromechanics Research Association, the Fluid Engineering Centre.

Christoph Schröder (presenter of paper 4.3 on behalf of Dr Kuntze) is Chief Engineer in the Department of Sewer Construction in the City of Hamburg.

Geoffrey Shiers has been a contractor for many years and has been particularly concerned with pipeline construction. He is now Contracts and Engineering Representative in the River Crossing Division of Reading and Bates Construction Company, specializing in directionally-controlled drilling for pipelines under rivers.

Michael Slipper has been working on the problems of underground mapping for utilities at the Water Research Centre.

James Thomson's experience in trenchless pipelaying goes back to the late 1950s, when he pioneered the use of pipe jacking in the UK. He developed a number of techniques and equipment for auger boring and pipe jacking, including the installation of bridge abutments and road underpasses by jacking.

Satoru Tohyama is Vice-President of the Japan Sewage Works Agency, in which he was previously Head of the Planning Division and, more recently, an Executive Director. Between 1975 and 1981 he served in the Department of Sewerage and Sewage Purification in the Ministry of Construction.

Dr Hans-Peter Uffmann is Technical Director of Krings International GmbH in Heinzburg, West Germany, and was previously Chief Engineer at the Institute for Construction Machinery and Construction Methods in Aachen.

John Wilson is a Senior Engineer, and now National Co-ordinator, with the National Joint Utility Group.

Hiroshi Yamamoto is a Senior Staff Engineer with the Nippon Telegraph and Telephone Public Corporation, Japan.

Dr Peter Zollman has been a Director of various instrument and machinery development firms, and is now Managing Director of Zed Instruments Ltd.

LIST OF DELEGATES

Abdullah bin Ali H Al-Harthy	Manager, Public Health Services, Ministry of Defence	Sultanate of Oman
Afschrift, P	Assistant Manager, S A Denys	Belgium
Akiyama, J	Assistant Manager, Tokyo Gas Co Ltd	Japan
Alstott, D	President, Edmands Engineering Inc	USA
Andersen, C	Engineer, Forenede Plast AS	Denmark
Anderson, TA	Depute Director, Dept of Sewerage, Strathclyde Regional Council	UK
Archbold, D	Asst Engr Sewerage Co-ordination, Northumbrian Water	UK
Archer, MFG	Director, Howard Humphreys & Partners	UK
Ashworth, M	Distribution Engineer, North Eastern Gas Board	UK
Aubry, J	Ingénieur, SADE	France
Audé, J	Ingénieur Principal, Spie Batignolles	France
Aylott, RFV	Deputy City Engineer, City of London	UK
Aylward, AR	Project Engineer, Water Research Centre Engineering	UK
Baillie, AM	Director (Marketing), William Tawse Ltd	UK
Bajkay, A	Head of Dept, Transinvest/Mining Shaft Sinking Co	Hungary
Belmore, KG	Deputy Engr & Surveyor, Waverley Borough Council	UK
Bielecki, R	Head of Design Dept, Sewer Constr, City of Hamburg	W Germany
Bjorklund, I	Principal Engineer, VBB Consulting Engineers	Sweden
Blackhurst, G	Manager, Electro/Mech Divn, UMO Plant Ltd	UK
Blake, JA	Postgraduate Engineer, Reading Borough Council	UK
Blake, LS	Director, Construction Industry Research and Information Association (Session Chairman)	UK
Blakey, WN	Technical Manager, Thorn EMI Cable Television Ltd	UK
Bland, A	Appraisal Engineer, Anglian Water Authority	UK
Bland, CEG	Director of Eng'g Services, Clay Pipe Development Association	UK
Blane, M	Asst Project Manager, Ministry of Defence	UK
Boak, JG	Regional Manager, South, Balfour Beatty Power Constr Ltd	UK
Bohman, A	Head of Dept, Byggnads & Industriservice AB BINAB	Sweden
Bonito, JDL	Engineer, Servicos Municipalizados de Oeiras	Portugal
Boon, J	Technical Officer, ARC Slimline Ltd	UK
Bosse, CB	Engineer, Carl B Bosse	W Germany
Bourgoin, F	Ing Serv Mat, SADE	France
Bower, RK	Project Manager, Kerr Construction Inc	USA
Boyd, HF	Consulting Engineer, Wavin (NI) Ltd	UK
Bradley, DE	Eastern Gas Board	UK
Bradley, RM	Project Manager, Engineering Science	W Malaysia
Braid, VE	British Gas Corporation	UK
Brazier, RO	Group Engr, Main Drainage, London Borough of Hackney	UK
Brennan, G	Technical Director, Concrete Pipe Association	UK
Briddon, K	Distribution Engineer, British Gas Corporation	UK
Brinch, T	Kristian Olimb A/S	Norway
Brome, JA	Marketing Director, Plumett Ltd	UK
Burger, AWM	Adj Director, Van Eijk Civil Technics BV	Netherlands
Burr, DG	Associate, SSC Consultants Ltd	UK
Burr, T	Liaison Engineer, Thames Water, Central Division	UK
Burt PN	Technical Representative, J Mastenbroek & Co	UK
Cairns, JJ	Representative, Hugh Pearl (Land Drainage) Ltd	UK
Caygill, G	Distribution Engineer, Wessex Water Authority	UK
Chapman, JR	Southern Water Authority	UK

Child, M	New Works Engineer, Anglian Water, Cambridge Division	UK
Cirak, E	Method of Constr. Eng, Sezai Turkes Feyzi Akkaya Construction Co	Turkey
Clark, RTF	Principal Engineer, Arun District Council	UK
Clough, P	Development Engineer, North West Gas Board	UK
Clow, DG	Head of Civil & Mech Eng'g Section, British Telecom	UK
Coe, AL	Distribution Engineer, Thames Water	UK
Coller, P	Iseki Poly-Tech Inc	UK
Collier, G	Principal Engineer (Constr), Eastern Electricity Board	UK
Connolly, F	Partner, Peter Brett Associates	UK
Cook, G	Senior Engineer, Sir William Halcrow & Partners	UK
Copeland, A	Principal Engineer, Sewers, Wandsworth Borough Council	UK
Core, D	Assistant Engineer, Wessex Water Authority	UK
Cornfield, G	Section Leader, Electricity Council	UK
Cottam, K	Senior Engineer, Southern Water Authority, Kent Division	UK
Cottrell, K	Area Direct Labour Manager, Welsh Water, SE Division	UK
Coulthart, A	Public Health Engineer, Property Services Agency	UK
Cowell, AR	Resident Engineer, Sir William Halcrow & Partners	UK
Cox, GC	Consultant, M McDowell Co-Partnership	UK
Craig, RN (Author)	Principal Engineer, Sir William Halcrow & Partners	UK
Crossingham, RA	District Engineer, North Thames Gas, Eastern Area	UK
Crouchman, F	UK Sales Manager, JTC International Ltd	UK
Cullen, N	Project Leader, Water Research Centre Engineering	UK
Daniels, DJ (Author)	Asst Division Manager, Eng'g Research Station, British Gas	UK
Dart, WW	Construction Specialist, Kerr Construction Inc	USA
Davidson, PJ	Asst Area Distribution Engineer, British Gas	UK
Dawson, DB	Design Assistant, British Rail, Western Region	UK
Dawson, LW	Senior Asst Distribution Engineer, North Thames Gas	UK
de Bijl, DJ	Consultant, Grontmij NV	Netherlands
de Vreede, LJT	Director, Waterschap 'De Dommel'	Netherlands
Dean, W	Distrib Engr, Wrexham & E. Denbighshire Water Co	UK
Decker, CT	Executive Vice-President, Carylon Corporation	USA
Delbaere, D	Chief Executive, SA Denys	Belgium
Dobb, K	Area Manager, Water Supply, Thames Water	UK
Donaldson, TP	Project Manager, Ministry of Defence	UK
Dudley, GSBG	Partner, Stewart Scott & Partners	UK
Duke, RG	Director, RS Kennedy & Co Ltd	UK
Dumrul, O	Project Manager, Sezai Turkes Feyzi Akkaya Constr Co	Turkey
Eberhard, G	Dow Chemicals GmbH, R & D	W Germany
Eccles, WW	Assistant Engineer, Buckinghamshire County Council	UK
Eckardt, H	Senior Engineer, Ed Zueblin AG	W Germany
Eckart, JE	Chief Engineer, Kerr Construction Inc	USA
Edgerton, DJ	Head of Group, Local Line Planning, British Telecom	UK
Edmondson, D	Marketing Officer, ARC Concrete Ltd	UK
Ellertson, D	Construction Specialist, Kerr Construction Inc	USA
Ellison, T	Graduate Engineer, AB BESAB	Sweden
Elvidge, AF	Manager, Methods Group, Water Research Centre Eng'g	UK
Encik Abdul Aziz bin Shah Man	Encik Abdul Aziz bin Shah Man	Egypt
Erdemgil, M	Kiska Construction Corporation	Turkey
Erickson, DM	Vice-President, Hydrogroup Inc	USA
Etherton, PT (Author)	Merstan Impact Moling Ltd	UK
Evans, DO	Area Distribution Manager, West Midlands Gas Board	UK

Everett, PJ	Executive Engineer, British Telecom	UK
Exelby, RE	Regional Co-ordinator, Yorkshire Water Authority,	UK
Fairhurst, RM (Author)	Higher Research Engineer, BHRA The Fluid Engineering Centre	UK
Faldager, I	Engineer, Teknologisk Institut, Tastrup	Denmark
Ferguson, EB	Director, Hatrick-Bruce (Construction) Ltd	UK
Ferguson, IK	Asst Distribution Co-ordinator (Operations), Scottish Gas	UK
Ferrero, G	Head of R & D, Gas Technology Division, Italgas	Italy
Fiddes, D	General Manager, Water Research Centre Engineering	UK
Field, DB	Director, Water Research Centre Engineering	UK
Filtenborg, O	Managing Director, Forenede Plast AS	Denmark
Finkenbusch, J	Manager, Reinhard Rohrbau	W Germany
Finney, VA	Principal Assistant Engineer, Manchester Met Borough Council	UK
Fleming RC	Executive Director, Para Resources Group Inc	UK
Forder, RE	Safety and Standards Manager, Press Construction Ltd	UK
Forrest, D	Kennedy Pipelining Services Ltd	UK
Fossen, T	Managing Director, Entroprenoservice A/S	Norway
Francis, GS	Distribution Technical Manager, Wales Gas Board	UK
Franks, A	Leading Instructor, Construction Industry Training Board	UK
Freer, CN	Partner, Pick Everard Keay & Gimson	UK
Frödin, KA	Managing Director, Vretmaskin AB	Sweden
Fry, MR	Technical Director, Land Drainage Contractors Association	UK
Fullerton, RA	Technical Marketing Executive, Stanley Power Tools	UK
Gale, J	Project Engineer, Water Research Centre Engineering	UK
Garnett, A (Author)	Managing Director, Electrolocation Ltd	UK
Garnham, A	Distribution Operations Engineer, Northern Gas Board	UK
Garton, HW	Civil Engineer, Drainage Dept, Isle of Man Local Govt Bd	UK
Geddes, JW	Area Distribution Engineer, Scottish Gas Board	UK
Gibbon, W	Area Distribution Engineer, West Midlands Gas Board	UK
Gibson, CR	Principal Liaison Engineer, Thames Water, South & West Divn,	UK
Glennie, EB (Author)	Water Research Centre Engineering	UK
Glover, JB	Director, Hepworth Iron Co Ltd	UK
Glover, TD	District Engineer, South Eastern Gas Board	UK
Glynn, M	Senior Engineer, North West Water Authority	UK
Godber, E	Assistant Divisional Engineer, Lee Valley Water Company	UK
Godwin, RJ	Director of Research & Development, Silsoe College	UK
Goff, AFF	Director, Galliford Pipeline Services Ltd	UK
Granger, AJ	Project Manager, Kerr Construction Inc	USA
Grayer, RJ	Divisional Manager, Eve Construction Ltd	UK
Greasley, KC	Divisional Director, Edmund Nuttall Ltd, Permaline Divn	UK
Green, D	Distribution Services Engineer, North West Water	UK
Greenwood, BJ	Project Engineer, Mainlaying, South Staffs Waterworks Co	UK
Guldbaek, E	Product Manager, Nordisk Wavin A/S	Denmark
Haines, GE	British Telecom Research	UK
Halderman, RG	Geologist, Baker Energy Resources Corporation	USA
Hallas, JB	Group Engineer, Cheshire County Council	UK
Halloran, M	Assistant Distribution Engineer, East Midlands Gas Board	UK
Hammond, RG	Technical Adviser, Komatsu Euro Africa NV/SA	UK

Hancox, RD	Eastern Gas Board	UK
Hardy, MF (Author)	County Surveyor, Hertfordshire County Council	UK
Harrison, NL	Manager, R & D, Humes Ltd, Concrete Division, Melbourne	Australia
Harvey, P	Principal Engineer, North Devon District Council	UK
Hashimoto, Y (Author)	Staff Engineer, Nippon Telegraph & Telephone Public Corpn	Japan
Haydock, JK	Project Engineer, Watson Hawksley	UK
Hazama, T	Assistant Manager, Osaka Gas Co Ltd	Japan
Heath, DC	Senior Civil Engineer, British Airports Authority	UK
Heavens, JW	Commercial Development Manager, ARC Slimline Ltd	UK
Heerkens, JCP	Head, Research & Advice Dept, Wegenboumij J Heijmans BV	Netherlands
Henderson, W	Construction Specialist, Kerr Construction Inc	USA
Henke, CF	University of Stuttgart	W Germany
Henry, K	Mains Engineer, North West Water (Western)	UK
Hewston, M	Assistant Distribution Manager, Rickmansworth Water Co	UK
Hindhaugh, GMA	Associate Director, Ward Ashcroft & Parkman	UK
Hoppe, F	Engineer, Kebaco Nord Rohr und Kanalsanierungs	W Germany
Howe, M (Author)	Asst Division Manager, Eng'g Research Stn, British Gas	UK
Hughes, K	Principal Assistant Engineer, North West Water	UK
Humphrey, RC	Director, Wastewater Collection, CH2M Hill Inc	USA
Hunter, P (Author)	Principal Engineer, Eng'g Research Stn, British Gas	UK
Imanaga, Y	Sada Kensetsu Ltd	Japan
Irvine, DJ (Author)	Technical Services Manager, Tarmac Construction Ltd	UK
Ishibashi, N	Japan Sewage Works Agency	Japan
Jackson, J	District Engineer, NW Area, North Thames Gas	UK
Jameson, DA	Senior Engineer (Development), Southern Gas Board	UK
Jellicoe, KCG	Civil Engineer, DoE Department of Transport	UK
Johansson, T	Civil Engineer, Uponor, Fristad	Sweden
John, H-J	Geschäftsführer, Meyer & John GmbH & Co	W Germany
Johnson, DB	Technical Manager, Oundle Division, Anglian Water	UK
Jones, DE	Assistant Chief Engineer, Dacorum Borough Council	UK
Jones, DG	Asst Distribution Manager, Rickmansworth Water Co	UK
Jones, MC	Principal Engineer, Wessex Water Authority	UK
Jørgensen, FA	Product Manager, Nordisk Wavin A/S	Denmark
Jukka, H	Manager, Work Department, Vesto Oy	Finland
Kabata, N	Suiken Ltd	Japan
Kawabata, A	Sokushin Koji Ltd	Japan
Kelly, J	PPTO, DoE (NI) Central Design Unit	UK
Kent, E	Head of Group, British Telecom	UK
Kiesling, E	President, L Berringer nach FKG GmbH & Co	W Germany
Kimura, K	Kido Kensetsu Kogyo Ltd	Japan
King, PJ	Executive Engineer, British Telecom	UK
Kiya, T	Watanabe-Gumi Ltd	Japan
Kleinman, AM	Construction Specialist, Kerr Construction Inc	USA
Koh Ah Teck	Director, Jurutera Konsultant (SEA) SDN BHD	W Malaysia
Kosova, H	Vice-President, Carylon Corporation	USA
Kramer, SR (Author)	Project Manager, Gas Research Institute	USA
Krämer-Wasserka, T	Chief Design Engr, Gewerkschaft Eisenhütte Westfalia	W Germany
Lammeramt, M	Works Manager, Eternit	Belgium
Langridge, RE	Partner, D Balfour (Consulting Engineers)	UK
Leach, G	British Gas Corporation	UK
Lee Chai Kam	Director, Syed Muhammad, Hooi & Binnie SDN BHD	W Malaysia
Lee, IS	Chief Engineer, McNicholas Construction Co Ltd	UK

Leinonen, J	Research Scientist, Technical Research Centre	Finland
Lim, BC	Managing Director, Mersing Constr & Eng'g SDN BHD	W Malaysia
Lloyd, DO	Managing Director, Halcrow Water	UK
Lloyd, FV	North West Gas	UK
Lofthouse, P	Director of Operations, Southern Water Authority	UK
Loo Sim Boon	Public Works Dept, Bandar Seri Begawan	Brunei
Lou Chi Tong	Manager, Associated Consultants	Brunei
Lowther, WJ	Civil Engineer, London Docklands Development Corpn	UK
Lundahl, B	Division Manager, Stabilator	Sweden
Lystbaek, JT	Division Manager, Per Aarsleff A/S	Denmark
Macdonald, ICB	Divisional Director, Tunnelling Divn, Fairclough Civil Eng'g Ltd	UK
McFarland, JAC	PPTO, DoE (NI) Central Design Unit	UK
McNally, CR	Director, C & M McNally Eng Inc	Canada
Manabe, S	Sumiko Kaihatsu Ltd	Japan
Manuel, WH	Head of Section, External Plant, British Telecom	UK
Masaka, I	Kurimoto Ltd	Japan
Matsui, H	Nippon Zenith Pipe Ltd	Japan
Matsuura, M	Meada Construction Ltd	Japan
Melsheimer, RF	Borzall Equipment Corporation	USA
Menzel, SWO	Chief Executive, Rib-Loc Systems BV	Australia
Mercer, J	Senior Project Manager, FlowMole Corporation	USA
Millar, AW	Senior Engineer, Peter Fraenkel Leslie & Reid	UK
Miller, PJ	Project Manager, Kerr Construction Inc	USA
Millican, RC	Eastern Gas Board	UK
Mills, P	Asst Distribution Engineer, Wessex Water Authority	UK
Moir, IWM	Camp Dresser & McKee Inc	UK
Moore, BL	Vice-President, Horizontal Boring & Tunnelling Co	USA
Moore, G	Civil Engineer, Bechtel GB Ltd	UK
Morgan, DJ	Senior Patents Adviser, British Gas Corporation	UK
Morgan, RH	Distribution Manager, Rickmansworth Water Company	UK
Moss, AF (Author)	Past Chairman, Pipe Jacking Association	UK
Moss, GF	Manager, Water Research Centre Engineering	UK
Mowl, J	Area Distribution Engineer, West Midlands Gas Board	UK
Müller, H	Manager, Kanal-Müller GmbH & Co	W Germany
Munro, AG (Author)	Acting Engr Met Works, Melbourne & Met Bd of Wks	Australia
Nakamoto, M	Mitsui Construction Ltd	Japan
Nakamura, M	Okumura Corporation	Japan
Nakano, H	Kido Kensetsu Kogyo Ltd	Japan
Nanninga, H	Senior Civil Engineer, City of Savannah, GA	USA
Neil, D	Assistant Director, Strathclyde Regional Council	UK
Newbon, V	Principal, Newbon Associates	UK
Nimmo, D	District Engineer, South Eastern Gas Board	UK
Noble, JW	Water Controller, Southern Water Authority, Kent Divn	UK
Norman, PW	Agency Liaison Engineer, Welsh Water, SW Division	UK
Nutaba, N	Nisan Kensetsu Ltd	Japan
O'Callaghan, J		
O'Reilly, MP	Head of Ground Eng'g Divn, Tpt & Road Research Laboratory	UK
O'Rourke, TD (Author)	Associate Professor, Cornell University	USA
O'Sullivan, JJ	Senior Engineer, Cork Corporation	Ireland
Oda, T	LS Planning Ltd	Japan
Olimb, K (Author)	Kristian Olimb A/S	Norway
Owen, JA	Assistant Distribution Engineer, British Gas Corpn	UK

Ozkaya, U	Engineer, Sezai Turkes Feyzi Akkaya Constr Co	Turkey
Parkinson, RW	Manager, Water Research Centre Engineering	UK
Parkyns, EH	Maintenance & Standards Engineer, South Eastern Gas	UK
Peach, AF	Senior Product Manager, Durapipe	UK
Pearson, A	Eastern Gas Board	UK
Phillips, P	Asst Mains Engr (Contracts), North West Water, N Divn	UK
Polen, B	NBM	Netherlands
Poole, AG (Author)	Divisional Manager, Avon Lippiatt Hobbs Ltd	UK
Poole, RW	Research Manager, Constr Industry Res & Info Assoc	UK
Powell, NT	Chief Engineer (Works), Cornwall County Council	UK
Price, JO	Engineer & Surveyor, South Pembrokeshire District Council	UK
Pullar, RA	Convenor, Water Services Committee, Tayside Reg Council	UK
Reed, K (Author)	Section Leader, Water Research Centre	UK
Rees, DF	Donald Rees Partnership	UK
Reid, JW	Kennedy Pipelining Services Ltd	UK
Rennet, RJ	Depute Director (New Works), Tayside Regional Council	UK
Renshaw, AW	Executive Engineer, British Telecom	UK
Reynolds, JH (Author)	Senior Engineer, Southern Water Authority, Hampshire Divn	UK
Robinson, B	District Engineer, Doncaster, East Midlands Gas Board	UK
Robinson, J	Partner, John Taylor & Sons	UK
Robinson, SE	Wavin (NI) Ltd	UK
Robson, JD	Principal Engineer, Allott & Lomax	UK
Rollinson, DJ	Manager, White Horse Contractors Ltd	UK
Rooke, A (Author)	Director, European Sales, Iseki Poly-Tech Inc	UK
Rosbrook, RB (Author)	Development Engineer, Southern Water Authority	UK
Rouse, T	New Works Engineer, Anglian Water Authy, Oundle Divn	UK
Russell, PJ	Industrial Eng'g Manager, Yorkshire Water Authority	UK
Rutherford, R	Executive Director, Para Resources Group Inc	UK
Ryan, DHJ	Senior Engineer (Masin Drainage), London Borough of Ealing	UK
Ryan, JTA	Marketing Director, DJ Ryan & Sons Ltd	UK
Sada, G	Sada Kensetsu Ltd	Japan
Sakai, K	Nakagawa Hume Pipe Ltd	Japan
Sano, J	Taibei Kogyo Ltd	Japan
Sasaki, S	Komatsu Kensetsu Ltd	Japan
Sato, Y	Komatsu Euro Africa NV/SA	UK
Saunders, DH (Author)	BHRA The Fluid Engineering Centre	UK
Saville, DR	Principal Public Health Engineer, Kyle Stewart Ltd	UK
Schaap, MJ	Adviesbureau Bongaerts, Kuyperen Huiswaard	Netherlands
Schlapp, H	Research Engineer, Seba-Dynatronic Mess und Ortungs Technik	W Germany
Schmidt	Krings International GmbH & Co	W Germany
Schröder, C	Chief Engr, Dept of Sewer Constr, City of Hamburg	W Germany
Schroeyers, J	Ingénieur, SADE	France
Scott, DA	Regional Projects Manager, Thames Water	UK
Senba, Y	Haneda Hume Pipe Ltd	Japan
Shaw, C	Executive Engineer, Warrington & Runcorn Devel. Corpn	UK
Sheridan, D	Contracts Manager, John Mowlem & Co plc	UK
Shiers, GE (Author)	Contracts & Eng'g Rep, Reading & Bates Construction Co	UK

Shigemune, K	Original Engineering Ltd	Japan
Shy, JB	McLaughlin Mfg, SE Division Inc	USA
Shy, MH	Vice-President, R & D McLaughlin Mfg SE Divn Inc	USA
Simmons, D	Distribution Operating Engineer, Wales Gas Board	UK
Simonds, B	Section Engineer (Divns & Maintenance), Bucks County Council	UK
Sloman, PR	Sales Director, Hugh Pearl (Land Drainage) Ltd	UK
Smith D	Eastern Gas Board	UK
Smith, K	Regional Manager, Cawthraw & Co Ltd	UK
Smith, RC	Main Engineer, Bristol Waterworks Company	UK
Snook, C	Assistant Engineer, Welsh Water, SE Division	UK
Speed, DHM	Chief Engineer, Newcastle & Gateshead Water Company	UK
Spiller, JH	Distribution Engineer, South Western Electricity Board	UK
Stafford, JS	Bridgwater Bros	UK
Stakes, T	Assistant Civil Engineer, Albright & Wilson	UK
Stanley, KG	Head of Group, British Telecom	UK
Starking, C	Eastern Gas Board	UK
Steele, MD	Marketing Manager, Pozament Ltd	UK
Stein, D	Associate Professor, Ruhr-Universität Bochum, (Session Chairman)	W Germany
Stephenson, WR	General Manager, Terratech Drilling	UK
Storah, BW	Asst Reg Distribution Engineer, North West Gas Board	UK
Stores, OJ	Owner, Duval Horizontal Boring Co	USA
Storm-Hansen, T	Civil Engineer, Entroprenoserve A/S	Norway
Stott, PF	Nash Professor, Kings College, U of London (Session Chairman)	UK
Stuber, LM	President, EMC Engineering	USA
Talaat, ZH	CWO	Egypt
Tan, I	Managing Director, Kenweld, Singapore Pte Ltd	Singapore
Tanaka, H	Hinode Water Engineering Equipment Ltd	Japan
Taylor, JEG	Engineering Chemist, South Eastern Gas Board	UK
Taylor, SJ	Group Engineer, Amber Valley District Council	UK
Teague, DJ	Senior Engineer, Revenue, S Pembrokeshire District Council	UK
Telford, PJ	Principal Engineer, Revenue, Severn Trent Water Authority	UK
Tennant, R	Group Engineer (Highways), Westminster City Council	UK
Tharby, R	Telecommunications Consultant (Session Chairman) Chairman, National Joint Utilities Group	UK
Thebock,	Technical Manager, Thebock & Feil GmbH	W Germany
Thirkettle, RJ	Eng'g & Marketing Consultant, Condor Im-Export GmbH	W Germany
Thomas, G	Senior Engr (Distribution), Thames Water, Chilterns	UK
Thomson, JC (Author)	Consultant	UK
Thoresson, O	Contracting Manager, Stockholms VA-verk	Sweden
Tohyama, S (Author)	Vice-President, Japan Sewage Works Agency	Japan
Tolan, L	Service Engineer, Wavin Ireland Ltd	Ireland
Tozzi, K	Operations Manager, Southern Water Authy, I of Wight	UK
Tranter, BL	Managing Director, Kenton Utilities & Developments Ltd	UK
Trim, D	President, Wade-Trim Group Inc	USA
Turner, J	Engineer, Distribution, Newcastle & Gateshead Water Co	UK
Turtle, D	Eng'g Assistant, Wrexham & E, Denbighshire Water Co	UK
Tyson, N	Company Accountant, Leicestershire Drainage Co	UK

Uffman, HP (Author)	Technical Director, Krings International GmbH & Co	W Germany
Ursell, A	Pedershaab Ltd	UK
Usami, S	Nippon Pressed Concrete Ltd	Japan
Usher, M	Distribution Engineer, Northumbrian Water Authority	UK
Vale, PG	Director, George Law Plant Ltd	UK
van Lohuizen, HPS	Professor, Technical University of Delft (Session Chairman)	Netherlands
Vits, L	Keramo NV	Belgium
Wade, R	Chairman of Board, Wade-Trim Group Inc	USA
Wadley, JJ	Partner, Babtie Shaw & Morton	UK
Wadsworth, JB	Eastern Gas Board	UK
Wagner, V	Kebaco Nord Rohr und Kanalsanierung	W Germany
Walker, N	Contracts Director, Tate Pipe Lining Processes Ltd	UK
Wallis, P	Area Distribution Engineer, West Midlands Gas Board	UK
Walsh, PM	P Walsh Civil Engineering Contractors	Ireland
Walton, MJ	Senior Engineer, West Midlands County Council	UK
Weerkens, JCP	Wegenbouwmij J Heijmans BV	Netherlands
Whinn, CJ	Chartered Engineer, Northumbrian Water Authority	UK
Wilkinson, B	Principal Engineer (Sewerage), Yorkshire Water Authority	UK
Williams, B	Director, Seer TV Surveys Ltd	UK
Willis, AL	Eastern Gas Board	UK
Wilson, D	Asst County Surveyor, Berkshire County Council	UK
Wilson, JL (Author)	Senior Engineer, National Joint Utilities Group	UK
Witte, C	Witte Bohrtechnik GmbH	W Germany
Wood, J	Senior Asst Engineer, Sheffield Met District Council	UK
Woods, G	Chief Civil Engineer, Naylor Bros (Clayware) Ltd	UK
Worger, S	Head of Group, British Telecom Reasearch	UK
Worst, WJP	Consultant, Grontmij NV	Netherlands
Yamauchi, M	Sato Kogyo Ltd	Japan
Yokohata, T	Toho Chikakoki Ltd	Japan
Yoshida, S	Manager, Tokyo Gas Co Ltd	Japan
Young, IM	General Sales Manager, Edwards & Jones Ltd	UK
Ywema, R	Heidemij Adviesbureau	Netherlands
Zollman, PM (Author)	Managing Director, ZED Instruments Ltd	UK

Press

Ashley, S	Editor, *Building Services Journal*
Cooper, RM	Institution of Mechanical & General Technician Engineers
Martin, D	Editor, *Tunnels & Tunnelling*
Tiratsoo, J	Editor, *Pipes & Pipelines International*
Wiseman, R	Editor, *World Water*

Secretariat

Cdr D J Dacam OBE	Secretary, IPHE
Miss J Little	Assistant Secretary, IPHE
Mrs M Minards	

LIST OF EXHIBITORS

Aliva Ltd	26/28 Beaconsfield Road, Hatfield, Herts	UK
All-Japan Trenchless Technology Transfer Study Group	c/o LS Planning Ltd, 102 Chiyoda-ku, Fujimi 1-7-8, Tokyo	Japan
Alphacrete Construction Linings (UK) Ltd	Hiview House, Highgate Road, London NW5 1TN	UK
AMK Pipe Technology Ltd	18 Chester Road, Castle, Northwich, Cheshire	UK
Aqua Pipes Ltd	Darlingscott Road, Shipston on Stour, Warks	UK
ARC Concrete Ltd	Mells Road, Mells, Frome, Somerset BA11 3PD	UK
Avent PRS (Pipe Renovation Systems)	Porte Marsh Road, Calne, Wiltshire SN11 9BW	UK
Avon Lippiatt Hobbs	Station Road, Westbury, Wiltshire BA13 3TN	UK
Biggs Wall & Company Ltd	Hampden House, Arlesey, Bedfordshire SG15 6RT	UK
Cementation Construction Ltd	Mitcham House, 681 Mitcham Road, Croydon CR9 3AP	UK
Century Oils Ltd	PO Box 2, Hanley, Stoke-on-Trent ST1 5HU	UK
Data Acquisition Ltd	Electron House, Higher Hillgate, Stockport SK1 3QD	UK
Decon Plant & Engineering	Wylds Road, Bridgwater, Somerset TA6 4BH	UK
Delta Construction Co	High Green, Great Shelford, Cambridge CB2 5EG	UK
Demco Ltd	Lindley Grange, Watling Street, Nuneaton, Warks CV10 0TT	UK
Drain Technology Ltd	Queen Street Warehouses, Queen Street, Northwich, Cheshire	UK
Edmund Nuttall Ltd, Permaline Division	Roundwood Industrial Estate, Ossett, West Yorkshire WF5 9SO	UK
Electrolocation Ltd	Western Drive, Bristol BS14 OAZ	UK
Europlast Rohrwerk GmbH	PO Box 13 01 60, D-4200 Oberhausen 11	W Germany
Horizonger Pipe Jacking Association	c/o Samwa Kizai KK, 103 Chiyoda-ku, Nihon-Bashi, Kayaba-Choo 2-4-9, Tokyo	Japan
Insight Vision Systems Ltd	Unit 1, Merebrook Industrial Units, Hanley Road, Malvern Worcs WR13 6NP	UK
Insituform Group Ltd	Borough House, Trinity Square, St Peter Port, Guernsey	Channel Islands
Johnston Construction Ltd	Watling Street, Shawell, Lutterworth, Leicestershire LE17 6AS	UK
Johnston Pipes Ltd	Doseley, Telford, Salop TF4 3BX	UK
Komatsu Ltd	Komatsu Building, 2-3-6 Akasaka, Minato-ku, Tokyo 107	Japan
Krings International (UK) Ltd	5 Elton Road, Clevedon, Avon	UK
Land & Marine Engineering Ltd	Port Causeway, Bromborough, Wirral, Merseyside L62 4TG	UK

19
14 NORTHERN INDUSTRIAL
& MARINE SERVICES CO LTD

A.L.H. SYSTEMS
SLIPLINING

Lilley Construction Ltd	Tunnel Road, Galley Common, Nuneaton CV10 9PP	UK
LS Planning Ltd	102 Chiyoda-ku, Fujimi 1-7-8, No 5 Nitto Building Suite 202, Tokyo	Japan
Lundby Systems AB	Box 7059, S-402 31 Gothenburg	Sweden
Mainline Renewal Systems Ltd	Low House, Pilgrims Way, Thurnham, Maidstone, Kent ME14 3LD	UK
Merstan Impact Moling Ltd	London House, 10/14 High Street, Milton, nr Sittingbourne, Kent	UK
Mini Tunnels International Ltd	Westminster House, High Street, Old Woking, Surrey GU22 9LF	UK
Mitsui Construction Co Ltd	101 Chiyoda-ku, Iwamoto-cho, 3-10-1, Tokyo	Japan
Morgan-Grampian (Construction Press) Ltd	30 Calderwood Street, Woolwich, London SE18 6QH	UK
Motherwell Bridge Engineering Group	Clough Green, Cawthorne, Barnsley S75 4AD	UK
MT Drain and Sewer Surveys Ltd	2-4 Mount Pleasant Road, Aldershot, Hampshire	UK
Naylor Bros (Clayware) Ltd	Cawthorne, Barnsley, South Yorkshire S75 4AD	UK
New Civil Engineer	Telford House, PO Box 101, 26-34 Old Street, London EC1P 1JH	UK
Nittoh Koji Co Ltd	101 Chiyoda-ku, Kaji-cho 2-3-12, Tokyo	Japan
Northern Industrial & Marine Services Co Ltd	Hockley Way, Alfreton Trading Estate, Alfreton, Derbyshire DE5 7FA	UK
Okumura Corporation	2-2-2 Matsuzaki-cho, Abeno-ku, Osaka 545	Japan
Pearpoint Ltd	32 Woolmer Trading Estate, Bordon, Hants GU35 9QF	UK
Pipe Jacking Association	Suite 411, London International Press Centre, Shoe Lane, London EC4A 3JB	UK
Pipeseal Systems Ltd	97 Dykehead Street, Queenslie Industrial Estate, Glasgow G33 4DQ	UK
Pneumatic Punchers Ltd	Unit 0, London Road Trading Estate, Sittingbourne, Kent ME10 1NQ	UK
Portable Factory Equipment Ltd	48 Smith Street, Hockley, Birmingham B19 3EW	UK
Press Construction Ltd (Utilities Division)	PO Box 35, Swindon, Wiltshire SN1 3AU	UK
Rees Instruments Ltd	Westminster House, High Street, Old Woking, Surrey	UK
DJ Ryan & Sons Ltd	Inglewhite Road, Longridge, Lancashire	UK
Seer TV Surveys Ltd	Westminster House, Old Woking, Surrey GU22 9LF	UK
Smole Association	c/o Watanabe Kumi KK, 106 Minato-ku, Minami-Azabu 1-18-4, Tokyo	Japan
Spun Concrete Ltd	Lichfield Road, Branston, Burton upon Trent, Staffordshire DE14 3HF	UK
Stanley Power Tools	Nelson Way, Cramlington, Northumberland NE23 9JS	UK
Stanton and Staveley Ltd	PO Box 72, Nottingham	UK
Steenbergeen	Holland Drain, PO Box 14083 35, 3508C Utrecht	Netherlands

Surveyor Public Works Weekly	Business Press International Ltd, Room 607, Surrey House, 1 Throwley Way, Sutton, Surrey SM1 4QQ	UK
Tate Pipe Lining Processes Ltd	Oldham Street, Denton, Manchester M34 3SR	UK
Telespec Ltd	Tannery House, Tannery Lane, Send, Woking, Surrey GU23 7HB	UK
Toho Chikakoki Co Ltd	2-2 Uchisaiwai-cho 1-chome, Chiyoda-ku, Tokyo 100	Japan
Tone Boring Co Ltd	6 Meguro, 1-Chome Meguro-ku, Tokyo	Japan
TT UK Ltd	10 Windsor Road, Bedford MK42 9SU	UK
Tube Headings	PO Box 22, Bentley House, Bentley Works, Doncaster DN5 OBT	UK
Tunnelling Accessories	16 Western Road, St Leonards-on-Sea, East Sussex	UK
Vretmaskin AB	Rissneleden 69, S-172 44 Sundbyberg	Sweden
Water Services Journal	Queensway House, 2 Queensway, Redhill, Surrey	UK
Wilfred Hartnagel GmbH	D-7401 Neustetten 1	W Germany